Lecture Notes in Physics

New Series m: Monographs

Springer
Berlin
Heidelberg
New York
Barcelona
Budapest
Hong Kong
London
Milan
Paris
Santa Clara
Singapore
Tokyo

The Editorial Policy for Monographs

The series Lecture Notes in Physics reports new developments in physical research and teaching - quickly, informally, and at a high level. The type of material considered for publication in the New Series m includes monographs presenting original research or new angles in a classical field. The timeliness of a manuscript is more important than its form, which may be preliminary or tentative. Manuscripts should be reasonably self-contained. They will often present not only results of the author(s) but also related work by other people and will provide sufficient motivation, examples, and applications.
The manuscripts or a detailed description thereof should be submitted either to one of the series editors or to the managing editor. The proposal is then carefully refereed. A final decision concerning publication can often only be made on the basis of the complete manuscript, but otherwise the editors will try to make a preliminary decision as definite as they can on the basis of the available information.
Manuscripts should be no less than 100 and preferably no more than 400 pages in length. Final manuscripts should preferably be in English, or possibly in French or German. They should include a table of contents and an informative introduction accessible also to readers not particularly familiar with the topic treated. Authors are free to use the material in other publications. However, if extensive use is made elsewhere, the publisher should be informed. Authors receive jointly 50 complimentary copies of their book. They are entitled to purchase further copies of their book at a reduced rate. As a rule no reprints of individual contributions can be supplied. No royalty is paid on Lecture Notes in Physics volumes. Commitment to publish is made by letter of interest rather than by signing a formal contract. Springer-Verlag secures the copyright for each volume.

The Production Process

The books are hardbound, and quality paper appropriate to the needs of the author(s) is used. Publication time is about ten weeks. More than twenty years of experience guarantee authors the best possible service. To reach the goal of rapid publication at a low price the technique of photographic reproduction from a camera-ready manuscript was chosen. This process shifts the main responsibility for the technical quality considerably from the publisher to the author. We therefore urge all authors to observe very carefully our guidelines for the preparation of camera-ready manuscripts, which we will supply on request. This applies especially to the quality of figures and halftones submitted for publication. Figures should be submitted as originals or glossy prints, as very often Xerox copies are not suitable for reproduction. For the same reason, any writing within figures should not be smaller than 2.5 mm. It might be useful to look at some of the volumes already published or, especially if some atypical text is planned, to write to the Physics Editorial Department of Springer-Verlag direct. This avoids mistakes and time-consuming correspondence during the production period.
As a special service, we offer free of charge LaTeX and TeX macro packages to format the text according to Springer-Verlag's quality requirements. We strongly recommend authors to make use of this offer, as the result will be a book of considerably improved technical quality.
Manuscripts not meeting the technical standard of the series will have to be returned for improvement.
For further information please contact Springer-Verlag, Physics Editorial Department II, Tiergartenstrasse 17, D-69121 Heidelberg, Germany.

Peter Bouwknegt Jim McCarthy
Krzysztof Pilch

The $\mathcal{W}_3$ Algebra

Modules, Semi-infinite Cohomology and BV Algebras

Springer

Authors

Peter Bouwknegt
Department of Physics and Mathematical Physics
University of Adelaide
Adelaide, SA 5005, Australia

Jim McCarthy
Department of Physics and Mathematical Physics
University of Adelaide
Adelaide, SA 5005, Australia

Krzysztof Pilch
Department of Physics and Astronomy
University of Southern California
Los Angeles, CA 90089-0484, USA

Cataloging-in-Publication Data applied for.

Die Deutsche Bibliothek - CIP-Einheitsaufnahme

Bouwknegt, Peter:
The W3 algebra modules, semi-infinite cohomology and BV algebras / Peter Bouwknegt ; Jim MacCarthy ; Krzysztof Pilch. - Berlin ; Heidelberg ; New York ; Barcelona ; Budapest ; Hong Kong ; London ; Milan ; Santa Clara ; Singapore ; Paris ; Tokyo : Springer, 1996
(Lecture notes in physics : N.s. M, Monographs ; 42)
ISBN 3-540-61528-8
NE: MacCarthy, Jim:; Pilch, Krzysztof:; Lecture notes in physics / M

ISSN 0075-8450 (Lecture Notes in Physics)
ISSN 0940-7677 (Lecture Notes in Physics, New Series m: Monographs)
ISBN 3-540-61528-8 Springer-Verlag Berlin Heidelberg New York

Printed in Germany

Typesetting: Camera-ready by authors
Cover design: *design & production* GmbH, Heidelberg
SPIN: 10481258   55/3142-543210 - Printed on acid-free paper

To Our Parents

Preface

The study of $\mathcal{W}$ algebras began in 1985 in the context of two-dimensional conformal field theories, the aim being to explore higher-spin extensions of the Virasoro algebra. Given the simultaneous growth in the understanding of two-dimensional metric gravity inspired by analyses of string models, it was inevitable that these algebras would be applied to give analogues of putative higher-spin gravity theories. This book is an exposition of the past few years of our work on such an application for the $\mathcal{W}_3$ algebra: in particular, the BRST quantization of the noncritical 4D $\mathcal{W}_3$ string. We calculate the physical spectrum as a problem in BRST cohomology. The corresponding operator cohomology forms a BV algebra, for which we provide a geometrical model.

The $\mathcal{W}_3$ algebra has one further generator, of spin three, in addition to the (spin two) energy-momentum tensor which generates the Virasoro algebra. Contrary to the Virasoro algebra, it is an algebra defined by nonlinear relations. In deriving our understanding of the resulting gravity theories we have had to develop a number of results on the representation theory of $\mathcal{W}$ algebras, to replace the standard techniques that were so successful in treating linear algebras.

The book is essentially self-contained, and as such it can be understood by the typical patient mathematician or mathematical physicist. For a full appreciation of the context for applications – beyond the outline given in the Introduction – the reader should consult the references. We expect the work to be of interest to readers with backgrounds in a number of different fields, such as vertex operator algebras, homological algebra, and string theory, or in applications of the above.

With this in mind, we have divided the presentation into three parts, which, to a large extent, may be read independently. In Part I we develop the machinery required to study $\mathcal{W}$ modules and apply it, in particular, to Verma modules and Fock modules of the $\mathcal{W}_3$ algebra at central charge $c = 2$. In Part II we use these results to compute the semi-infinite cohomology of the $\mathcal{W}_3$ algebra with values in the tensor product of a $c = 2$ Fock module and a $c = 98$ Fock module. In Part III, after developing some general results about BV algebras and their modules, and discussing some examples, we show how the corresponding operator cohomology can be given the structure of a BV algebra. This BV structure has a natural geometric model.

May 1996 | Peter Bouwknegt
Adelaide | Jim McCarthy
Los Angeles | Krzysztof Pilch

Acknowledgements

The authors would like to thank each others' physics departments, as well as the Theory Division at CERN, for hospitality at various times during this work. We have enjoyed discussions with I. Bars, K. de Vos, E. Frenkel, E. Getzler, W. Lerche, M. Varghese, G. Moore, I. Penkov, C. Pope, A. Schwarz, P. van Driel, N. Warner and C.-J. Zhu. We would especially like to thank G. Zuckerman for several discussions on the BV algebra structure of 2D gravity, and, in particular, for suggesting the description of polyvectors on the base affine space in Sect. 4.6.2.

We made extensive use of Mathematica^{TM} routines developed by L. Romans in deriving some of the results in Sects. 2.3.2 and 2.4.1, and of the OPEdefs package of C. Thielemans in Sects. 3.5.3 and 5.4 and Appendix J.

P.B. acknowledges the support of the Packard Foundation during his time at the University of Southern California, during which most of this work was done. P.B. and J.M. acknowledge the support of the Australian Research Council and of the Green Hurst Institute for Theoretical Physics, and K.P. acknowledges the partial support by the U.S. Department of Energy Contract #DE-FG03-84ER-40168.

Contents

Appendices

1 Introduction and Preliminaries

1.1 General Introduction

This book is an exposition of our work, over the past few years, on the $\mathcal{W}_3$ algebra: the representation theory; the corresponding semi-infinite cohomology for special modules; the operator algebra of related $\mathcal{W}_3$ string models and its BV algebra interpretation. This study has several motivations from different directions, which we will briefly indicate before returning to an outline of the results.

1.1.1 Physical Context and Motivation

In $\mathcal{W}$ gravities[1] the vector-generated diffeomorphism symmetry of metric gravity is extended by higher tensor structures. The resulting gauge theory then involves massless higher-spin tensor fields in addition to the spin two field corresponding to metric deformations. It is an intriguing possibility that there should be some corresponding generalization of geometry which will allow a natural description of these $\mathcal{W}$ generalizations. A number of groups have made preliminary studies of the subject (see, e.g., [Bi,Gv,GvMa,Hu1,Ma]), but as yet it has not been developed to an elegant theory. One can hope that a better understanding of the models themselves will aid in this development. Recent progress in constructing $\mathcal{W}$ gravities in two dimensions, where the quantization may be carried through, suggests this as the promising avenue for exploration. Given that a first quantized description of a propagating string must be independent of the parametrization of the two-dimensional string world-sheet, a study of 2D $\mathcal{W}$ gravity is further motivated as a possible extension of string theory. We will follow the string motivation here by restricting our attention to models for which the matter content is a conformal field theory.

Let us briefly recall the corresponding situation for 2D metric gravity. In a conformal gauge quantization using the DDK ansatz [Da,DiKa], a well-defined BRST quantization of the model exists for a restricted range of the central charge of the matter conformal field theory; namely, $c^M \leq 1$. The model then splits into almost-decoupled left- and right-moving "chiral" sectors, and the physical states can be computed [BMP3,LiZu1] from the BRST cohomology (also known as the semi-infinite cohomology) of the Virasoro algebra with values in a tensor product of two scalar field Fock modules. In [Wi2], Witten instigated the study of the algebra of the corresponding physical operators for the case of a single free matter scalar ($c^M = 1$), the 2D (Virasoro) string. He found a rich structure and

[1] For a general review see, e.g., [dBGo1,Hu2,Po,SSvN] and references therein.

an interesting, but incomplete, geometrical interpretation. Further study showed how this structure could be implemented in deriving the physical consequences of the model [Kl,Vc,WiZw], and how it extends to $c^M \leq 1$ [KMS]. Finally, Lian and Zuckerman [LiZu2] identified the underlying mathematical structure as a Batalin-Vilkovisky (BV) algebra – a special case of a Gerstenhaber (G) algebra (see also [Gt,PeSc]). They found that 2D gravity models generically have this BV algebra structure, and they further showed how the geometrical nature of Witten's results arise through a homomorphism from the BV algebra of operator cohomology to the BV algebra of regular polyvector fields on the (complex) plane. The remainder of the physical operator algebra was then interpreted, through a nondegenerate pairing, as a module of this BV algebra.

To describe $\mathcal{W}$ gravity one must quantize a gauge theory based on a nonlinear algebra of constraints; namely, a $\mathcal{W}$ algebra extension of the Virasoro algebra (see [BoSc1,BoSc2] and references therein). It is this algebraic structure which allows a definition of certain $\mathcal{W}$ gravity models through BRST quantization even though the associated $\mathcal{W}$ geometry is not yet well understood in general. In fact, working by analogy with metric 2D gravity, there exists a well-motivated BRST quantization of $\mathcal{W}$ gravity coupled to conformal matter with a restricted range for the central charge in the matter conformal field theory [BLNW1]. For technical reasons, the most complete discussion of noncritical $\mathcal{W}$ strings to date has been given for $c^M \leq 2$ in models based on the $\mathcal{W}_3$ algebra (see, e.g., [BBRT,BLNW1-2,BMP5-7,LeSe,LPWX]). A convenient representation[2] for this class of models is the system of $\mathcal{W}_3$ gravity coupled to a matter sector consisting of two free scalar fields. It is presumed that a similar treatment may be given in the general rank N case for $c^M \leq N$, and indeed many of the results we present are completely general with this in mind. The models with $c^M < 2$ are obtained by choosing appropriate background charges in the free field matter system, and a rather well-known projection maps these to the minimal models of the $\mathcal{W}_3$ algebra [BMP2,FKW].

The analysis of the 2D string recalled above has been extended to $\mathcal{W}_3$ gravity models over a number of years [BMP5-7]. In this book we will discuss the special case of $c^M = 2$. In the corresponding string interpretation the matter scalar fields would embed the world sheet of the string into a two-dimensional spacetime. But, moreover, since this is a noncritical theory there are dynamical gravitational degrees of freedom – under the DDK-type ansatz these are described by a pair of scalar fields of "wrong sign" with a background charge, the so-called Liouville sector. Thus, in this string language, the model describes a $(2+2)$-dimensional string in a nontrivial background. We call it the 4D $\mathcal{W}_3$ string.

[2] This representation is directly relevant to the situation in which all "cosmological constant" terms are tuned to zero. Indirect arguments make these results relevant to the generic case as well.

1.1.2 Mathematical Context and Motivation

$\mathcal{W}$ algebras are nonlinear extensions of the Virasoro algebra (see [BoSc1,BoSc2] and references therein). In the class of $\mathcal{W}$ algebras the simplest one is the so-called $\mathcal{W}_3$ algebra, which possesses – apart from the Virasoro generators L_n, $n \in \mathbb{Z}$ – one additional (infinite) set of generators W_n, $n \in \mathbb{Z}$. The $\mathcal{W}_3$ algebra, being the simplest infinite dimensional algebra with nonlinear defining relations, is a useful laboratory to see which properties, constructions and techniques from the Lie algebra case extend, or do not extend as the case may be, to the nonlinear case. Specifically, in this book, we study the structure theory of a suitable category of $\mathcal{W}_3$ modules (e.g., composition series) and various aspects of homological algebra (e.g., resolutions, semi-infinite cohomology).

There are two immediately obvious differences with the Lie algebra case. First, the adjoint action of the Cartan subalgebra on $\mathcal{W}_3$ is not diagonalizable, and similarly its action on most interesting $\mathcal{W}_3$ modules is not diagonalizable.[3] As a consequence, we are led to incorporate so-called generalized Verma modules – i.e., modules induced from an indecomposable module of the Cartan subalgebra by the negative root operators – into our framework. Secondly, the tensor product of two $\mathcal{W}_3$ modules does not, in general, carry the structure of a $\mathcal{W}_3$ module. This necessitates a generalization of what we mean by semi-infinite cohomology of an algebra with values in a tensor product of modules. Moreover, it prevents (at least a straightforward) application of many standard techniques in calculating such a cohomology.

Besides the interest from a purely technical point of view, studying the semi-infinite cohomology of $\mathcal{W}_3$ algebras is interesting because it provides us with beautiful, yet highly nontrivial, examples of so-called BV algebras [Gt,Ko,LiZu2,PeSc]. BV algebras are $\mathbb{Z}$-graded, supercommutative, associative algebras with a second order derivation Δ of degree -1, satisfying $\Delta^2 = 0$. They naturally possess the structure of a G algebra [Gs]; i.e., a $\mathbb{Z}$-graded, supercommutative, associative algebra under a product, as well as a $\mathbb{Z}$-graded Lie superalgebra under a bracket, and such that the bracket acts as a superderivation of the product. G algebras show up in many different areas of mathematics. An important example of a G algebra is the set of polyderivations $\mathcal{P}(\mathcal{R})$ of a commutative algebra $\mathcal{R}$. It is an interesting question to determine for which algebras $\mathcal{R}$, the set of polyderivations $\mathcal{P}(\mathcal{R})$ possesses, in fact, the structure of a BV algebra. In this book we present some examples where this turns out to be the case, namely, the free algebra $\mathcal{C}_N$ on N generators and the algebra $\mathcal{R}_N$ obtained from $\mathcal{C}_{2N}$ by dividing out the ideal generated by a quadratic relation.

A special class of BV algebras is that for which the homology of the BV operator Δ vanishes. It turns out that the semi-infinite cohomology of the $\mathcal{W}_3$ algebra with coefficients in the tensor product of two $\mathcal{W}_3$ Fock modules can be equipped with the structure of a BV algebra and contains, in a sense, the algebra of polyderivations $\mathcal{P}(\mathcal{R}_3)$ in exactly such a way as to make the homology of Δ trivial. In fact, at least superficially, the cohomology looks like a specific

[3] This is very reminiscent of the Lie superalgebra case.

"patching" of a set of G modules $\mathcal{P}(\mathcal{R}_3, M_w)$, i.e., polyderivations with coefficients in an $\mathcal{R}_3$ module M_w, labeled by elements of the Weyl group of $\mathfrak{sl}_3$ ($\mathcal{P}(\mathcal{R}_3, M_{w=1}) \cong \mathcal{P}(\mathcal{R}_3)$). This patching is nontrivial. However, there is another approach using the fact that $\mathcal{P}(\mathcal{R}_3)$ can be identified with the set of polyvector fields (with polynomial coefficients) on the so-called base affine space of $SL(3, \mathbb{C})$ – which gives a precise description of this patching. This latter description gives immediate conjectures for the BV structure corresponding to an arbitrary $\mathcal{W}$ algebra, together with interesting insights into the geometrical aspects of these theories. The main ingredient of this geometric description of the BV structure of $\mathcal{W}$ strings is a BV algebra BV[$\mathfrak{g}$] associated with a simple, simply-laced Lie algebra $\mathfrak{g}$.

1.2 Outline and Summary of Results

A semi-infinite cohomology may be defined for the $\mathcal{W}_3$ algebra by analogy with that for the Virasoro case. Corresponding to the two sets of generators L_m and W_m, $m \in \mathbb{Z}$, introduce two sets of ghost oscillators $(b_m^{[i]}, c_m^{[i]})$, $i = 2, 3$, generating the Fock space F^{gh}. For any two positive energy[4] $\mathcal{W}_3$ modules V^M and V^L, such that $c^M + c^L = 100$, there exists a complex $(V^M \otimes V^L \otimes F^{\mathrm{gh}}, d)$, graded by ghost number, and with a differential (BRST operator) d of degree one [BLNW1,TM], with leading terms

$$d = \sum_m c_{-m}^{[3]} \left(\frac{1}{\sqrt{\beta^M}} W_m^M - \frac{i}{\sqrt{\beta^L}} W_m^L \right) + c_{-m}^{[2]} \left(L_m^M + L_m^L \right) + \dots , \tag{1.1}$$

where $\beta = 16/(22 + 5c)$. The cohomology of d at degree n will be denoted by $H^n(\mathcal{W}_3, V^M \otimes V^L)$ and called the BRST cohomology of the $\mathcal{W}_3$ algebra on $V^M \otimes V^L$.

The central problem motivating the present study was the computation of this cohomology for Fock modules, particularly the case $c^M = 2$ which is the case of interest for the 4D $\mathcal{W}_3$ string. The result is given in Chap. 3. In turn, this problem spawned several other studies of mathematical and possibly physical interest. In particular, the calculations of Chap. 3 require a detailed knowledge of the representation theory of $\mathcal{W}_3$, which is discussed in Chap. 2. General results show that the corresponding operator cohomology, $\mathfrak{H}$, forms a BV algebra [LiZu2]. In Chap. 4 we discuss examples of BV algebras which allow us to develop a detailed understanding of $\mathfrak{H}$ in Chap. 5. The results there go a long way towards verifying those of Chap. 3.

In the remainder of this introduction we summarize the main results of each chapter. A glossary of notation is included at the end of the book.

---

[4] A positive energy module is L_0 diagonalizable with finite dimensional eigenspaces, and with the spectrum bounded from below. The category $\mathcal{O}$ of relevant positive energy modules is defined more precisely in Chap. 2.

The modules of interest in Chap. 2 are Fock modules (typically denoted by F) and (generalized) Verma modules (typically M). The structure of a given module $V \in \mathcal{O}$ can be exhibited in part through its composition series, JH(V) (Theorem 2.3). In fact (Lemma 2.15) there is a 1–1 correspondence between primitive vectors in V of (generalized) weight (h, w) and irreducible modules $L(h, w, c)$ appearing in JH(V). Denote the multiplicity with which a given irreducible module L appears in JH(V) by $(V : L)$. Linear independence of the characters of irreducible modules (Theorem 2.2), gives the result (Theorem 2.29) that

$$(M(\Lambda, \alpha_0) : L) \;=\; (F(\Lambda, \alpha_0) : L)\,. \tag{1.2}$$

Thus by studying Fock modules one obtains detailed information about related Verma modules, and *vice versa*. The crucial result here is Corollary 2.28, which shows that generically the two are isomorphic as $\mathcal{W}_3$ modules. In particular, one finds that if $-i(\Lambda + \alpha_0\rho) \in D_+$, with $\alpha_0{}^2 < -4$, then $F(\Lambda, \alpha_0) \cong \overline{M}(\Lambda, \alpha_0)$.

For $c = 2$ ($\alpha_0 = 0$), the case of most interest in this book, the Fock module is unitary with respect to a Hermitian inner product and we show in Theorem 2.31 that $F(\Lambda, 0)$ is completely reducible. Moreover, for $\Lambda \in P$,

$$(F(\Lambda, 0) : L(\Lambda', 0)) \;=\; m_\Lambda^{\Lambda'}\,, \tag{1.3}$$

where $m_\Lambda^{\Lambda'}$ is the multiplicity of the weight Λ in the finite-dimensional irreducible $\mathfrak{sl}_3$ module $\mathcal{L}(\Lambda')$. This is a proof of the Kazhdan-Lusztig conjecture for this special case.[5] Further, for $c = 2$, we apply this understanding to derive (generalized) Verma module resolutions for irreducible $\mathcal{W}_3$ modules.

The calculation of the semi-infinite cohomology is detailed in Chap. 3. We begin, in Sect. 3.2, by finding a lattice of momenta for the matter Fock modules so that the cohomology complex lifts to a Vertex Operator Algebra (VOA), $\mathfrak{C}$, on which the differential acts as the charge of a spin-1 current. By means of the usual state/operator mapping of conformal field theory, the main results of this section are summarized as the cohomology of the complex $(\mathfrak{C}, d)$, denoted $H(\mathcal{W}_3, \mathfrak{C})$ (with a slight abuse of notation). When considered as an algebra with the induced VOA structure, we will denote this operator cohomology by $\mathfrak{H}$. There exists a bilinear form on $\mathfrak{C}$ which induces a nondegenerate pairing between the cohomology at gh $= n$ and gh $= 8 - n$ (Theorem 3.12).

The fundamental result which allows the techniques developed in Chap. 2 to be applied is the reduction theorem (Theorem 3.8):

For an arbitrary generalized Verma module $M^{(\kappa)}(\Lambda^M, \alpha_0^M)$ *and a contragredient Verma module* $\overline{M}(\Lambda^L, \alpha_0^L)$, $c^M + c^L = 100$, *the cohomology* $H(\mathcal{W}_3, M^{(\kappa)}(\Lambda^M, \alpha_0^M) \otimes \overline{M}(\Lambda^L, \alpha_0^L))$ *is nonvanishing if and only if*

$$-i(\Lambda^L + \alpha_0^L\rho) \;=\; w(\Lambda^M + \alpha_0^M\rho)\,, \tag{1.4}$$

for some $w \in W$, *in which case it is spanned by the states*

[5] For a general discussion of the Kazhdan-Lusztig conjecture for $\mathcal{W}$ algebras, see [dVvD1-2]).

$$v_0\,,\quad c_0^{[2]}v_0\,,\quad c_0^{[3]}v_{\kappa-1}\,,\quad c_0^{[3]}c_0^{[2]}v_{\kappa-1}\,, \tag{1.5}$$

where $v_i = v_i^M \otimes \overline{v}^L \otimes |0\rangle_{\text{gh}}$, $i=0,\ldots,\kappa-1$, *span the highest weight space.*

For $c^M = 2$ and $-i\Lambda^L + 2\rho \in P_+$, combining this theorem with the Verma module resolutions of Chap. 2 computes $H(\mathcal{W}_3, L(\Lambda,0)\otimes F(\Lambda^L,2i))$ (Theorem 3.14). The Fock space decomposition theorem (Theorem 2.31) then computes (Theorem 3.17) the desired cohomology, $H^n(\mathcal{W}_3, F(\Lambda^M,0)\otimes F(\Lambda^L,2i))$. The resulting operator cohomology for this sector of Liouville momenta may be decomposed under $\mathfrak{sl}_3 \oplus (\mathfrak{u}_1)^2$ into cones of finite-dimensional irreducible modules at different ghost numbers (Theorem 3.19). For the remaining sectors, i.e., for $w(-i\Lambda^L+2\rho)\in P_+$, $w\in W$, we are able to derive the full result from the assumption of a kind of Weyl group symmetry – the result being that up to ghost number shifts the cones are essentially reflected to the other Weyl chambers. Thus, the complete cohomology for the $c^M = 2$ case is summarized in Theorem 3.25.

General results [LiZu2] imply that the operator cohomology forms a BV algebra, $(\mathfrak{H}, \cdot\,, b_0)$, graded by ghost number, with the dot product given by the operator product expansion and with the BV operator identified with $b_0 \equiv b_0^{[2]}$. To prepare for a thorough analysis of the operator algebra, we develop in Chap. 4 some general machinery as well as explicit examples of G and BV algebras.

Given an Abelian ring, $\mathcal{R}$, the archetypal example of a G algebra is the algebra of polyderivations of the ring $\mathcal{R}$, $(\mathcal{P}(\mathcal{R}), \cdot\,, [-,-]_S)$, equipped with the Schouten bracket. For any BV algebra, $(\mathfrak{A}, \cdot\,, \Delta)$, the subspace $\mathfrak{A}^0$ is an Abelian ring with respect to the dot product. In fact, there is a natural G algebra homomorphism, π, from $(\mathfrak{A}, \cdot\,, \Delta)$ to $(\mathcal{P}(\mathfrak{A}^0), \cdot\,, [-,-]_S)$. If there is a compatible BV operator on the space of polyderivations, then we show under what conditions π lifts to a BV homomorphism (Theorem 4.12).

We illustrate the above in examples based on the ring $\mathcal{R}_N = \mathcal{C}_{2N}/\mathcal{I}$, where $\mathcal{C}_{2N} \cong \mathbb{C}[x^1,\ldots,x^{2N}]$ is a free Abelian algebra, and $\mathcal{I}$ is the ideal generated by a quadratic vanishing relation (Sect. 4.2.1). The natural $\mathfrak{so}_{2N}$ action by derivations of $\mathcal{C}_{2N}$ descends to $\mathcal{R}_N$. Using it, we construct an explicit basis for the space of polyderivations, $\mathcal{P}(\mathcal{R}_N)$, which is summarized in Theorem 4.16. Moreover, in Theorem 4.17, we find a finite set of generators and relations which characterize $\mathcal{P}(\mathcal{R}_N)$ as a dot algebra. These results allow us to demonstrate that $\mathcal{P}(\mathcal{R}_N)$ is actually a BV algebra (Theorem 4.24), and to explicitly calculate the homology of the corresponding BV operator (Theorem G.5). For comparison with the operator cohomology the most relevant case is $N=3$, so we present this case quite explicitly. The ground ring, $\mathcal{R}_3$, is a model space for $\mathfrak{sl}_3 \subset \mathfrak{so}_6$.

The theory of G and BV modules can be developed along the same lines, as we consider in Sect. 4.4. Given an $\mathcal{R}$ module, M, one may construct the polyderivations, $\mathcal{P}(\mathcal{R}, M)$, of $\mathcal{R}$ with values in M. Then in Theorem 4.31 we show under which conditions $\mathcal{P}(\mathcal{R}, M)$ will be a G module of $\mathcal{P}(\mathcal{R})$.

We restrict our example of this construction to the most relevant case of $N=3$. As a remnant of the $\mathfrak{so}_6$ (or, in fact, a "hidden" $\mathfrak{so}_8$) structure of $\mathcal{R}_3$, we see in Sect. 4.5.2 that there are six natural $\mathcal{R}_3$ module structures, M_w, $w\in W$.

The main result of this section is Theorem 4.33, where we show that for each $w \in W$, the space of twisted polyderivations $\mathfrak{P}_w \equiv \mathcal{P}(\mathcal{R}_3, M_w)$ is a G module of $\mathfrak{P} \equiv \mathcal{P}(\mathcal{R}_3)$. In Sect. 4.5.3 we are then able to give quite explicit $\mathfrak{sl}_3 \oplus (\mathfrak{u}_1)^2$ decompositions for these G modules.

It is well known that the model space of $\mathfrak{sl}_3$ can also be realized as the space of polynomial functions on the algebraic variety $A(SL(3,\mathbb{C})) = N_+\backslash SL(3,\mathbb{C})$, where N_+ is the complex subgroup (of $SL(3,\mathbb{C})$) generated by the positive root generators [BGG2]. The space A is called the base affine space of $SL(3,\mathbb{C})$. Thus the algebra of polyderivations $\mathcal{P}(\mathcal{R}_3)$ has a geometric realization as the algebra of polynomial polyvector fields on $A(SL(3,\mathbb{C}))$. This result generalizes to the complex Lie group, G, of any (simply-laced) Lie algebra $\mathfrak{g}$ with maximal nilpotent subalgebra $\mathfrak{n}_+ \subset \mathfrak{g}$. In Sect. 4.6 we introduce the base affine space $A = A(G)$, and characterize the polynomial polyvectors $\mathcal{P}(A)$ as the $\mathfrak{n}_+$-invariant elements in $\mathcal{E}(G)\otimes\bigwedge\mathfrak{b}_-$, where $\mathcal{E}(G)$ is the space of regular functions on G and we identify $\mathfrak{b}_- = \mathfrak{n}_+\backslash\mathfrak{g}$

The fact that $\mathcal{P}(A)$ is a BV algebra is a classical result [Ko]. Given the discussion above it is natural to consider an extension of this algebra. We show in Sect. 4.6 that, in the context outlined above, to the Lie algebra $\mathfrak{g}$ one may associate a BV algebra, BV[$\mathfrak{g}$], whose underlying graded commutative algebra is given by the cohomology, $H(\mathfrak{n}_+, \mathcal{E}(G) \otimes \bigwedge\mathfrak{b}_-)$. The BV algebra $\mathcal{P}(A)$ is thus incorporated as the zeroth order cohomology. In Sect. 4.6.4 we construct a one parameter family of BV operators on BV[$\mathfrak{g}$] which reduce to that corresponding to the Schouten bracket on $\mathcal{P}(A)$. We show that, except at one point, all BV operators define equivalent BV structures on BV[$\mathfrak{g}$] and have trivial homology.

This study is not just a mathematical curiosity since it turns out that BV[$\mathfrak{g}$] provides a model for the $\mathcal{W}$ cohomology. For the $\mathfrak{sl}_2$ and $\mathfrak{sl}_3$ cases this result is based on Theorem 4.43, an exact computation of the cohomology $H(\mathfrak{n}_+, \mathcal{E}(G) \otimes \bigwedge\mathfrak{b}_-)$. For the general case we have been content to compute the cohomology for all weights $\Lambda \in P_+$ sufficiently deep inside the fundamental Weyl chamber (Theorem 4.42), where one finds

$$H(\mathfrak{n}_+, \mathcal{L}(\Lambda) \otimes \bigwedge\mathfrak{b}_-) \;\cong\; H(\mathfrak{n}_+, \mathcal{L}(\Lambda)) \otimes \bigwedge\mathfrak{b}_- \,. \tag{1.6}$$

In Sect. 4.6.5 this result is compared, for the $\mathfrak{sl}_3$ case, with the modules of generalized polyderivations. It is found that BV[$\mathfrak{sl}_3$] precisely glues these modules into a BV algebra.

In Chap. 5 we put the above results together to obtain a description of the BV algebra of operator cohomology, $\mathfrak{H}$. We first observe in Theorem 5.4 that $\mathfrak{H}^0 \cong \mathcal{R}_3$, thus bringing into play all the results of Chap. 4.[6] In particular, we prove (Theorem 5.13):

i. *There exists a natural map* $\pi : \mathfrak{H} \to \mathfrak{P}$ *that is a BV algebra homomorphism between* $(\mathfrak{H}, \cdot, b_0)$ *and* $(\mathfrak{P}, \cdot, \Delta)$.
ii. *Let* $\mathfrak{I} \equiv \operatorname{Ker} \pi$ *be a BV ideal of* $\mathfrak{H}$. *We have an exact sequence of BV algebras*

[6] The description of the ground ring of $\mathcal{W}_N$ gravity in terms of the $\mathfrak{sl}_N$ model space was anticipated in [MMMO].

$$0 \longrightarrow \mathfrak{I} \longrightarrow \mathfrak{H} \xrightarrow{\pi} \mathfrak{P} \longrightarrow 0.$$

There exists a dot algebra homomorphism $\imath : \mathfrak{P} \to \mathfrak{H}$, such that $\pi \circ \imath = \mathrm{id}$, i.e., the sequence splits as a sequence of $\imath(\mathfrak{P})$ dot modules.

Similarly, we show that there are $\mathfrak{H}^0$ modules, $\widehat{M}_w \subset \mathfrak{I}$, which are isomorphic to M_w as $\mathcal{R}_3$ modules (Theorem 5.18). Indeed, up to some subtleties at the boundaries of the different Weyl chambers, we deduce that the bulk of $\mathfrak{I}$ admits a description in terms of the twisted polyderivations $\mathfrak{P}_w$ (Theorem 5.20).

To be more precise, one can try to understand how these different sectors of twisted polyderivations are patched together. We first show in Theorem 5.22 that the homology of b_0 on $\mathfrak{H}$ is trivial. Thus $(\mathfrak{H}, \cdot, b_0)$ is an extension of $\mathfrak{P}$ which is acyclic with respect to the BV operator. Next, we construct a further projection on $\mathfrak{I}$, $\pi' : \mathfrak{I}^n \longrightarrow \mathcal{P}^{n-1}(\mathcal{R}_3, M_{r_1} \oplus M_{r_2})$, which is the identity on $\mathfrak{I}^1 \cong \widehat{M}_{r_1} \oplus \widehat{M}_{r_2}$. Then (Theorem 5.24) the map π' is a G morphism between the G module $\mathfrak{I}$ of $\mathfrak{H}$ and the G module $\mathfrak{P}_{r_1} \oplus \mathfrak{P}_{r_2}$ of $\mathfrak{P}$. Together with the nondegenerate pairing on cohomology, this gives a fairly explicit description of the dot module structure of $\mathfrak{H}$ over $\imath(\mathfrak{P})$.

In fact, from the results in Sect. 4.6 we know precisely such an extension, namely BV[$\mathfrak{sl}_3$]! Thus we have a remarkable characterization of the BV algebra of chiral operator cohomology in the $\mathcal{W}_3$ string. Furthermore, this result may be immediately extended to the study of $\mathcal{W}_N$ strings[7] for general N, and leads to the following conjectures:

There is an isomorphism of BV algebras

$$H(\mathcal{W}[\mathfrak{g}], \mathbb{C}) \cong H(\mathfrak{n}_+, \mathcal{E}(G) \otimes \bigwedge \mathfrak{b}_-) \equiv \mathrm{BV}[\mathfrak{g}], \qquad (1.7)$$

where $\mathfrak{g} = \mathfrak{sl}_N$, or, more generally, a simple, simply-laced Lie algebra.

The work outlined in this book checks the conjecture for $\mathfrak{sl}_3$, while for $\mathfrak{sl}_2$ it is also correct and provides a new understanding for that model. Indeed, in the last section of this introduction we summarize the present understanding of the $c^M = 1$ Virasoro case, the 2D string, including this new result. The reader who is familiar with that case, but interested to appreciate the new insights of this book, will hopefully find it a useful bridge.

Finally, there are a number of appendices which typically either record results of explicit calculations, or give technical details of particular derivations.

Short summaries of some of the results may be found in [BMP8,BoPi,McC].

---

[7] There is a systematic construction of the BRST operator for such $\mathcal{W}_N$ strings, though the procedure has not been carried out explicitly beyond $N = 4$; see [BLNW2,RSS].

1.3 The BV Algebra of the 2D $\mathcal{W}_2$ String

To complete the introduction we summarize the main features of the BV algebra of 2D gravity coupled to $c = 1$ matter, the so-called 2D $\mathcal{W}_2$ (Virasoro) string. In doing so we note both the similarities and differences with the $\mathcal{W}_3$ string, and to this end we follow the notation introduced later in that context. If the notation is not familiar the interested reader should consult the appropriate later section. We also use this opportunity to demonstrate the remarkable characterization of the BV algebra of operator cohomology embodied in (1.7). This last result is new, for the remainder the reader should consult [BMP3,LiZu1,MMMO,Wi2,WiZw,WuZh] and especially [LiZu2] for additional discussion and detailed proofs.

In this section, and only here, P and Q denote the $\mathfrak{sl}_2$ weight and root lattices, respectively, while $F^{\rm gh}$ is the Fock space of the Virasoro ghosts, i.e., the $j = 2$ (bc) system.

1.3.1 The Cohomology Problem

Recall that a Feigin-Fuchs module, $F(\Lambda, \alpha_0)$, of the Virasoro algebra is parametrized by the momentum, Λ, of the underlying Fock space of a single scalar field and a background charge α_0. If we interpret Λ as an $\mathfrak{sl}_2$ weight, then the highest weight of the Virasoro module is $h = \frac{1}{2}(\Lambda, \Lambda + 2\alpha_0\rho)$. The central charge is given by $c = 1 - 6\alpha_0^2$.

An important problem in the study of 2D Virasoro string is to compute the operator algebra, $\mathfrak{H}$, obtained as the semi-infinite (BRST) cohomology of the VOA associated with tensor products, $F(\Lambda^M, 0) \otimes F(\Lambda^L, 2i)$, of Feigin-Fuchs modules with $c = 1$ and $c = 25$, respectively. More precisely, we consider the $\mathcal{W}_2$ analogue of the complex $(\mathfrak{C}, d)$ in Sect. 3.2, defined by a lattice of momenta $L = \{(\Lambda^M, -i\Lambda^L) \in P \times P \,|\, \Lambda^M + i\Lambda^L \in Q\}$.

The vertex operator realization of $\widehat{\mathfrak{sl}_2}$ on the $c = 1$ Fock spaces, together with the Liouville momentum operator, $-ip^L$, give rise to an $\mathfrak{sl}_2 \oplus \mathfrak{u}_1$ symmetry on $\mathfrak{C}$, that commutes with the BRST operator and yields a decomposition of the cohomology, $H(\mathcal{W}_2, \mathfrak{C})$, into a direct sum of finite-dimensional irreducible modules. The complete description of the cohomology space is given by the following theorem, due to [BMP3,LiZu1,Wi2].

Theorem 1.1 *The relative cohomology $H_{\rm rel}(\mathcal{W}_2, \mathfrak{C})$ is isomorphic, as an $\mathfrak{sl}_2 \oplus \mathfrak{u}_1$ module, to the direct sum of finite-dimensional irreducible modules with highest weights in a set of disjoint "cones" $\{(\Lambda, \Lambda') + (\lambda, w\lambda) \,|\, \lambda \in P_+\}$, i.e.*

$$H^n_{\rm rel}(\mathcal{W}_2, \mathfrak{C}) \cong \bigoplus_{w \in W} \bigoplus_{(\Lambda, \Lambda') \in S^n_w} \bigoplus_{\lambda \in P_+} \mathcal{L}(\Lambda + \lambda) \otimes \mathbb{C}_{\Lambda' + w\lambda}\,, \tag{1.8}$$

where the sets S^n_w are given in Table 1.1.

Table 1.1. The tips S^n_w in the decomposition of $H_{\rm rel}(\mathcal{W}_2, \mathfrak{C})$

n	w	S^n_w
0	1	$(0,0)$
1	1	$(\rho,-\rho)$
	-1	$(0,-2\rho)$
2	-1	$(0,-4\rho)$

Remark. The required cohomology, $H(\mathcal{W}_2, F(\Lambda^M, 0) \otimes F(\Lambda^L, 2i))$, may be determined either directly, see, e.g., [BMP3], or by decomposing $F(\Lambda^M, 0)$ into irreducible modules, and then computing $H_{\rm rel}(\mathcal{W}_2, L(\Lambda, 0) \otimes F(\Lambda^L, 2i))$, see [LiZu1]. The latter cohomology is nonvanishing if and only if the weights Λ and Λ^L satisfy $-i\Lambda^L + 2\rho = w(\Lambda + \rho - \sigma\rho)$ for some $w, \sigma \in W$. This is different than in the $\mathcal{W}_3$ case, summarized in Corollary 4.38, where σ runs over an extension of the Weyl group of $\mathfrak{sl}_3$. Thus, the $\mathcal{W}_3$ cohomology displays qualitatively new features (which, in this particular case, can be explained by the more complex embedding pattern of primitive vectors in Verma modules of a higher rank $\mathcal{W}$ algebra).

The result of Theorem 1.1 may be conveniently summarized pictorially.

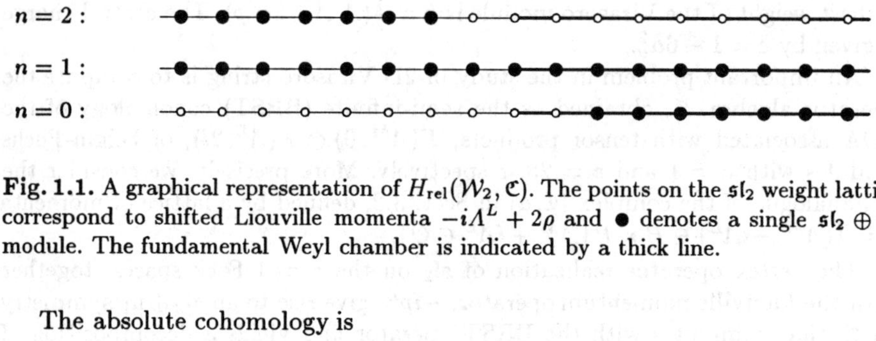

Fig. 1.1. A graphical representation of $H_{\rm rel}(\mathcal{W}_2, \mathfrak{C})$. The points on the $\mathfrak{sl}_2$ weight lattice correspond to shifted Liouville momenta $-i\Lambda^L + 2\rho$ and $\bullet$ denotes a single $\mathfrak{sl}_2 \oplus \mathfrak{u}_1$ module. The fundamental Weyl chamber is indicated by a thick line.

The absolute cohomology is

$$H^\bullet(\mathcal{W}_2, \mathfrak{C}) \;\cong\; H^\bullet_{\rm rel}(\mathcal{W}_2, \mathfrak{C}) \oplus H^{\bullet-1}_{\rm rel}(\mathcal{W}_2, \mathfrak{C})\,, \tag{1.9}$$

i.e., it has a doublet structure with the resulting pattern of $\mathfrak{sl}_2 \oplus \mathfrak{u}_1$ modules as in Fig. 1.2.

1.3.2 The BV Algebra and Structure Theorems

A BV algebra $(\mathfrak{A}, \cdot, \Delta)$ consists of a graded, graded commutative algebra $(\mathfrak{A}, \cdot)$ with an operator Δ, called the BV operator, which is a graded second order derivation of degree -1 on $\mathcal{A}$ satisfying $\Delta^2 = 0$ (see Chap. 4).

The ground ring of the BV algebra $(\mathfrak{H}, \cdot, b_0)$, where $\mathfrak{H} \equiv H(\mathcal{W}_2, \mathfrak{C})$ and $b_0 = b_0^{[2]}$, is defined to be the ring of operator cohomology at zero ghost number. From [Wi2] this ring is known to be isomorphic with the algebra of polynomial

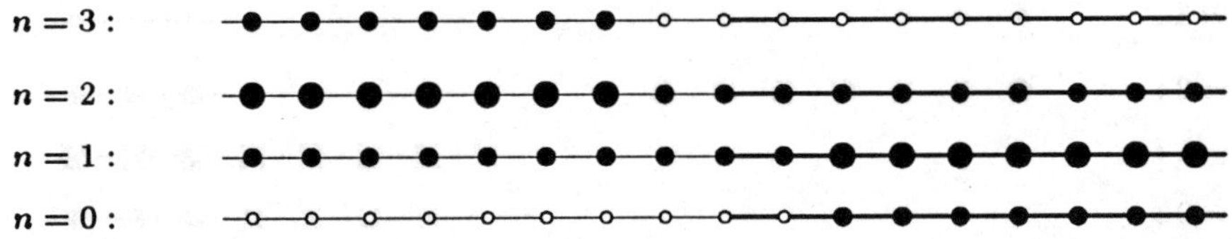

Fig. 1.2. A graphical representation of $H(\mathcal{W}_2, \mathbb{C})$, adopting the conventions of Fig. 1.1. The dots • and ● denote a degeneracy of one and two $\mathfrak{sl}_2 \oplus \mathfrak{u}_1$ modules, respectively.

functions on the complex plane $\mathbb{C}^2$, i.e., $\mathfrak{H}^0 \cong \mathcal{C}_2$. Let us denote the generators of $\mathfrak{H}^0$ by $\widehat{x}_i$, and those of $\mathcal{C}_2$ by x_i, for $i = 1, 2$. The space of polyvectors on $\mathbb{C}^2$ or, equivalently, the set of polyderivations of the commutative algebra $\mathcal{C}_2$, is denoted by $\mathcal{P}(\mathcal{C}_2)$. $\mathcal{P}(\mathcal{C}_2)$ carries the structure of a BV algebra where the BV operator Δ_S induces the Schouten bracket on polyvector fields.

A large part of the higher ghost number cohomology is described by the following structure theorem.

Theorem 1.2 [LiZu2]
i. The map[8] $\pi : \mathfrak{H} \to \mathcal{P}(\mathcal{C}_2)$, *introduced in Sect. 4.1.4, is a BV algebra homomorphism onto the BV algebra of polyderivations* $(\mathcal{P}(\mathcal{C}_2), \cdot, \Delta_S)$.
ii. There exists an embedding $\imath : \mathcal{P}(\mathcal{C}_2) \to \mathfrak{H}$ *that preserves the dot product and satisfies* $\pi \circ \imath = \mathrm{id}$.

In addition to $\imath(x_i) = \widehat{x}_i$, the embedding $\imath$ is completely characterized by the image $\imath(\Omega) = \widehat{\Omega}$ of the "volume element" (cf. (4.23)) $\Omega = \frac{\partial}{\partial x_1} \wedge \frac{\partial}{\partial x_2} - \frac{\partial}{\partial x_2} \wedge \frac{\partial}{\partial x_1}$, where $\widehat{\Omega}$ is the unique operator of ghost number two at the weight $(0, -2\rho)$, see Fig. 1.2. Let us denote $\mathfrak{H}_+ = \imath(\mathcal{P}(\mathcal{C}_2))$ and $\mathfrak{H}_- = \ker \pi$. The decomposition $\mathfrak{H} \cong \mathfrak{H}_+ \oplus \mathfrak{H}_-$ is shown in Fig. 1.3.

A straightforward calculation reveals that $b_0 \widehat{\Omega} \notin \imath(\mathcal{P}(\mathcal{C}_2))$, although $\Omega \in \mathfrak{H}_+$. In fact, as is easily seen from Fig. 1.3, $\widehat{\Omega}$ is the only element in $\mathfrak{H}_+$ with this property. By contrast, $\mathfrak{H}_-$ is a BV ideal in which all dot product and brackets vanish, and from a study of the action of $\mathfrak{H}$ on this ideal, one concludes:

Theorem 1.3 [LiZu2] *The BV algebra* $(\mathfrak{H}, \cdot, b_0)$ *is generated by* 1, *the ground ring generators* $\widehat{x}_i$, *and* $\widehat{\Omega}$.

For this Virasoro case the remainder of the cohomology is well described by "generalized polyderivations" of $\mathcal{C}_2$ (i.e., polyvectors with coefficients which are not in $\mathcal{C}_2$ but a module thereof, see Sect. 4.5.2). We may introduce on $\mathcal{C}_2$ two

[8] In [LiZu2] this map is denoted by ψ.

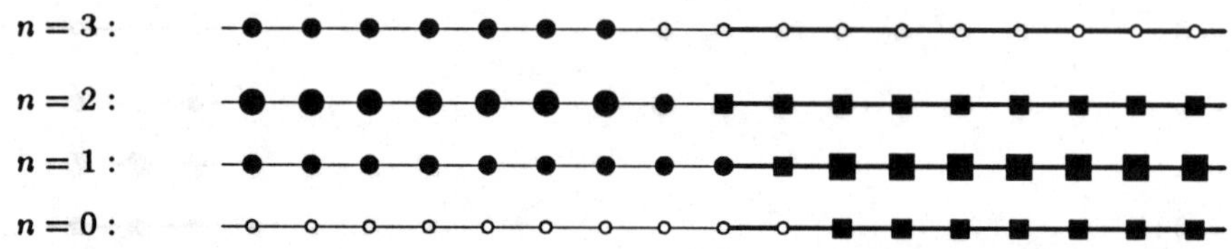

Fig. 1.3. The decomposition $\mathfrak{H} \cong \mathfrak{H}_+ \oplus \mathfrak{H}_-$. The modules in $\mathfrak{H}_+$ are denoted by the squares while those in $\mathfrak{H}_-$ by the dots. Again, degeneracies are indicated by size.

structures of a ground ring ($\mathcal{C}_2$) module: M_1, isomorphic to the ground ring itself, and the twisted module M_{-1} defined by the "dual" realization of the ground ring generators, $x_1 \to -\frac{\partial}{\partial x_2}$ and $x_2 \to -\frac{\partial}{\partial x_1}$. As shown in [LiZu2], the ghost number one cohomology in the negative Weyl chamber, i.e., $\mathfrak{H}^1_-$, is isomorphic, as a ground ring module, to M_{-1}. Furthermore, $\mathfrak{H}_-$ has a natural structure of a G module of $\mathcal{P}(\mathcal{C}_2)$, with respect to which one may identify it with the generalized polyderivations $\mathcal{P}(\mathcal{C}_2, M_{-1})$.

It should be emphasized here that the "gluing" of $\mathfrak{H}_+$ and $\mathfrak{H}_-$, accomplished by the BV operator b_0, is underlined by a simple algebraic principle.

Theorem 1.4 *The BV algebra* $(\mathfrak{H}, \cdot, b_0)$ *is an extension of the BV algebra* $(\mathcal{P}(\mathcal{C}_2), \cdot, \Delta_S)$ *of polyvectors on the base affine space of* $\mathfrak{sl}_2$, *for which the homology of the BV operator* b_0 *is trivial.*

1.3.3 The BV Algebra of $\mathfrak{n}_+$ Cohomology

The ring $\mathcal{C}_2$ may also be identified with the ring of regular functions on the base affine space of $SL(2,\mathbb{C})$ (the complex Lie group of $\mathfrak{sl}_2$). Let us quickly recall how this is done. In the following we use the notation $\mathcal{E}(X)$ for the space of regular functions on X.

Fix a Cartan decomposition $\mathfrak{sl}_2 \cong \mathfrak{n}_- \oplus \mathfrak{h} \oplus \mathfrak{n}_+ \cong \mathfrak{b}_- \oplus \mathfrak{n}_+$ and denote the corresponding Chevalley generators by $\{f, h, e\}$. Following [BGG2], define the base affine space of $SL(2,\mathbb{C})$ as the quotient $A = A(SL(2,\mathbb{C})) = N_+ \backslash SL(2,\mathbb{C})$, where N_+ is the subgroup generated by $\mathfrak{n}_+$. Under the geometric action of $(\mathfrak{sl}_2)_L \oplus (\mathfrak{sl}_2)_R$ by left- and right-invariant vector fields, $\mathcal{E}(SL(2,\mathbb{C}))$ decomposes as

$$\mathcal{E}(SL(2,\mathbb{C})) \cong \bigoplus_{\Lambda \in P_+} \mathcal{L}(\Lambda^*) \otimes \mathcal{L}(\Lambda)\,, \tag{1.10}$$

where $\mathcal{L}(\Lambda)$ is the irreducible finite dimensional $\mathfrak{sl}_2$ module with highest (dominant integral) weight Λ. It immediately follows that

$$\mathcal{E}(A) \cong \bigoplus_{\Lambda \in P_+} \mathbb{C}_{\Lambda^*} \otimes \mathcal{L}(\Lambda) . \tag{1.11}$$

The identification of $\mathcal{E}(A)$ with $\mathcal{C}_2$ is obvious, and using the construction in Appendix G we may make it explicit. Thereto, identify $\mathcal{E}(SL(2,\mathbb{C}))$ with polynomials in the matrix elements g_{ij}, $g_{11}g_{22} - g_{12}g_{21} = 1$, on which $(\mathfrak{sl}_2)_R$ acts as

$$\begin{aligned} \Pi^R(e) &= g_{11}\frac{\partial}{\partial g_{12}} + g_{21}\frac{\partial}{\partial g_{22}} , \\ \Pi^R(h) &= g_{11}\frac{\partial}{\partial g_{11}} - g_{22}\frac{\partial}{\partial g_{22}} + g_{21}\frac{\partial}{\partial g_{21}} - g_{12}\frac{\partial}{\partial g_{12}} , \\ \Pi^R(f) &= g_{12}\frac{\partial}{\partial g_{11}} + g_{22}\frac{\partial}{\partial g_{21}} . \end{aligned} \tag{1.12}$$

The expressions for Π^L, the realization of $(\mathfrak{sl}_2)_L$, are identical but for $g_{11} \leftrightarrow g_{22}$ and an overall minus sign in all generators. The elements $\Phi_\Lambda(g) = (g_{21})^{2j} \in \mathcal{E}(SL(2,\mathbb{C}))$, $(\Lambda, \alpha) = 2j = 0, 1, 2, \ldots$, correspond to the highest weight states (which we will denote by v_Λ) in the decomposition (1.10). The remaining states of the representation are obtained by the action of the lowering operators given above. Then $\mathcal{E}(A) \cong \mathcal{C}_2$ is explicit through the identification $x \equiv g_{21}$ and $y \equiv g_{22}$.

The polyvectors discussed above have a natural geometric realization as polyvectors on A. The space of regular polyvectors on A, $\mathcal{P}(A)$, can be identified as the regular sections of the homogenous vector bundle $SL(2,\mathbb{C}) \times_{N_+} \bigwedge \mathfrak{b}_-$, where we consider $\bigwedge \mathfrak{b}_-$ as an $\mathfrak{n}_+$ module through the identification $\bigwedge \mathfrak{b}_- \cong \bigwedge(\mathfrak{n}_+ \backslash \mathfrak{sl}_2)$. Put another way, if $\mathcal{E}(SL(2,\mathbb{C})) \otimes \bigwedge \mathfrak{b}_-$ denotes the space of regular functions on $SL(2,\mathbb{C})$ with values in $\bigwedge \mathfrak{b}_-$, then $\mathcal{P}(A)$ is simply given by the $\mathfrak{n}_+$-invariant elements under the natural $\mathfrak{n}_+$ action; i.e.,

$$\mathcal{P}(A) \cong (\mathcal{E}(SL(2,\mathbb{C})) \otimes \bigwedge \mathfrak{b}_-)^{\mathfrak{n}_+} . \tag{1.13}$$

Clearly there is a natural grading on $\mathcal{P}(A) = \bigoplus_n \mathcal{P}^n(A)$, induced from the decomposition

$$\bigwedge \mathfrak{b}_- \cong \bigoplus_{n=0}^{2} \bigwedge^n \mathfrak{b}_- , \tag{1.14}$$

and it has a natural structure of a graded commutative algebra induced from that of $\mathcal{E}(SL(2,\mathbb{C})) \otimes \bigwedge \mathfrak{b}_-$. In fact, $\mathcal{P}(A)$ carries a BV algebra structure – as is clear since it is isomorphic to that of $\mathcal{P}(\mathcal{C}_2)$. Again it may be convenient to use an explicit realization, identifying $\mathfrak{b}_-$ as the Fock space of the "ghost oscillators" $\{b^a, c_a\}$, $a \in \{0, -\}$, with vacuum $|bc\rangle$ satisfying $b^a|bc\rangle = 0$. The action of $\mathfrak{n}_+$ is then realized by

$$\Pi^{bc}(e) = c_0 b^- , \tag{1.15}$$

and the condition of $\mathfrak{n}_+$ invariance for a given $\Phi \in \mathcal{E}(SL(2,\mathbb{C})) \otimes \bigwedge \mathfrak{b}_-$ is simply

$$\Pi^L(x)\Phi = -\Pi^{bc}(x)\Phi , \qquad x \in \mathfrak{n}_+ . \tag{1.16}$$

At this point we have just obtained a geometric rewriting of previously known results. But now observe that in the present context the polyvectors can be identified as the zeroth order cohomology of $\mathfrak{n}_+$ with coefficients in $\mathcal{E}(SL(2,\mathbb{C}))\otimes \bigwedge\mathfrak{b}_-$. Thus it is natural to consider the algebra, $\mathrm{BV}[\mathfrak{sl}_2]$, defined by the full cohomology,

$$\mathrm{BV}[\mathfrak{sl}_2] \equiv H(\mathfrak{n}_+,\mathcal{E}(SL(2,\mathbb{C}))\otimes \textstyle\bigwedge\mathfrak{b}_-). \tag{1.17}$$

The main result of this section is that this algebra provides a model of the chiral operator algebra of the 2D string.

Theorem 1.5 $\mathrm{BV}[\mathfrak{sl}_2]$ *is a graded, graded commutative algebra with the product "$\cdot$" induced from the product on the underlying complex. Further, it has a BV algebra structure and is acyclic with respect to the corresponding BV operator, Δ. Moreover,* $(\mathrm{BV}[\mathfrak{sl}_2],\,\cdot\,,\Delta)$ *is isomorphic to* $(\mathfrak{H},\,\cdot\,,b_0)$ *as a BV algebra.*

Remark. For a complete calculation of the cohomology and a complete discussion of the BV algebra structure, the reader should refer to Chap. 4. For this introduction we content ourselves with a sketch of the proof of Theorem 1.5, with view to making it somewhat more explicit.

The computation of $\mathrm{BV}[\mathfrak{sl}_2]$. Introduce the BRST ghost oscillators corresponding to $\mathfrak{n}_+$, $\{\sigma^+,\omega_+\}$, and the associated ghost Fock space $F^{\sigma\omega}$ with vacuum $|\sigma\omega\rangle$ satisfying $\omega_+|\sigma\omega\rangle=0$. The cohomology $H(\mathfrak{n}_+,\mathcal{E}(SL(2,\mathbb{C}))\otimes\bigwedge\mathfrak{b}_-)$ may now be computed as the cohomology of the differential

$$d_B = \sigma^+\left(\Pi^L(e)+\Pi^{bc}(e)\right) \tag{1.18}$$

acting on the complex $\mathcal{C}(SL(2,\mathbb{C}))\equiv\mathcal{E}(SL(2,\mathbb{C}))\otimes F^{bc}\otimes F^{\sigma\omega}$. Explicitly,

$$d_B = \sigma^+\left(-g_{21}\frac{\partial}{\partial g_{11}}-g_{22}\frac{\partial}{\partial g_{12}}+c_0b^-\right). \tag{1.19}$$

The complex $\mathcal{C}(SL(2,\mathbb{C}))$ is bi-graded by the $\sigma\omega$ and the bc ghost numbers, with d_B of degree $(1,0)$. Clearly, this bi-degree passes to the cohomology. We will write $H^n(\mathfrak{n}_+,\mathcal{E}(SL(2,\mathbb{C}))\otimes\bigwedge\mathfrak{b}_-)$ for the cohomology in *total* ghost number n. Similarly, we will write $\mathrm{BV}^n[\mathfrak{sl}_2]$. When it is important to distinguish the bi-grading, we will write, e.g., $\mathrm{BV}^{(m,n)}[\mathfrak{sl}_2]$.

Since $\mathcal{E}(SL(2,\mathbb{C}))$ decomposes into irreducible $\mathfrak{sl}_2$ modules under the right action of $\mathfrak{sl}_2$, as in (1.10), we need just compute the cohomology on a $\mathcal{L}(\Lambda)$. A straightforward calculation yields the eight cohomology states listed in Table 1.2 for $j>0$, while for $j=0$ the states in the second position in the $m=1$ column are missing. Each state in Table 1.2 corresponds to a $2j+1$-dimensional $(\mathfrak{sl}_2)_R$ multiplet obtained by acting with $\Pi^R(f)$. The full cohomology is then found as the direct sum of this result over all irreducible modules, $\mathcal{L}(\Lambda)$. It is straightforward to write explicit formulae for the result in terms of the matrix elements g_{ij}.

The BV structure of $\mathrm{BV}[\mathfrak{sl}_2]$. The BV structure of interest is an extension of that of $\mathcal{P}(A)$ to a BV operator, Δ with trivial homology. The explicit Δ we take is

Table 1.2. The highest weight cohomology states for $j > 0$

$n\backslash m$	0	1	2
0	v_Λ	$c_0 v_\Lambda, (c_- - \frac{1}{2j}c_0 f) v_\Lambda$	$c_0 c_- v_\Lambda$
1	$\sigma^+ f^{2j} v_\Lambda$	$\sigma^+ c_- f^{2j} v_\Lambda, \sigma^+ c_0 f^{2j} v_\Lambda$	$\sigma^+ c_0 c_- f^{2j} v_\Lambda$

$$\Delta = -b^0(\Pi^L(h) - 2\sigma^+\omega_+ + 2) - b^- \Pi^L(f) + 2c_- b^- b^0 + 2\sigma^+ b^- b^0. \quad (1.20)$$

Then we have

Lemma 1.6 *The operator Δ is acyclic on* $\mathrm{BV}[\mathfrak{sl}_2]$.

Proof. We first decompose Δ under the assignment of degree zero to $\{g_{ij}, c_0, c_-\}$, and degree one to σ^+ (conjugate operators have opposite degree). Thus $\Delta = \Delta_0 + \Delta_1$, where the subscript denotes the additive degree of the operator,

$$\begin{aligned}
\Delta_0 &= b^0(-\Pi^L(h) + 2\sigma^+\omega_+ + 2c_- b^- - 2) - b^- \Pi^L(f), \\
\Delta_1 &= 2\sigma^+ b^- b^0.
\end{aligned} \quad (1.21)$$

Now note that

$$[\Delta_0, c_0]_+ = -\Pi^L(h) + 2\sigma^+\omega_+ + 2c_- b^- - 2, \quad (1.22)$$

and observe from Table 1.2 that the right hand side only vanishes for the two states $\Omega = c_0 c_-$ (the volume element) and $\Omega_{-1} = \sigma^+$. Thus, by the standard argument, these are the only nontrivial homology for Δ_0 (that they are nontrivial is clear by explicit computation). Since Δ_1 is nilpotent and $\Delta_1 \Omega = \Delta\Omega = 2\Omega_{-1}$, the homology of Δ is therefore trivial on $\mathrm{BV}[\mathfrak{sl}_2]$ (cf. Sect. 4.6.4). □

The BV isomorphism. As discussed above, $\mathrm{BV}^{(0,0)}[\mathfrak{sl}_2] \sim \mathcal{C}_2$, polynomials in x and y identified as group elements. From the calculation of cohomology and the G algebra structure induced from the BV operator, one easily shows that

$$V_x \equiv -c_- g_{22} - c_0 g_{12}, \quad V_y \equiv c_- g_{21} + c_0 g_{11}. \quad (1.23)$$

are cohomology states identified with $\frac{\partial}{\partial x}$ and $\frac{\partial}{\partial y}$, respectively. The dot algebra isomorphism $(\bigoplus_m \mathrm{BV}^{(0,m)}[\mathfrak{sl}_2], \cdot) \cong (\mathcal{P}(\mathcal{C}_2), \cdot)$ is then manifest from the explicit computation of cohomology. Further, the BV algebra homomorphism π discussed in Sect. 1.3.2 is clearly identified as the projection to zero $\sigma\omega$ ghost number.

It is natural to describe $\mathfrak{I} \equiv \mathrm{Ker}\,\pi$, the cohomology at nonzero $\sigma\omega$ ghost number, in terms of $\mathcal{P}(\mathcal{C}_2)$ modules. We will first identify the twisted ground ring module in terms of the zero bc ghost number cohomology. The state $\Omega_{-1} = \sigma^+$ is the singlet state in $\mathrm{BV}^{(1,0)}[\mathfrak{sl}_2]$. The remaining states in this subspace are obtained by taking multiple brackets with $\frac{\partial}{\partial x}$ and $\frac{\partial}{\partial y}$. The identification

$\mathrm{BV}^{(1,0)}[\mathfrak{sl}_2] \cong M_{-1}$ is completed by the observation that $x \cdot \Omega_{-1}$ and $y \cdot \Omega_{-1}$ are trivial in d_B cohomology. Then it is again manifest from the explicit calculation that $\bigoplus_m \mathrm{BV}^{(1,m)}[\mathfrak{sl}_2] \cong \mathcal{P}(C_2, M_{-1})$.

Thus the structure of $(\mathrm{BV}[\mathfrak{sl}_2], \cdot)$ is precisely isomorphic to that of $(\mathfrak{H}, \cdot)$, with

$$\mathfrak{H}_+ \cong \bigoplus_{m=0}^{2} \mathrm{BV}^{(0,m)}[\mathfrak{sl}_2], \qquad (1.24)$$

and

$$\mathfrak{H}_- \cong \bigoplus_{m=0}^{2} \mathrm{BV}^{(1,m)}[\mathfrak{sl}_2]. \qquad (1.25)$$

Note, in particular, that the vanishing of dot products and brackets within $\mathfrak{H}_-$ is a consequence of the ghost structure. The final identification as BV algebras uses Lemma 1.6 to explicitly decompose $\mathrm{BV}[\mathfrak{sl}_2]$ into doublets under the action of Δ. This can be directly transcribed to the corresponding decomposition of $\mathfrak{H}$.

2 $\mathcal{W}$ Algebras and Their Modules

2.1 $\mathcal{W}$ Algebras

2.1.1 Introduction to $\mathcal{W}$ Algebras

$\mathcal{W}$ algebras are certain nonlinear, higher spin extensions of the 2-dimensional conformal algebra, i.e., the Virasoro algebra. They were first introduced by Zamolodchikov [Za] and have subsequently been investigated by many people (see, e.g., [BoSc1,BoSc2], and references therein). The proper mathematical setting is that of "Vertex Operator Algebras" (VOAs) [FLM] or, equivalently, "Meromorphic Conformal Field Theory" [Go].

The simplest $\mathcal{W}$ algebras are the algebras $\mathcal{W}[\mathfrak{g}]$ associated to a simple, simply-laced Lie algebra $\mathfrak{g}$, either by Drinfel'd-Sokolov reduction or by a coset-construction. They have generators of conformal dimension equal to the orders of the independent Casimir operators of $\mathfrak{g}$. In particular $\mathcal{W}_N \equiv \mathcal{W}[\mathfrak{sl}_N]$ has $N-1$ generators of dimension $2, 3, \ldots, N$.

In this paper we restrict our attention to the simplest nonlinear $\mathcal{W}$ algebra, namely $\mathcal{W}_3$, although most of the results continue to hold for the more general algebras $\mathcal{W}[\mathfrak{g}]$. We formulate many of our results using generic Lie algebra notation so that the generalization to $\mathcal{W}[\mathfrak{g}]$ should be obvious.

2.1.2 The $\mathcal{W}_3$ Algebra

The $\mathcal{W}_3$ algebra with central charge $c \in \mathbb{C}$ can be defined as the quotient of the universal enveloping algebra of the free Lie algebra generated by L_m, W_m, $m \in \mathbb{Z}$, by the ideal generated by the following commutation relations

$$\begin{aligned}
[L_m, L_n]_- &= (m-n)L_{m+n} + \tfrac{c}{12}m(m^2-1)\delta_{m+n,0}\,, \\
[L_m, W_n]_- &= (2m-n)W_{m+n}\,, \\
[W_m, W_n]_- &= (m-n)\left(\tfrac{1}{15}(m+n+3)(m+n+2) - \tfrac{1}{6}(m+2)(n+2)\right)L_{m+n} \\
&\quad + \beta(m-n)\Lambda_{m+n} + \tfrac{c}{360}m(m^2-1)(m^2-4)\delta_{m+n,0}\,,
\end{aligned} \tag{2.1}$$

where $\beta = 16/(22+5c)$ and

$$\Lambda_m = \sum_{n\leq -2} L_n L_{m-n} + \sum_{n>-2} L_{m-n}L_n - \tfrac{3}{10}(m+3)(m+2)L_m\,. \tag{2.2}$$

Equivalently, one can introduce fields $T(z)$ and $W(z)$ (i.e., formal power series in $\mathcal{W}_3[[z, z^{-1}]]$) by

$$T(z) = \sum_{m\in\mathbb{Z}} L_m z^{-m-2}\,, \qquad W(z) = \sum_{m\in\mathbb{Z}} W_m z^{-m-3}, \tag{2.3}$$

in terms of which (2.1) can be translated into so-called "Operator Product Expansions" (OPEs) (see, e.g., [BPZ] for an early discussion of the use of OPEs in conformal field theory, and [Br,FHL,FLM] for the mathematical theory)

$$\begin{aligned}
T(z)T(w) &= \frac{c/2}{(z-w)^4} + \frac{2T(w)}{(z-w)^2} + \frac{\partial T(w)}{z-w} + \ldots\,, \\
T(z)W(w) &= \frac{3W(w)}{(z-w)^2} + \frac{\partial W(w)}{z-w} + \ldots\,, \\
W(z)W(w) &= \frac{c/3}{(z-w)^6} + \frac{2T(w)}{(z-w)^4} + \frac{\partial T(w)}{(z-w)^3} \\
&\quad + \frac{1}{(z-w)^2}\left(2\beta\Lambda(w) + \tfrac{3}{10}\partial^2 T(w)\right) \\
&\quad + \frac{1}{(z-w)}\left(\beta\partial\Lambda(w) + \tfrac{1}{15}\partial^3 T(w)\right) + \ldots\,,
\end{aligned} \tag{2.4}$$

where

$$\Lambda(z) = \sum_{m\in\mathbb{Z}} \Lambda_m z^{-m-3} = (TT)(z) - \tfrac{3}{10}\partial^2 T(z)\,. \tag{2.5}$$

It is useful to split the $\mathcal{W}_3$ generators into three groups according to their modings. Let

$$\mathcal{W}_{3,\pm} \equiv \{L_m, W_m \mid \pm m > 0\}\,, \qquad \mathcal{W}_{3,0} \equiv \{L_0, W_0\}\,. \tag{2.6}$$

Note that while the generators in $\mathcal{W}_{3,0}$ form an (Abelian) subalgebra of $\mathcal{W}_3$, the so-called Cartan subalgebra, the generators in $\mathcal{W}_{3,\pm}$ do not form a subalgebra. The action of $\mathcal{W}_{3,0}$ on the generators of $\mathcal{W}_3$ is explicitly given by

$$\begin{aligned}
[L_0, L_n]_- &= -n\, L_n\,, \\
[L_0, W_n]_- &= -n\, W_n\,, \\
[W_0, L_n]_- &= -2n\, W_n\,, \\
[W_0, W_n]_- &= -\tfrac{1}{15}n(n^2-4)L_n - \beta n \Lambda_n\,.
\end{aligned} \tag{2.7}$$

From here we easily see that it is impossible to diagonalize the action of $\mathcal{W}_{3,0}$ on $\mathcal{W}_3$, i.e., we do not have a root space decomposition with respect to the Cartan subalgebra. It should, however, be possible to find a generalized root space decomposition, i.e., a basis of $\mathcal{W}_3$ in which the action of $\mathcal{W}_{3,0}$ is in Jordan normal form. To our knowledge no such basis has been explicitly constructed yet, and we consider this an important open problem. Later, when we discuss modules of $\mathcal{W}_3$, we encounter the same problem of nondiagonalizability of the Cartan subalgebra, and we will discuss this issue in more detail. Here it suffices to remark that this nondiagonalizability is one of the major differences with, for example, affine Lie algebras or the Virasoro algebra, and is reflected in a much more subtle and complicated submodule structure. As a consequence, both the construction of resolutions and the calculation of semi-infinite cohomology are correspondingly more difficult.

2.2 $\mathcal{W}_3$ Modules

2.2.1 The Category $\mathcal{O}$

To prove properties of $\mathcal{W}_3$ modules in some generality, we first need to define a proper category of modules – henceforth referred to as the category $\mathcal{O}$. This category should be small enough to allow for certain "nice" properties, e.g., the existence of Jordan-Hölder series and the existence of a semi-infinite cohomology. On the other hand the category should be big enough to incorporate the (physically) interesting modules, such as free field Fock spaces and Verma modules, and to allow for the existence of certain homological constructions, e.g., resolutions of irreducible modules, within the category. In addition one usually requires that the category is closed under certain basic operations such as taking direct sums, tensor products and contragredients.

For (affine) Lie algebras $\mathfrak{g}$ the category $\mathcal{O}$ is, loosely speaking, the category of $\mathfrak{h}$-diagonalizable modules with finite-dimensional weight spaces and weights bounded from above [BGG2,Ka]. To require $\mathcal{W}_{3,0}$ diagonalizability for $\mathcal{W}_3$ modules is too strong a requirement, however. As we will see later, as a direct consequence of (2.7), in general not even Verma modules are W_0 diagonalizable. We thus only require the modules to be L_0 diagonalizable.[1] Moreover, since L_0 is identified with the energy operator, we require the L_0 eigenvalues to be bounded from below. For an L_0 diagonalizable module V, let $V_{(h)} = \{v \in V \,|\, L_0\, v = h\, v\}$ be its eigenspaces, such that $V = \coprod_{h\in\mathbb{C}} V_{(h)}$. Let $P(V) = \{h \in \mathbb{C} \,|\, V_{(h)} \neq 0\}$ denote the set of L_0 eigenvalues of V. We refer to the L_0 eigenvalue of a state $v \in V$ as its "L_0 level."

Definition 2.1 [FKW] *The category $\mathcal{O}$, of "positive-energy $\mathcal{W}$ modules," is the set of L_0 diagonalizable modules V such that each L_0 eigenspace $V_{(h)}$ is finite-dimensional, and for which there exist a finite set $h_1, \ldots, h_s \in \mathbb{C}$ ($h_i \not\equiv h_j \bmod 1$ for all i, j), such that $P(V) \subset \bigcup_{i=1}^{s} \bigcup_{N\in\mathbb{Z}_{\geq 0}} \{h \in \mathbb{C} \,|\, h = h_i + N\}$.*

It is clear that (finite) direct sums, submodules and quotients of modules in $\mathcal{O}$ are again in $\mathcal{O}$. Generally, however, contrary to the Lie algebra case, tensor products of modules in $\mathcal{O}$ are not in $\mathcal{O}$ for the simple reason that – due to the nonlinear nature of the $\mathcal{W}_3$ algebra – they do not carry the structure of a $\mathcal{W}_3$ module.

[1] Modules with a nondiagonalizable action of the Virasoro generator L_0 also exist (see, e.g., [Gu,RoSa]), and have some important applications. However, we do not need them for the purpose of this paper, so we do not include them in the definition of the category $\mathcal{O}$.

Despite the fact that W_0 need not be diagonalizable on $V \in \mathcal{O}$, we do of course have a generalized eigenspace decomposition (Jordan normal form) of W_0 on each $V_{(h)}$. We have $V_{(h)} = \coprod_{w \in \mathbb{C}} V_{(h,w)}$, where, for $w \in \mathbb{C}$, we have denoted by $V_{(h,w)} = \{v \in V_{(h)} \,|\, \exists N \in \mathbb{N} : (W_0 - w)^N v = 0\}$ the generalized eigenspaces of W_0.

Within each Jordan block of $V_{(h,w)}$ we may choose a basis $\{v_0, \ldots, v_{\kappa-1}\}$ such that $(W_0 - w)\, v_i = v_{i-1}$ for $i = 1, \ldots, \kappa - 1$ and $(W_0 - w)\, v_0 = 0$. That is, with respect to this basis

$$W_0 = \begin{pmatrix} w & 1 & & & \\ & w & 1 & & \\ & & \ddots & \ddots & \\ & & & w & 1 \\ & & & & w \end{pmatrix}. \tag{2.8}$$

For such a basis, we also use the notation

$$v_{\kappa-1} \stackrel{W_0 - w}{\longrightarrow} v_{\kappa-2} \stackrel{W_0 - w}{\longrightarrow} \cdots \stackrel{W_0 - w}{\longrightarrow} v_0 . \tag{2.9}$$

The (generalized) character ch_V of a module $V \in \mathcal{O}$ is now defined as

$$\mathrm{ch}_V(q,p) = \mathrm{Tr}_V \left(q^{L_0} p^{W_0}\right) = \sum_{(h,w)\in\mathbb{C}^2} \dim_{\mathbb{C}} \left(V_{(h,w)}\right) q^h p^w . \tag{2.10}$$

We have

Theorem 2.2 *The characters ch_L, where L runs over the set of irreducible modules in $\mathcal{O}$ are linearly independent over $\mathbb{C}$, in particular $\mathrm{ch}_L = \mathrm{ch}_{L'}$ iff $L \cong L'$.*

Proof. The proof of this theorem is given after Theorem 2.14. □

Unfortunately, no explicit expression for the generalized character is known for any $\mathcal{W}_3$ module (with the exception of the trivial module, of course). For most purposes it suffices, however, to consider the specialization $\mathrm{ch}_V(q,1)$ of the character. Expressions for these specialized characters are known for most interesting modules in $\mathcal{O}$.

Clearly, since $\mathcal{W}_3$ is $\mathbb{Z}$ graded, any $V \in \mathcal{O}$ decomposes into a direct sum $V = \bigoplus_i V(h_i)$ of $\mathcal{W}_3$ modules, where h_i are the elements in Definition 2.1, and the L_0 eigenvalues of $V(h_i)$ are concentrated on strings $h_i + N$, $N \in \mathbb{Z}_{\geq 0}$. For all practical purposes, we may thus equally well consider modules built on a single h_i only. Let us denote this subcategory by $\mathcal{O}_{(h_i)}$.

An important ingredient in unraveling the structure of a module $V \in \mathcal{O}$ is to consider filtrations of V by submodules, in particular so-called "composition series" or "Jordan-Hölder series" that are characterized by the condition that quotients of subsequent terms in the filtration should be irreducible. As opposed to the finite-dimensional setting, such composition series do not, in general, exist for infinite-dimensional algebras. We need a slight modification of the usual construction; namely, a "cut-off" which renders the filtrations finite. In complete analogy with the affine Lie algebra case we have (see [Ka])

Theorem 2.3 *Every module $V \in \mathcal{O}_{(h)}$ possesses a composition series (or Jordan-Hölder series)* JH(V)*; i.e., for all $N \in \mathbb{Z}_{\geq 0}$, there exists a (finite) filtration by submodules of V (denoted by* $\mathrm{JH}_N(V)$*)*

$$V = V_0 \supset V_1 \supset \ldots \supset V_s = 0, \tag{2.11}$$

and a subset $I \subset \{0, \ldots, s-1\}$, such that

i. V_i/V_{i+1} is irreducible for $i \in I$,
ii. $\coprod_{M \leq N} (V_i/V_{i+1})_{(h+M)} = 0$ for $i \notin I$.

Proof. The proof (which requires some results on highest weight modules to be discussed in Sect. 2.2.2) parallels the one given in [Ka] with one minor modification; namely, the maximal element has to be chosen to be an eigenvector of W_0. That this can always be done is obvious. See [Ka] for more details. □

For any $V \in \mathcal{O}_{(h)}$, only irreducible modules $L \in \mathcal{O}_{(h)}$ appear as quotients in the composition series of V. So, for any such $L \in \mathcal{O}_{(h)}$, choose $M \in \mathbb{Z}_{\geq 0}$ such that $P(L) \subset h + M + \mathbb{Z}_{\geq 0}$ and $V_{(h+M)} \neq 0$. Then, choose any $N \geq M$ and denote by $(V:L)$ the multiplicity with which the irreducible module L appears in the composition series $\mathrm{JH}_N(V)$. Clearly, $(V:L)$ is independent of the choice of $N \geq M$. Moreover,

Theorem 2.4 *We have*

$$\mathrm{ch}_V = \sum_L (V:L)\, \mathrm{ch}_L \,. \tag{2.12}$$

Proof. As in [Ka]. □

Remarks

i. The characters ch_L of the irreducible modules L are not only independent (Theorem 2.2), but they also span the space of characters ch_V, $V \in \mathcal{O}$.
ii. The fact that we would like quotients of subsequent terms in the composition series to be irreducible forced us to introduce additional terms whose quotients do not contribute states up to the level we are interested in. This can be avoided by the following modification of Theorem 2.3, which is easily seen to be equivalent to the original. Let us say that a $\mathcal{W}_3$ module $V \in \mathcal{O}_{(h)}$ is "irreducible up to level N" if for all proper submodules $W \subset V$ we have $W \bigcap (\coprod_{k \leq N} V_{(h+k)}) = 0$. Then, for every $V \in \mathcal{O}_{(h)}$ and $N \in \mathbb{Z}_{\geq 0}$, we have a filtration (2.11) of V such that each V_i/V_{i+1} is irreducible up to level N. We give examples of such filtrations in Sect. 2.3.2.

Definition 2.5 *Let $V \in \mathcal{O}$.*

i. A vector $v \in V$ is called "primitive" if there exists a proper submodule $U \subset V$ such that $\mathcal{W}_{3,+} \cdot v \subset U$ while $v \notin U$.
ii. A vector $v \in V$ in called "pseudo-singular" (or p-singular, for short) if $\mathcal{W}_{3,+} \cdot v = 0$.
iii. A vector $v \in V$ is called "singular" if it is p-singular and a $\mathcal{W}_{3,0}$ eigenvector.

Let us denote the set of singular, p-singular and primitive vectors in $V \in \mathcal{O}$ by $\mathrm{Sing}(V)$, $\mathrm{pSing}(V)$ and $\mathrm{Prim}(V)$, respectively, and let SV be the module generated by all p-singular vectors in V. Clearly, $\mathrm{Prim}(V) \supset \mathrm{pSing}(V) \supset \mathrm{Sing}(V)$. Also, a primitive vector $v \in V$ becomes p-singular in the quotient module V/U.

While the notion and use of a singular vector is probably well-known from the Virasoro analogue, the notion of a primitive vector might be less familiar. It is, nevertheless, rather useful. In particular, after examining in more detail the collection of irreducible modules in $\mathcal{O}$, we establish a 1–1 correspondence between primitive vectors in a module $V \in \mathcal{O}$ and the set of irreducible modules L occuring in the composition series $\mathrm{JH}(V)$ of V (see Lemma 2.15).

The last property of modules in $\mathcal{O}$ that we discuss before proceeding to more explicit examples is that to every module $V \in \mathcal{O}$ there is associated a contragredient module $\overline{V} \in \mathcal{O}$. Hereto, let $\omega_{\mathcal{W}}$ be the $\mathbb{C}$-linear anti-involution of $\mathcal{W}_3$ defined by[2]

$$\omega_{\mathcal{W}}(L_n) = L_{-n}\,, \qquad \omega_{\mathcal{W}}(W_n) = W_{-n}\,. \tag{2.13}$$

For future use, note that (2.13) is equivalent to the anti-involution acting on fields in $\mathcal{W}_3[[z, z^{-1}]]$ defined by

$$\omega_{\mathcal{W}}(T(z)) = z^{-4}\,T(z^{-1})\,, \qquad \omega_{\mathcal{W}}(W(z)) = z^{-6}\,W(z^{-1})\,. \tag{2.14}$$

Definition 2.6 *Consider $V \in \mathcal{O}$. The "contragredient module" $\overline{V}$ is defined, as a vector space, to be $\overline{V} = \coprod_{N\in\mathbb{Z}} \mathrm{Hom}_{\mathbb{C}}(V_{(h+N)}, \mathbb{C})$. The $\mathcal{W}_3$ module structure is given by*

$$x\,f(v) \equiv f(\omega_{\mathcal{W}}(x)\,v)\,, \tag{2.15}$$

where $f \in \overline{V}$, $v \in V$ and $x \in \mathcal{W}_3$.

Clearly, for $V \in \mathcal{O}$, the contragredient module $\overline{V}$ is again in $\mathcal{O}$. For comparison, note that the module dual to V has L_0 eigenvalues bounded from *above* and thus is not in $\mathcal{O}$ [FeFu].

Lemma 2.7 *Let $V, W \in \mathcal{O}$ and suppose $\pi \in \mathrm{Hom}_{\mathcal{W}}(V, W)$. Then, there exists a $\overline{\pi} \in \mathrm{Hom}_{\mathcal{W}}(\overline{W}, \overline{V})$ defined by $\overline{\pi}(\overline{w})(v) = \overline{w}(\pi(v))$ for all $v \in V, \overline{w} \in \overline{W}$.*

---

[2] In later sections we also need the anti-linear anti-involution defined by these relations. We denote it by $\overline{\omega}_{\mathcal{W}}$.

2.2.2 (Generalized) Verma Modules

Important examples of modules in $\mathcal{O}$ are the so-called "Verma modules" or, more generally, "highest weight modules." In this section we recall their definitions and some important properties. It turns out that we will need modules which are slightly more general than Verma modules, the so-called "generalized Verma modules." However, their structure theory can be developed along the same lines as for Verma modules, so we just restrict ourselves to stating the analogous theorems.

Definition 2.8 *A $\mathcal{W}_3$ module $V \in \mathcal{O}$ is called a highest weight module with highest weight $(h, w) \in \mathbb{C}^2$ if there exists a nonzero vector $v_V \in V$, the so-called "highest weight vector," such that*

$$\mathcal{W}_{3,+} \cdot v_V = 0\,, \qquad L_0\, v_V = h\, v_V\,, \qquad W_0\, v_V = w\, v_V\,, \tag{2.16}$$

and

$$V \cong \mathcal{W}_3 \cdot v_V\,. \tag{2.17}$$

Remark. Note that a "module with highest weight" is a module that satisfies (2.16), but not necessarily (2.17). For example, Fock spaces – to be discussed in Sect. 2.2.3 – are modules with highest weight, but are *not* highest weight modules in general.

Verma modules are highest weight modules which are, in a sense, maximal. Namely,

Definition 2.9 *A Verma module $M(h, w, c)$ is the module "induced" by the action of $\mathcal{W}_{3,-}$ from a highest weight vector v_M of highest weight (h, w), i.e., the $\mathcal{W}_3$ module $\mathcal{W}_3 \cdot v_M$, divided by the ideal generated by the relations*

$$\mathcal{W}_{3,+} \cdot v_M \;=\; 0\,, \qquad L_0\, v_M \;=\; h\, v_M\,, \qquad W_0\, v_M \;=\; w\, v_M\,. \tag{2.18}$$

The action of $\mathcal{W}_3$ on (2.18) is determined by the commutation relations (2.1) and equation (2.18).

One of the most important properties of Verma modules is their co-universality.

Lemma 2.10 *Let $V \in \mathcal{O}$ and let $v_0 \in V$ be a singular vector of weight (h, w). Then there exists a unique $\mathcal{W}_3$ homomorphism $\pi \in \mathrm{Hom}_{\mathcal{W}_3}(M(h, w, c), V)$ such that $\pi(v_M) = v_0$ where v_M is the highest weight vector of $M(h, w, c)$.*

Proof. Clearly, π is uniquely defined by $\pi(x\, v_M) = x\, \pi(v_M) = x\, v_0$ for all $x \in \mathcal{W}_3$. □

And, as an immediate consequence, we have

Corollary 2.11 *Every highest weight module $V \in \mathcal{O}$ with highest weight $(h,w) \in \mathbb{C}^2$ is a quotient of the Verma module $M(h,w,c)$.*

Proof. Let v_M be the highest weight vector of $M(h,w,c)$, and let v_V be the highest weight vector of V. Lemma 2.10 provides us with a unique $\mathcal{W}_3$ homomorphism $M(h,w,c) \xrightarrow{\pi} V$ such that $\pi(v_M) = v_V$. Clearly, because of (2.17), π is an epimorphism. Let $K = \operatorname{Ker}\pi$, then we have $V \cong M(h,w,c)/K$. □

For many purposes it is useful to have an explicit basis of the Verma module. Clearly, the set of all monomials

$$e_{m_1\ldots m_K;n_1\ldots n_L} = L_{-m_1}\cdots L_{-m_K}W_{-n_1}\cdots W_{-n_L}\,v\,, \tag{2.19}$$

with $m_i, n_j \in \mathbb{Z}_{>0}$, $K, L \in \mathbb{Z}_{\geq 0}$, form a basis of $M(h,w,c)$, but this basis is overcomplete. One may, instead, find a linearly independent set.

Theorem 2.12 [Wa2] (Poincaré-Birkhoff-Witt) *The vectors $e_{m_1\ldots m_K;n_1\ldots n_L}$ with $m_1 \geq \ldots \geq m_{K-1} \geq m_K > 0$, $n_1 \geq \ldots \geq n_{L-1} \geq n_L > 0$ constitute a set of independent basis vectors of the Verma module $M(h,w,c)$.*

Remark. Similarly, the vectors

$$\overline{e}_{m_1\ldots m_K;n_1\ldots n_L}\,, \qquad m_1 \geq \ldots \geq m_K > 0\,, \quad n_1 \geq \ldots \geq n_L > 0\,, \tag{2.20}$$

dual to $e_{m_1\ldots m_K;n_1\ldots n_L}$, constitute a basis for the contragredient Verma module $\overline{M}(h,w,c)$.

The idea of the proof is quite standard. Note that the theorem would be straightforward if the algebra was Abelian. The idea is thus to try to reduce the problem to that of an Abelian algebra and then show that the correction terms are immaterial. We present the details of the proof since similar ideas are most crucial in later sections; e.g., in the computation of a certain semi-infinite cohomology using the Koszul complex of the Abelianized algebra as a starting point.

Proof of Theorem 2.12. We define a grading of $\mathcal{W}_3$ by $\deg(L_n) = 1$ and $\deg(W_n) = 2$.[3] Note that, with respect to this grading, the degrees of terms on the right hand side of the commutator (2.1) are strictly less than the degree of the left hand side. Similarly, we associate a degree to the monomials (2.19) by $\deg\left(e_{m_1\ldots m_K;n_1\ldots n_L}\right) = K + 2L$. Now observe that for any permutation $\sigma \in S_K$ and $\sigma' \in S_L$ we have

$$e_{m_{\sigma(1)}\ldots m_{\sigma(K)};n_{\sigma'(1)}\ldots n_{\sigma'(L)}} = e_{m_1\ldots m_K;n_1\ldots n_L} + \ldots\,, \tag{2.21}$$

[3] Evidently, other choices of degree are possible. A choice that works for any $\mathcal{W}$ algebra is to put $\deg(W_n^{(\Delta)}) = \Delta$, where $W^{(\Delta)}(z)$ is a $\mathcal{W}$ generator of conformal dimension Δ (see, e.g., [Wa2]).

where the dots stand for a (finite) sum of monomials (2.19) of degrees strictly less than $K+2L$. We can choose σ and σ' such that we obtain the lexicographical ordering $m_1 \geq \ldots \geq m_{K-1} \geq m_K > 0$, $n_1 \geq \ldots \geq n_{L-1} \geq n_L > 0$. The theorem is now proved by induction on the degree. □

The above proof can be formalized. Note that, upon defining $M_d = \{v \in M \mid \deg v \leq d\}$, we obtain an increasing filtration

$$\mathbb{C} \cdot v \equiv M_0 \subset M_1 \subset M_2 \subset \cdots \subset M\,. \tag{2.22}$$

Similarly, we obtain an increasing filtration of $\mathcal{W}_3$. Now, obviously, the associated graded space

$$\mathrm{Gr}\, M \equiv \coprod_{d \geq 0} (M_{d+1}/M_d)\,, \tag{2.23}$$

becomes a module of the Abelian algebra $\mathrm{Gr}\, \mathcal{W}_3$. In fact,

$$e_{m_{\sigma(1)},\ldots,m_{\sigma(K)};n_{\sigma'(1)},\ldots,n_{\sigma'(L)}} = e_{m_1,\ldots,m_K;n_1,\ldots,n_L}\,, \tag{2.24}$$

in $\mathrm{Gr}\, M$.

A useful observation is the following. Since every highest weight module V is the image of a Verma module M under a $\mathcal{W}_3$ homomorphism (see Corollary 2.11), V inherits the increasing filtration of M upon defining $V_d = \pi(M_d)$. Under this grading $\mathrm{Gr}\, V$ becomes a $\mathrm{Gr}\, \mathcal{W}_3$ module.

Even though $\mathcal{W}_{3,\pm}$ does not define a subalgebra of $\mathcal{W}_3$, and thus, strictly speaking, the universal enveloping algebra $U(\mathcal{W}_{3,\pm})$ is not defined, for practical purposes and motivated by Theorem 2.12 it is useful to define $U(\mathcal{W}_{3,\pm})$ not as an algebra but merely as a subspace of $\mathcal{W}_3$ as follows

Definition 2.13 *The universal envelope $U(\mathcal{W}_{3,-})$ of $\mathcal{W}_{3,-}$ is defined to be the subspace of $\mathcal{W}_3$ spanned by the vectors*

$$L_{-m_1} \cdots L_{-m_K} W_{-n_1} \cdots W_{-n_L}\,, \tag{2.25}$$

where $m_1 \geq \ldots \geq m_{K-1} \geq m_K > 0\,, n_1 \geq \ldots \geq n_{L-1} \geq n_L > 0$, and similarly for $U(\mathcal{W}_{3,+})$.

From Theorem 2.12 it follows that every $v \in M(h,w,c)$ can be written as $v = u\, v_M$ for some $u \in U(\mathcal{W}_{3,-})$, and that for $v = u_1 u_2\, v_M$ with $u_1, u_2 \in U(\mathcal{W}_{3,-})$ we can find $u \in U(\mathcal{W}_{3,-})$ such that $v = u\, v_M$ by using the $\mathcal{W}_3$ commutation relations (2.1) and the defining relations (2.16) for v_M.

We now return to the study of Verma modules and a particular class of quotient modules, namely the irreducible modules.

Theorem 2.14

i. $M(h,w,c)$ contains a maximal submodule $I(h,w,c)$ so that the quotient $L(h,w,c) \equiv M(h,w,c)/I(h,w,c)$ is irreducible. Conversely, every irreducible module $L \in \mathcal{O}$ is isomorphic to some $L(h,w,c)$.

ii. $I(h,w,c) \cong PM(h,w,c)$*, where* $PM(h,w,c)$ *is the submodule of* $M(h,w,c)$ *generated by all (proper) primitive vectors in* $M(h,w,c)$*.*
iii. Every (nonzero) $\mathcal{W}_3$ *homomorphism of Verma modules is injective.*

Proof. (i) Standard.
(ii) Clearly, by maximality of $I(h,w,c)$ we have $PM(h,w,c) \subset I(h,w,c)$. Now suppose $I(h,w,c) \not\cong PM(h,w,c)$. Take a vector $v \in I(h,w,c)\backslash PM(h,w,c)$ of minimal level. Since $\mathcal{W}_{3,+}$ lowers the level, $\mathcal{W}_{3,+} \cdot v$ vanishes in the quotient $I(h,w,c)\backslash PM(h,w,c)$, so $\mathcal{W}_{3,+} \cdot v \in PM(h,w,c)$. But this implies that $v \in PM(h,w,c)$, which is a contradiction.
(iii) Suppose $\pi \in \mathrm{Hom}_{\mathcal{W}_3}(M,M')$. Denote by v_M and $v_{M'}$ the highest weight vectors of the Verma modules M and M', respectively. Clearly, $\pi(v_M) = u_2\, v_{M'}$ for some $u_2 \in U(\mathcal{W}_{3,-})$. To show injectivity of π we have to prove that $u_1 u_2\, v_{M'} = 0$ with $u_1 \in U(\mathcal{W}_{3,-})$ implies $u_1 = 0$ or $u_2 = 0$. This is obvious for Abelian algebras, so we use the fact that the $\mathcal{W}_3$ homomorphism of Verma modules induces a $\mathrm{Gr}\,\mathcal{W}_3$ homomorphism of the associated graded Verma modules (see the discussion after Theorem 2.12). □

Proof of Theorem 2.2. Follows from the fact that, because of Theorem 2.14 (i), L is determined up to isomorphism by (h,w,c), i.e., by the leading term in the character $\mathrm{ch}_L(q,p)$. □

After having determined that all irreducible modules in the category $\mathcal{O}$ are of the type $L(h,w,c)$ (Theorem 2.14 (i)) we now have

Lemma 2.15 *There is a 1–1 correspondence between elements in* $\mathrm{Prim}(V)$ *of generalized weight* (h,w) *and irreducible modules* $L(h,w,c)$ *occurring in* $\mathrm{JH}(V)$*.*

Proof. Let $v \in \mathrm{Prim}(V)$ and have generalized weight (h,w). By definition there exists a proper submodule $U \subset V$ such that $v \notin U$ while $\mathcal{W}_{3,+} \cdot v \subset U$. Consider the module V/U. Clearly $L(h,w,c)$ occurs in $\mathrm{JH}(V/U)$. By merging the JH series for V/U with the JH series for U we immediately obtain that $L(h,w,c) \in \mathrm{JH}(V)$. Conversely, suppose $L(h,w,c) \in \mathrm{JH}(V)$, then $L(h,w,c) \cong V_s/V_{s+1}$ for some s (and N sufficiently large). Let v be a representative of the highest weight vector of $L(h,w,c)$ in $V_s \subset V$. Clearly, $\mathcal{W}_{3,+} \cdot v \in V_{s+1}$ while $v \notin V_{s+1}$, i.e., $v \in \mathrm{Prim}(V)$. □

A very convenient ingredient in the study of the submodule structure of Verma modules is the determinant of a certain bilinear form defined on Verma modules, the so-called Shapovalov form. Let us first briefly recall the definition and properties of the Shapovalov form.

First, applying Lemma 2.10 to the highest weight vector $\bar{v}_M$ of the contragredient Verma module $\overline{M}(h,w,c)$ yields a (unique) $\mathcal{W}_3$ homomorphism

$$\imath \in \mathrm{Hom}_{\mathcal{W}_3}(M(h,w,c), \overline{M}(h,w,c))\,, \tag{2.26}$$

such that $\imath(v_M) = \bar{v}_M$. This, in turn, immediately provides us with a bilinear form $\langle -|-\rangle_M : M(h,w,c) \times M(h,w,c) \longrightarrow \mathbb{C}$ by

$$\langle u|v\rangle_M = \imath(u)(v), \qquad u,v \in M(h,w,c), \tag{2.27}$$

which is such that $\langle v_M|v_M\rangle_M = \imath(v_M)(v_M) = \bar{v}_M(v_M) = 1$. Moreover, the fact that $\imath$ is a $\mathcal{W}_3$ homomorphism translates into the property that the form (2.27) is contravariant with respect to $\omega_{\mathcal{W}}$; namely, for all $u,v \in M(h,w,c)$,

$$\begin{aligned} \langle x\,u|v\rangle_M &= \imath(x\,u)(v) \\ &= (x\,\imath(u))(v) \\ &= \imath(u)(\omega_{\mathcal{W}}(x)\,v) \\ &= \langle u|\omega_{\mathcal{W}}(x)\,v\rangle_M . \end{aligned} \tag{2.28}$$

Conversely, every contravariant bilinear form $\langle -|-\rangle$ on $M(h,w,c) \times M(h,w,c)$ such that $\langle v_M|v_M\rangle = 1$ leads to a $\mathcal{W}_3$ homomorphism $\bar{\imath} \in \mathrm{Hom}_{\mathcal{W}_3}(M(h,w,c), \overline{M}(h,w,c))$, satisfying $\bar{\imath}(v_M) = \bar{v}_M$, by defining $\bar{\imath}(v) = \langle v|-\rangle_M$. Moreover, because of the uniqueness of $\imath$, we necessarily have $\bar{\imath} = \imath$.

To get a more explicit form for $\langle -|-\rangle_M$, define for any $v \in M(h,w,c)$, the "vacuum expectation value" $\langle v\rangle$ as the coefficient of the highest weight vector v_M of $M(h,w,c)$ in v. Now, for $u,v \in M(h,w,c)$ and $u = x\,v_M$ for some $x \in U(\mathcal{W}_{3,-})$, the formula

$$\langle u|v\rangle \equiv \langle \omega_{\mathcal{W}}(x)v\rangle \tag{2.29}$$

clearly defines such a contravariant bilinear form, and hence equals $\langle -|-\rangle_M$ above.

Now, upon recalling that the "radical" $\mathrm{Rad}\,\langle -|-\rangle_M$ of the form $\langle -|-\rangle_M$ is defined as

$$\mathrm{Rad}\,\langle -|-\rangle_M \equiv \{u \in M(h,w,c) \mid \langle u|v\rangle_M = 0,\ \forall\, v \in M(h,w,c)\}, \tag{2.30}$$

we can formulate the main properties of the Shapovalov form

Theorem 2.16

i. $M(h,w,c)$ *carries a unique contravariant bilinear form* $\langle -|-\rangle_M$ *such that* $\langle v_M|v_M\rangle_M = 1$, *where* v_M *is the highest weight vector of* $M(h,w,c)$. *This form is symmetric.*

ii. The generalized eigenspaces of $\mathcal{W}_{3,0} = \{L_0, W_0\}$ *are pairwise orthogonal.*

iii. Ker $\imath \cong$ Rad $\langle -|-\rangle_M \cong I(h,w,c)$, *hence* $L(h,w,c) \cong M(h,w,c)/I(h,w,c)$ *carries a unique contravariant bilinear form* $\langle -|-\rangle_L$ *such that* $\langle v_L|v_L\rangle_L = 1$, *where* v_L *is the highest weight vector of* $L(h,w,c)$. *This form is nondegenerate.*

Proof. (i) It remains to show that $\langle -|-\rangle_M$ is symmetric. This is, however, evident from (2.29), and the fact that $\omega_{\mathcal{W}}{}^2 = 1$.
(ii) Consider two generalized eigenspaces spanned by $\{u_0, \ldots, u_{\kappa-1}\}$ and $\{v_0, \ldots, v_{\kappa'-1}\}$ corresponding to eigenvalues (h,w) and (h',w'), respectively. That is,

$L_0\, u_i = h\, u_i$ for all i, and $L_0\, v_j = h'\, v_j$ for all j. Further, $(W_0 - w)^M\, u_i = 0$ for all $M \geq i+1$ and $(W_0 - w')^M\, v_j = 0$ for all $M \geq j+1$. Then, as usual, $(h-h')\langle v_j|u_i\rangle_M = \langle v_j|L_0\, u_i\rangle_M - \langle L_0\, v_j|u_i\rangle_M = 0$, and so $\langle v_j|u_i\rangle_M = 0$ for $h \neq h'$. Moreover, for $M \geq i+1$, we have

$$0 = \langle v_j|(W_0-w)^M\, u_i\rangle_M = \sum_{k=0}^{M} \binom{M}{k} (w'-w)^{M-k} \langle (W_0-w')^k\, v_j|u_i\rangle_M \,. \quad (2.31)$$

Consider (2.31) for $j=0$ and arbitrary i. It follows from (2.31) that $\langle v_0|u_i\rangle_M = 0$ for $w \neq w'$. Now proceed by induction to j to conclude that $\langle v_j|u_i\rangle_M = 0$, for all i, j, if $w \neq w'$.
(iii) By definition, Ker $\imath \cong$ Rad$\langle -|-\rangle_M$. Furthermore, Rad$\langle -|-\rangle_M$ is clearly a (proper) submodule of $M(h,w,c)$, so it remains to be shown that $I(h,w,c) \subset$ Rad$\langle -|-\rangle_M$. By Theorem 2.14 (ii) we have $I(h,w,c) \cong PM(h,w,c)$, so suppose $v \in PM(h,w,c)$. Clearly $v_M \notin U(\mathcal{W}_{3,+}) \cdot v$, and, in view of (2.29), this immediately implies that $v \in$ Rad$\langle -|-\rangle_M$. □

It turns out to be convenient to parametrize the Verma modules $M(h,w,c)$ by an $\mathfrak{sl}_3$ weight $\Lambda \in \mathfrak{h}^*_{\mathbb{C}}$ and a complex scalar $\alpha_0 \in \mathbb{C}$ (called the "background charge") as follows

$$\begin{aligned} c(\alpha_0) &= 2 - 24{\alpha_0}^2 \,, \\ h(\Lambda,\alpha_0) &= -(\theta_1\theta_2 + \theta_1\theta_3 + \theta_2\theta_3) = \tfrac{1}{2}(\Lambda, \Lambda + 2\alpha_0\rho) \,, \\ w(\Lambda,\alpha_0) &= \sqrt{3\beta}\theta_1\theta_2\theta_3 \,, \end{aligned} \quad (2.32)$$

where

$$\theta_i = (\Lambda + \alpha_0\rho, \epsilon_i) \,, \quad (2.33)$$

and ϵ_i, $i=1,2,3$ are the weights of the 3-dimensional representation of $\mathfrak{sl}_3$ with highest weight Λ_1, i.e.,

$$\epsilon_1 = \Lambda_1 \,, \qquad \epsilon_2 = \Lambda_2 - \Lambda_1 \,, \qquad \epsilon_3 = -\Lambda_2 \,. \quad (2.34)$$

The origin of this parametrization will become apparent in Sect. 2.2.3. Clearly,

Lemma 2.17 *We have*

$$h(\Lambda,\alpha_0) \;=\; h(\Lambda',\alpha_0) \,, \qquad w(\Lambda,\alpha_0) \;=\; w(\Lambda',\alpha_0) \,, \quad (2.35)$$

if and only if $\Lambda' = w(\Lambda + \alpha_0\rho) - \alpha_0\rho$ *for some Weyl group element* $w \in W$.

For convenience, we define, for fixed background charge α_0, the shifted (or "dotted") action of the Weyl group W of $\mathfrak{g}$ on P by

$$w \cdot \Lambda \;=\; w(\Lambda + \alpha_0\rho) - \alpha_0\rho \,, \qquad w \in W \,. \quad (2.36)$$

Henceforth, we simply write $M(\Lambda,\alpha_0)$ for $M(h(\Lambda,\alpha_0), w(\Lambda,\alpha_0), c(\alpha_0))$, and similarly $L(\Lambda,\alpha_0)$, for the irreducible quotient. Since we will mostly restrict our attention to a specific value of the central charge c (or background charge α_0) we will, in fact, often write $M(\Lambda)$ etc., if no confusion can arise. Note that, because of Lemma 2.17, we have $M(\Lambda, \alpha_0) \cong M(\Lambda', \alpha_0)$ if $\Lambda' = w \cdot \Lambda$ for some $w \in W$; so to parametrize Verma modules one may choose $\Lambda + \alpha_0 \rho$ to lie in a specific Weyl chamber if one so desires.

We denote the subspace of $M(\Lambda, \alpha_0)$ of L_0 eigenvalue $h(\Lambda, \alpha_0) + N$ by $M(\Lambda, \alpha_0)_N$.[4] Now, let $\imath_N : M(\Lambda, \alpha_0)_N \longrightarrow \overline{M}(\Lambda, \alpha_0)_N$ denote the restriction of $\imath$ to $M(\Lambda, \alpha_0)_N$. Clearly, $\imath_N$ is a linear map between two vector spaces of equal dimension; so, after choosing bases for $M(\Lambda, \alpha_0)_N$ and $\overline{M}(\Lambda, \alpha_0)_N$, the determinant of $\imath_N$ (the so-called Kac determinant) is well-defined. Put

$$\mathcal{S}(\Lambda, \alpha_0)_N \equiv \det \imath_N . \tag{2.37}$$

Equivalently, for any basis $\{u_i\}$ of $M(\Lambda, \alpha_0)_N$,

$$\mathcal{S}(\Lambda, \alpha_0)_N \sim \det(\langle u_i | u_j \rangle_M) , \tag{2.38}$$

where $\sim$ means proportionality with a factor independent of Λ and α_0. Obviously, $\operatorname{Ker} \imath_N = 0$ if and only if $\mathcal{S}(\Lambda, \alpha_0)_N \neq 0$. Moreover, $\mathcal{S}(\Lambda, \alpha_0)_N \neq 0$ clearly implies $\mathcal{S}(\Lambda, \alpha_0)_k \neq 0$ for all $k \leq N$. So, we conclude from Theorem 2.16,

Lemma 2.18 *The Verma module $M(\Lambda, \alpha_0)$ is irreducible up to level N if and only if $\mathcal{S}(\Lambda, \alpha_0)_N \neq 0$.*

The following explicit result for the Kac determinant is well-known (see, e.g., [BoSc1] and references therein)

Theorem 2.19

$$\mathcal{S}(\Lambda, \alpha_0)_N \sim \prod_{\alpha \in \Delta} \prod_{\substack{r,s \in \mathbb{N} \\ rs \leq N}} \Big((\Lambda + \alpha_0 \rho, \alpha) - (r\alpha_+ + s\alpha_-) \Big)^{p_2(N-rs)} , \tag{2.39}$$

where we have introduced $\alpha_\pm$ such that $\alpha_0 = \alpha_+ + \alpha_-$, $\alpha_+ \alpha_- = -1$. The (two-color) partition function $p_2(N)$ is defined by $\prod_{n \geq 1} (1 - q^n)^{-2} = \sum_{N \geq 0} p_2(N)\, q^N$.

Proof. The proof can be given either by constructing a sufficient number of singular vectors explicitly (for example through a free field construction), or, as in [KaRa], by determining a sufficient number of vanishing lines using the fact that we have a realization of $\mathcal{W}_3$ on the coset module $(\widehat{\mathfrak{sl}_3})_1 \oplus (\widehat{\mathfrak{sl}_3})_k / (\widehat{\mathfrak{sl}_3})_{k+1}$ (see, e.g., [BoSc1]). □

An immediate consequence of Theorem 2.19 is the $\mathcal{W}_3$ analogue of the so-called Kac-Kazhdan condition

Corollary 2.20 *The Verma module $M(\Lambda, \alpha_0)$ is irreducible if and only if $(\Lambda + \alpha_0 \rho, \alpha) \notin (\mathbb{N}\alpha_+ + \mathbb{N}\alpha_-)$ for all roots $\alpha \in \Delta$.*

[4] Note that $M(\Lambda, \alpha_0)_N = M(\Lambda, \alpha_0)_{(h(\Lambda)+N)}$ in the notation introduced in Sect. 2.2.1.

In the case of the Virasoro algebra the Kac determinant (2.39) suffices to determine the complete structure of submodules of a Verma module [FeFu]. In particular, one finds that $\mathrm{Prim}(M) = \mathrm{Sing}(M)$, and that the weights of all singular vectors are concentrated on the orbit of the highest weight under the affine Weyl group of $\widehat{\mathfrak{sl}_2}$. For $\mathcal{W}_3$, however, knowledge of just the Kac determinant (2.39) is not enough to ascertain the submodule structure of a Verma module, essentially because (2.39) only carries information about the L_0 weight. The full submodule structure could probably be deduced if, for instance, we could find the (nonzero) generalized eigenspaces of $\mathcal{W}_{3,0}$ and were able to compute the determinant of the bilinear form on these generalized eigenspaces. It is generally believed that, also for the $\mathcal{W}_3$ algebra, the weights of all primitive vectors in a Verma module are concentrated on a certain orbit of the highest weight vector under the affine Weyl group of $\widehat{\mathfrak{sl}_3}$. To our knowledge, no general proof of this "linkage principle" is available, but there is ample evidence for it coming from the Quantum Drinfel'd-Sokolov reduction. Remarkably, in the case of our immediate interest, namely $c = 2$ Verma modules, it can indeed be proved – see Theorem 2.33. Clearly, knowing the weights of all primitive vectors in Verma modules is of utmost importance, since these not only determine the possible irreducible modules in the composition series of a Verma module (by Lemma 2.15), but also the possible nontrivial $\mathcal{W}_3$ homomorphisms between Verma modules.

We have already emphasized that for generic modules $V \in \mathcal{O}$ the action of the Cartan subalgebra $\mathcal{W}_{3,0}$ need not be diagonalizable. In fact, this phenomenon already occurs in Verma modules. Consider thereto the following example (see [Wal]). The level $h+1$ eigenspace in the Verma module $M(h,w,c)$ is two-dimensional and spanned by $\{L_{-1}\,v_M, W_{-1}\,v_M\}$. The action of W_0 on this two-dimensional space is given by

$$\begin{aligned} W_0(L_{-1}\,v_M) &= (w\,L_{-1} + 2\,W_{-1})\,v_M\,, \\ W_0(W_{-1}\,v_M) &= \left(\tfrac{1}{5}\,(-1 + 2\beta(5h+1))\;L_{-1} + w\,W_{-1}\right)\,v_M\,. \end{aligned} \tag{2.40}$$

Clearly, W_0 is not diagonalizable iff the following equation holds

$$2\beta(5h+1) - 1 \;=\; 0\,. \tag{2.41}$$

In that case, we have a two-dimensional generalized eigenspace corresponding to a W_0 eigenvalue w. Upon defining[5]

$$\begin{aligned} v_0 &= W_{-1}\,v_M\,, \\ v_1 &= \tfrac{1}{2}L_{-1}\,v_M\,, \end{aligned} \tag{2.42}$$

we find from (2.40)

$$\begin{aligned} W_0\;v_1 &= w\,v_1 + v_0\,, \\ W_0\;v_0 &= w\,v_0\,, \end{aligned} \tag{2.43}$$

i.e., the vectors $\{v_0, v_1\}$ form a Jordan basis for $M(h,w,c)$ at L_0 level $h+1$.

[5] Note that v_1 is determined by (2.43) up to the addition of an arbitrary multiple of v_0.

For example, (2.41) is satisfied for $h = w = 0$ and $c = 2$ (i.e., $\Lambda = 0$, $\alpha_0 = 0$). In this specific case it is easily seen that, in fact, v_0 is a singular vector whilst v_1 is p-singular. Thus, the module generated by v_1, which we denote by $M(v_1)$, is a proper submodule of $M(0,0,2)$ (note that $v_0 = W_0 v_1 \in M(v_1)$). Moreover, $M(v_1)$ is entirely contained in the maximal ideal and therefore has to be projected out in (the first step of) a resolution of the irreducible module $L(0,0,2)$ (see Sect. 2.4 for more details). We elaborate on this example in later sections.

To summarize, we have seen in the simple example above that modules where the highest weight space corresponds to some indecomposable representation of the Cartan subalgebra $\mathcal{W}_{3,0}$, such as $M(v_1)$, naturally occur as submodules of Verma modules, and, moreover, that these modules are required in the construction of (Verma module type) resolutions for the irreducible modules.

After this lengthy discussion, let us now introduce generalized Verma modules.

Let $V^{(\kappa)}$ denote a κ-dimensional indecomposable representation of $\mathcal{W}_{3,0}$ of generalized weight (h, w). As explained in Sect. 2.2.1, we may choose a basis $\{v_0, \ldots, v_{\kappa-1}\}$ of $V^{(\kappa)}$ such that (see (2.9))

$$v_{\kappa-1} \xrightarrow{W_0 - w} \ldots \xrightarrow{W_0 - w} v_1 \xrightarrow{W_0 - w} v_0 \xrightarrow{W_0 - w} 0 . \tag{2.44}$$

Definition 2.21 *The generalized Verma module $M^{(\kappa)}(h, w, c)$, with generalized highest weight (h, w), is defined as the $\mathcal{W}_3$ module "induced" from $V^{(\kappa)}$ by the action of $\mathcal{W}_{3,-}$, i.e., the $\mathcal{W}_3$ module $\mathcal{W}_3 \cdot V^{(\kappa)}$ modded out by the ideal generated by the relations*

$$\mathcal{W}_{3,+} \cdot V^{(\kappa)} = 0 , \tag{2.45}$$

as well as

$$\begin{aligned} L_0\, v_i &= h\, v_i \quad \text{for } i = 0, \ldots, \kappa - 1 , \\ W_0\, v_i &= \begin{cases} w\, v_i + v_{i-1} & \text{for } i = 1, \ldots, \kappa - 1 \\ w\, v_0 & \text{for } i = 0 . \end{cases} \end{aligned} \tag{2.46}$$

The action of $\mathcal{W}_3$ on $M^{(\kappa)}(h, w, c)$ is defined by means of the commutation relations (2.1) and the relations (2.45) and (2.46).

Most of the theorems as well as their proofs that have been discussed for Verma modules have a straightforward analogue for generalized Verma modules. We refrain from going into details. Let us just remark that a basis of $M^{(\kappa)}(h, w, c)$ is provided by the vectors

$$e^{(i)}_{m_1 \ldots m_K; n_1 \ldots n_L} = L_{-m_1} \cdots L_{-m_K} W_{-n_1} \cdots W_{-n_L}\, v_i , \tag{2.47}$$

where, $m_1 \geq \ldots \geq m_{K-1} \geq m_K > 0$ and $n_1 \geq \ldots \geq n_{L-1} \geq n_L > 0$, while $i = 0, \ldots, \kappa - 1$.

There exists a unique bilinear symmetric form $\langle - | - \rangle_M$ on $M^{(\kappa)}(h, w, c)$, contravariant with respect to $\omega_{\mathcal{W}}$, and such that $\langle v_i | v_j \rangle_M = \delta_{ij}$.

Since a generalized Verma module $M^{(\kappa)}(h, w, c)$ is generated (over $\mathcal{W}_3$) by the vector $v_{\kappa-1}$, each $\pi \in \mathrm{Hom}_{\mathcal{W}_3}(M^{(\kappa)}(h, w, c), V)$ with $V \in \mathcal{O}$ is uniquely determined by the image of $v_{\kappa-1}$ under π (cf., Lemma 2.10). Clearly, $\pi(v_{\kappa-1}) \in \text{pSing}(V)$ and has generalized weight (h, w). Moreover, $(W_0 - w)^{\kappa}\, \pi(v_{\kappa-1}) = 0$. Conversely, for every $v \in \text{pSing}(V)$ of generalized weight (h, w) such that $(W_0 - w)^{\kappa}\, v = 0$ the map $\pi(v_{\kappa-1}) = v$ uniquely extends to a $\mathcal{W}_3$ homomorphism $\pi : M^{(\kappa)}(h, w, c) \to V$.

In particular, if we apply the above to the case that V itself is a generalized Verma module, we find a sequence of $\mathcal{W}_3$ homomorphisms

$$0 \longrightarrow M(h,w,c) \xrightarrow{\pi_0} \ldots \xrightarrow{\pi_{\kappa-3}} M^{(\kappa-1)}(h,w,c) \xrightarrow{\pi_{\kappa-2}} M^{(\kappa)}(h,w,c)\,, \qquad (2.48)$$

defined by $\pi_i(v_i) = v_i$ for $i = 0, \ldots, \kappa - 2$. Even more so, since every π_i is injective, we obtain a decreasing filtration of $M^{(\kappa)}(h, w, c)$ by generalized Verma modules, i.e.,

$$M^{(\kappa)}(h,w,c) \;\supset\; M^{(\kappa-1)}(h,w,c) \;\supset\; \ldots \;\supset\; M(h,w,c)\,, \qquad (2.49)$$

such that all quotients are isomorphic to $M(h, w, c)$. This filtration is very useful in relating properties of generalized Verma modules to those of ordinary Verma modules and, in particular, for relating generalized Verma module cohomology to Verma module cohomology by means of the spectral sequence associated to the filtration. As an example, it follows from the filtration (2.49) that the weights of primitive vectors in $M^{(\kappa)}(\Lambda, \alpha_0)$ coincide with the weights of those in $M(\Lambda, \alpha_0)$.

More generally, for $\kappa_1 < \kappa_2$, we have an injective $\pi \in \text{Hom}_{\mathcal{W}_3}(M^{(\kappa_1)}(h, w, c), M^{(\kappa_2)}(h, w, c))$ such that the quotient is isomorphic to $M^{(\kappa_2-\kappa_1)}(h, w, c)$; i.e., we have exact sequences

$$0 \longrightarrow M^{(\kappa_1)}(h,w,c) \xrightarrow{\pi} M^{(\kappa_2)}(h,w,c) \longrightarrow M^{(\kappa_2-\kappa_1)}(h,w,c) \longrightarrow 0\,. \qquad (2.50)$$

Although it is still true that any $\mathcal{W}_3$ homomorphism of a Verma module to a generalized Verma module is injective (cf., Theorem 2.14 (iii)), this property does not hold for $\mathcal{W}_3$ homomorphisms between arbitrary generalized Verma modules. Consider, for example, the $\mathcal{W}_3$ automorphism π of $M^{(2)}(h, w, c)$ defined by $\pi(v_1) = v_0$. Clearly π is not injective. For a more complicated example, consider the Verma module $M = M(0, 0, 2)$. We have already seen that there exists a $\mathcal{W}_3$ homomorphism $M^{(2)}(1, 0, 2) \xrightarrow{\pi} M(0, 0, 2)$ defined by $\pi(v_1') = v_1$ (with v_1 as defined in (2.42)), whose image is $M(v_1)$. Explicit computation shows that $M(v_1) \not\cong M^{(2)}(1, 0, 2)$. We come back to this example in more detail in Sect. 2.3.2.

2.2.3 Fock Spaces

In this section we define an extremely useful realization of the $\mathcal{W}_3$ algebra known as the "free field realization." The corresponding modules are known as Fock modules or Feigin-Fuchs modules.

Let $\mathcal{A}$ be the oscillator algebra (Heisenberg algebra) with basis $\{\alpha^i_m \mid m \in \mathbb{Z}, i = 1, 2\}$ and commutation relations

$$[\alpha^i_m, \alpha^j_n]_- \;=\; m\delta_{m+n,0}\delta^{ij}\,. \qquad (2.51)$$

In terms of formal power series

$$i\partial\phi^i(z) = \sum_{n\in\mathbb{Z}} \alpha_n^i \, z^{-n-1} \quad \in \mathcal{A}[[z,z^{-1}]]\,, \tag{2.52}$$

also referred to as "free scalar fields," the commutation relations (2.51) are encoded in the following OPEs

$$i\partial\phi^i(z)\; i\partial\phi^j(w) = \frac{\delta^{ij}}{(z-w)^2} + \ldots . \tag{2.53}$$

Also, for convenience, we often use vector notation by introducing an orthonormal basis (with respect to the Euclidean inner product) $\{e_1, e_2\}$ of $\mathbb{C}^2$, e.g.,

$$\boldsymbol{\alpha}_n \equiv \sum_i \alpha_n^i \, e_i\,, \tag{2.54}$$

and identify $\mathbb{C}^2$ with the weight space $\mathfrak{h}_{\mathbb{C}}^*$ of $\mathfrak{sl}_3$. The algebra $\mathcal{A}$ has a Cartan decomposition $\mathcal{A} \cong \mathcal{A}_- \oplus \mathcal{A}_0 \oplus \mathcal{A}_+$, where

$$\mathcal{A}_\pm = \{\alpha_n^i \mid \pm n > 0\}\,, \qquad \mathcal{A}_0 = \{\alpha_0^i\}\,. \tag{2.55}$$

The universal enveloping algebra $U(\mathcal{A})$ as well as its local completion $U(\mathcal{A})_{\rm loc}$ [FeFr3] are defined as usual.

For any $\Lambda \in \mathfrak{h}_{\mathbb{C}}^*$ and $\alpha_0 \in \mathbb{C}$ we have an (irreducible) $\mathcal{A}$ module $F(\Lambda, \alpha_0)$, i.e., a Fock module, with basis

$$f_{m_1,\ldots,m_M;n_1,\ldots n_N} = \alpha_{-m_1}^2 \cdots \alpha_{-m_M}^2 \alpha_{-n_1}^1 \cdots \alpha_{-n_N}^1 |\Lambda\rangle\,, \tag{2.56}$$

where $n_1 \geq n_2 \geq \ldots \geq n_N$, $m_1 \geq m_2 \geq \ldots \geq m_M$ and the "vacuum vector" $|\Lambda\rangle$ satisfies

$$\begin{aligned} \boldsymbol{\alpha}_m\,|\Lambda\rangle &= 0\,, \qquad \text{for} \quad m > 0\,, \\ \boldsymbol{\alpha}_0\,|\Lambda\rangle &= \Lambda\,|\Lambda\rangle\,. \end{aligned} \tag{2.57}$$

The action of $\mathcal{A}$ on $F(\Lambda, \alpha_0)$ is defined by (2.51) and (2.57).[6] It is usual to extend the representation by the operator $\boldsymbol{q}$ such that $[q^i, \alpha_n^j]_- = i\delta_{n,0}\delta^{ij}$. In canonical quantization, $\boldsymbol{q}$ is simply the zero mode of the free scalar field,

$$\phi^i(z) = q^i - i\alpha_0^i \log z + i \sum_{n\neq 0} \frac{\alpha_n^i}{n}\, z^{-n}\,. \tag{2.58}$$

The vacuum vector of different Fock modules, i.e., $|\Lambda\rangle$ for different Λ, can be generated from $|0\rangle$ (the so-called $SL(2,\mathbb{C})$ invariant vacuum) via $|\Lambda\rangle = e^{i\Lambda\cdot q}|0\rangle$.

It is useful to state more precisely the relation between operators and states. First, observe that for fixed Λ there is an isomorphism between the states in

---

[6] Clearly, at this point, the parameter α_0 does not play a role, since all modules $F(\Lambda, \alpha_0)$ for different values of α_0 are isomorphic as $\mathcal{A}$ modules. The reader should easily distinguish, in context, the complex number α_0 from the vector of "momentum" operators $\boldsymbol{\alpha}_0$.

(2.56) and the space of fields obtained by a finite number of normal products of a finite number of derivatives of the basic fields $i\partial\phi^j(z)$. Moreover, using the normal ordering prescription, we have $\lim_{z\to 0} e^{i\Lambda\cdot\phi(z)}|0\rangle = |\Lambda\rangle$. So, the isomorphism can be extended to arbitrary Λ as follows: Introduce the space, $\mathfrak{V}$, of normal-ordered fields of the form $P(i\partial\phi^j(z))e^{i\Lambda\cdot\phi(z)}$, where P is a polynomial in $i\partial\phi^j(z)$ and its derivatives, and $\Lambda \in \mathfrak{h}^*_{\mathbb{C}}$. Then, for any state $|\mathcal{O}\rangle \in F(\Lambda,\alpha_0)$, there is a corresponding field $\mathcal{O}(z) \in \mathfrak{V}$ such that

$$|\mathcal{O}\rangle = \lim_{z\to 0} \mathcal{O}(z)|0\rangle\,. \tag{2.59}$$

When the space of allowed Λ in $\mathfrak{V}$ is restricted to a lattice L such that the OPEs of all fields are meromorphic, we call the space $\mathfrak{V}$ a chiral algebra. Under certain further conditions we call the chiral algebra a Vertex Operator Algebra (VOA). The strongest of these is the requirement that, for any two fields in the chiral algebra, the OPE in one order is related to that in the other order by analytic continuation. For further discussion, and a complete list of the defining relations for a VOA, see, e.g., [Br,FHL,FLM]. To impose these conditions generally requires that we extend the construction of operator fields to include phase-cocycles. An example which will be required later is discussed in Appendix B.

For any given $\alpha_0 \in \mathbb{C}$, let $\omega_{\mathcal{A}}$ be the $\mathbb{C}$-linear anti-involution of $\mathcal{A}$ defined by (recall that w_0 is the Coxeter element of the Weyl group)

$$\omega_{\mathcal{A}}(\alpha_n) \;=\; w_0(\alpha_{-n}) - 2\alpha_0\rho\delta_{n,0}\,, \tag{2.60}$$

which we may equivalently specify on fields, i.e., $\mathcal{A}[[z,z^{-1}]]$, by

$$\omega_{\mathcal{A}}(i\partial\phi(z)) \;=\; z^{-2}\,w_0(i\partial\phi(z^{-1})) - 2\alpha_0\rho\,z^{-1}\,. \tag{2.61}$$

Clearly, $\omega_{\mathcal{A}}$ extends to a $\mathbb{C}$-linear anti-involution on $U(\mathcal{A})_{\rm loc}$.[7] Since

$$\omega_{\mathcal{A}}(\alpha_0)|\Lambda\rangle \;=\; (w_0(\Lambda) - 2\alpha_0\rho)|\Lambda\rangle \;=\; (w_0\cdot\Lambda)|\Lambda\rangle\,, \tag{2.62}$$

the anti-involution $\omega_{\mathcal{A}}$ provides us with a map

$$F(w_0\cdot\Lambda,\alpha_0)\,\times\,F(\Lambda,\alpha_0)\longrightarrow F(\Lambda,\alpha_0)\,, \tag{2.63}$$

defined by

$$(x\,v_{F'},y\,v_F)\;\mapsto\;\omega_{\mathcal{A}}(x)y\,v_F\,, \tag{2.64}$$

where $x,y \in U(\mathcal{A}_-)_{\rm loc}$, and we have denoted, for convenience, $v_F = |\Lambda,\alpha_0\rangle$ and $v_{F'} = |w_0\cdot\Lambda,\alpha_0\rangle$.

Furthermore, we define the "vacuum expectation value" $\langle - \rangle\,:\,F(\Lambda,\alpha_0)\longrightarrow\mathbb{C}$ as the coefficient of v_F in the expansion of $v \in F(\Lambda,\alpha_0)$ in the basis (2.56).

---

[7] There exist many other anti-involutions on $\mathcal{A}$ and consequently many different bilinear forms contravariant with respect to the chosen anti-involution. The one we have chosen here is the most natural with regards to the $\mathcal{W}_3$ module structure of $F(\Lambda,\alpha_0)$ (see Theorem 2.24). We will, however, introduce a different anti-involution and the associated form, needed for the proof of Theorem 2.31, shortly.

By combining the bilinear map (2.63) with $\langle - \rangle$ we obtain a bilinear form

$$\langle -|-\rangle_F \; : \; F(w_0 \cdot \Lambda, \alpha_0) \times F(\Lambda, \alpha_0) \longrightarrow \mathbb{C}, \tag{2.65}$$

i.e.,

$$\langle v|w\rangle \equiv \langle \omega_{\mathcal{A}}(x)w\rangle, \tag{2.66}$$

where $v = x\, v_{F'}$ with $x \in U(\mathcal{A}_-)_{\rm loc}$. We have

Theorem 2.22 *There exists a unique bilinear form $\langle -|-\rangle_F : F(w_0 \cdot \Lambda, \alpha_0) \times F(\Lambda, \alpha_0) \longrightarrow \mathbb{C}$, contravariant with respect to $\omega_{\mathcal{A}}$, such that $\langle v_{F'}|v_F\rangle_F = 1$, where v_F and $v_{F'}$ are the highest weight vectors of $F(\Lambda, \alpha_0)$ and $F(w_0 \cdot \Lambda, \alpha_0)$, respectively. This form is nondegenerate.*

Another useful anti-involution of $\mathcal{A}$, to be denoted by $\overline{\omega}_{\mathcal{A}}$, is defined by

$$\overline{\omega}_{\mathcal{A}}(\alpha_n^i) = \alpha_{-n}^i, \tag{2.67}$$

or, equivalently, by

$$\overline{\omega}_{\mathcal{A}}(i\partial\phi^i(z)) = z^{-2}\, i\partial\phi^i(z^{-1}). \tag{2.68}$$

We extend $\overline{\omega}_{\mathcal{A}}$ to an anti-linear anti-involution on $U(\mathcal{A})_{\rm loc}$. In complete analogy with (2.66) we now obtain a sesquilinear form

$$(-|-)_F \; : \; F(\Lambda, \alpha_0) \times F(\Lambda, \alpha_0) \longrightarrow \mathbb{C}, \tag{2.69}$$

by

$$(x\, v_F|y\, v_F)_F = \langle \overline{\omega}_{\mathcal{A}}(x) y\, v_F\rangle. \tag{2.70}$$

In fact,

Theorem 2.23 *There exists a unique sesquilinear form $(-|-)_F : F(\Lambda, \alpha_0) \times F(\Lambda, \alpha_0) \longrightarrow \mathbb{C}$, contravariant with respect to $\overline{\omega}_{\mathcal{A}}$, such that $(v_F|v_F)_F = 1$. This form is Hermitian, i.e., $(v|w)_F = \overline{(w|v)_F}$, and positive definite. The basis (2.56) is orthogonal with respect to $(-|-)_F$.*

Proof. Standard (see, e.g., [KaRa]). □

The following theorem is the reason for the term "free field realization."

Theorem 2.24 *For any $\alpha_0 \in \mathbb{C}$ such that $c = 2 - 24{\alpha_0}^2$, we have a homomorphism of algebras $\varrho : \mathcal{W}_3 \longrightarrow U(\mathcal{A})_{\rm loc}$ defined by*

$$\varrho(T(z)) = -\tfrac{1}{2}\partial\phi \cdot \partial\phi - i\alpha_0\rho \cdot \partial^2\phi, \tag{2.71}$$

$$\begin{aligned} \varrho(W(z)) = {} & \tfrac{-i}{3}\sqrt{\tfrac{\beta}{2}}\,(\partial\phi^1\partial\phi^1\partial\phi^2 - \partial\phi^2\partial\phi^2\partial\phi^2) \\ & + \alpha_0\sqrt{\beta}(\tfrac{\sqrt{3}}{2}\partial\phi^1\partial^2\phi^1 + \partial\phi^2\partial^2\phi^1 - \tfrac{\sqrt{3}}{2}\partial\phi^2\partial^2\phi^2) \\ & + \tfrac{i}{2}\sqrt{\tfrac{3\beta}{2}}\alpha_0^2\,(\partial^3\phi^1 - \tfrac{1}{\sqrt{3}}\partial^3\phi^2)\,. \end{aligned} \tag{2.72}$$

Furthermore,

$$\varrho\,\omega_{\mathcal{W}} = \omega_{\mathcal{A}}\,\varrho\,, \qquad \text{for all } \alpha_0 \in \mathbb{C}, \tag{2.73}$$

and

$$\varrho\,\overline{\omega}_{\mathcal{W}} = \overline{\omega}_{\mathcal{A}}\,\varrho\,, \qquad \text{for } \alpha_0 = 0\,, \tag{2.74}$$

such that, in particular, the form $\langle - | - \rangle_F$ *is contravariant with respect to* $\omega_{\mathcal{W}}$ *for all* $\alpha_0 \in \mathbb{C}$, *and* $(-|-)_F$ *is contravariant with respect to* $\overline{\omega}_{\mathcal{W}}$ *for* $\alpha_0 = 0$.

Proof. By a straightforward, albeit tedious, calculation. □

Remark. The homomorphism ϱ was first discussed in [FaZa]. In [FaLu1,FaLu2] a systematic method to derive ϱ, the so-called Quantum Drinfel'd-Sokolov (QDS) reduction, was first presented. In (2.72) we have chosen an orthonormal basis with respect to which the simple roots of $\mathfrak{sl}_3$ are $\alpha_1 = (\sqrt{2}, 0)\,,\ \alpha_2 = (-1/\sqrt{2}, \sqrt{3/2})$.

By means of the homomorphism ϱ we can equip the $\mathcal{A}$ module $F(\Lambda, \alpha_0)$ with the structure of a $\mathcal{W}_3$ module. We denote this module by $F(\Lambda, \alpha_0)$ as well. Clearly, $F(\Lambda, \alpha_0) \in \mathcal{O}$; the highest weight space is one-dimensional and spanned by $|\Lambda, \alpha_0\rangle$. The central charge of this representation, along with the weight of $|\Lambda, \alpha_0\rangle$, are parametrized exactly as in (2.32) by the background charge α_0. The module contragredient to $F(\Lambda, \alpha_0)$, as a $\mathcal{W}_3$ module, is determined by the following theorem

Theorem 2.25 *We have a* $\mathcal{W}_3$ *isomorphism*

$$\imath_F \;:\; F(w_0 \cdot \Lambda, \alpha_0) \;\xrightarrow{\;\cong\;}\; \overline{F}(\Lambda, \alpha_0)\,, \tag{2.75}$$

where $\imath_F$ *is explicitly given by* $\imath_F(v) = \langle v | - \rangle_F$.

Proof. The fact that $\imath_F \in \mathrm{Hom}_{\mathcal{W}_3}(F(w_0 \cdot \Lambda, \alpha_0), \overline{F}(\Lambda, \alpha_0))$ follows from the contravariance of $\langle - | - \rangle_F$ with respect to $\omega_{\mathcal{A}}$ and (2.73). That $\imath_F$ is, in fact, an isomorphism follows from the fact that $\langle - | - \rangle_F$ is nondegenerate (see Theorem 2.22). □

To determine the structure of $F(\Lambda, \alpha_0)$ as a $\mathcal{W}_3$ module it turns out to be useful to "compare" $F(\Lambda, \alpha_0)$ to a (contragredient) Verma module. We have

Theorem 2.26 *Let* v_M, v_F *and* $\bar{v}_M$ *denote, respectively, the highest weight vectors of* $M(\Lambda, \alpha_0)$, $F(\Lambda, \alpha_0)$ *and* $\overline{M}(\Lambda, \alpha_0)$.

i. There exist unique $\mathcal{W}_3$ *homomorphisms*

$$M(\Lambda, \alpha_0) \;\xrightarrow{\;\imath'\;}\; F(\Lambda, \alpha_0) \;\xrightarrow{\;\imath''\;}\; \overline{M}(\Lambda, \alpha_0)\,, \tag{2.76}$$

such that $\imath'(v_M) = v_F$ *and* $\imath''(v_F) = \bar{v}_M$.

ii. $\imath = \imath''\,\imath'$.

iii. $\imath'(x\,v_M) \;=\; \varrho(x)\,v_F$ *and* $\overline{\imath''}(x\,v_M) \;=\; \varrho(x)\,v_{F'}$.

iv. We have a commutative diagram

$$\begin{array}{ccc} M(\Lambda,\alpha_0) \times M(\Lambda,\alpha_0) & \stackrel{\overline{\imath''} \times \imath'}{\longrightarrow} & F(w_0 \cdot \Lambda, \alpha_0) \times F(\Lambda,\alpha_0) \\ \Big\downarrow \langle - | - \rangle_M & & \Big\downarrow \langle - | - \rangle_F \\ \mathbb{C} & \stackrel{\text{id}}{\longrightarrow} & \mathbb{C} \end{array} \tag{2.77}$$

Proof. (i) The existence and definition of $\imath'$ follows from Lemma 2.10. Similarly, Lemma 2.10 gives a $\mathcal{W}_3$ homomorphism $\overline{\imath''} : M(\Lambda,\alpha_0) \longrightarrow F(w_0 \cdot \Lambda, \alpha_0) \cong \overline{F}(\Lambda,\alpha_0)$ which, by Lemma 2.7, is contragredient to the map $\imath''$ sought for.
(ii) Follows from the uniqueness of $\imath$.
(iii) Follows from the uniqueness of $\imath'$ and $\imath''$.
(iv) Let $x, y \in U(\mathcal{W}_{3,-})$, then

$$\begin{aligned} \langle \overline{\imath''}(x\, v_M), \imath'(y\, v_M)\rangle_F &= \langle (\omega_{\mathcal{A}}\, \rho)(x)\rho(y)\, v_F\rangle_F \\ &= \langle (\rho\, \omega_{\mathcal{W}})(x)\rho(y)\, v_F\rangle_F \\ &= \langle \rho(\omega_{\mathcal{W}}(x) y)\, v_F\rangle_F \\ &= \langle \omega_{\mathcal{W}}(x) y\, v_M\rangle_M \\ &= \langle x\, v_M, y\, v_M\rangle_M\,. \end{aligned} \tag{2.78}$$

□

Let, $\imath'_N$ and $\imath''_N$ denote the restrictions of $\imath'$ and $\imath''$ to $M(\Lambda,\alpha_0)_N$ and $F(\Lambda,\alpha_0)_N$, respectively. Since $\imath'_N$ and $\imath''_N$ are linear maps between vector spaces of equal dimension we can define

$$\mathcal{S}'(\Lambda,\alpha_0)_N \equiv \det \imath'_N\,, \qquad \mathcal{S}''(\Lambda,\alpha_0)_N \equiv \det \imath''_N\,, \tag{2.79}$$

where the determinants are defined by means of bases $\{v_i\}$ and $\{w_i\}$ of $F(\Lambda,\alpha_0)_N$ and $M(\Lambda,\alpha_0)_N$, respectively. We have

$$\mathcal{S}'(\Lambda,\alpha_0)_N \sim \det\left(\langle v_i | \imath'(w_j)\rangle_F\right). \tag{2.80}$$

In addition, it follows from the proof of Theorem 2.26 (i) that

$$\mathcal{S}''(\Lambda,\alpha_0)_N = \mathcal{S}'(w_0 \cdot \Lambda, \alpha_0)_N\,. \tag{2.81}$$

We have

Theorem 2.27 [BMP5]

$$\begin{aligned} \mathcal{S}'(\Lambda,\alpha_0)_N &\sim \prod_{\alpha\in\Delta_+} \prod_{\substack{r,s\in\mathbb{N} \\ rs\le N}} \Big((\Lambda+\alpha_0\rho,\alpha) - (r\alpha_+ + s\alpha_-)\Big)^{p_2(N-rs)}, \\ \mathcal{S}''(\Lambda,\alpha_0)_N &\sim \prod_{\alpha\in\Delta_+} \prod_{\substack{r,s\in\mathbb{N} \\ rs\le N}} \Big((\Lambda+\alpha_0\rho,\alpha) + (r\alpha_+ + s\alpha_-)\Big)^{p_2(N-rs)}. \end{aligned} \tag{2.82}$$

Sketch of proof. The proof is based on the explicit construction of a sufficient number of singular vectors in $F(\Lambda,\alpha_0)$ in terms of multi-contour integrals over products of screeners. Note that, by Theorem 2.26 (ii), the Kac determinant $\mathcal{S}(\Lambda,\alpha_0)$ of Theorem 2.20 factorizes, up to a proportionality factor, as $\mathcal{S}'(\Lambda,\alpha_0)\mathcal{S}''(\Lambda,\alpha_0)$. □

As an immediate consequence of Theorem 2.27 we have

Corollary 2.28 [BMP5]

i. *The Fock module $F(\Lambda,\alpha_0)$ is isomorphic with $M(\Lambda,\alpha_0)$ or $\overline{M}(\Lambda,\alpha_0)$ provided the following condtions hold for all $\alpha \in \Delta_+$*

$$F(\Lambda,\alpha_0) \cong \begin{cases} M(\Lambda,\alpha_0) & \text{if } (\Lambda+\alpha_0\rho,\alpha) \notin (\mathbb{N}\alpha_+ + \mathbb{N}\alpha_-) \\ \overline{M}(\Lambda,\alpha_0) & \text{if } (\Lambda+\alpha_0\rho,\alpha) \notin -(\mathbb{N}\alpha_+ + \mathbb{N}\alpha_-)\,. \end{cases} \tag{2.83}$$

In particular, if $(\Lambda+\alpha_0\rho,\alpha) \notin (\mathbb{N}\alpha_+ + \mathbb{N}\alpha_-)$ for all $\alpha \in \Delta$, then $M(\Lambda,\alpha_0) \cong F(\Lambda,\alpha_0) \cong \overline{M}(\Lambda,\alpha_0)$ are irreducible.

ii. *For ${\alpha_0}^2 \in \mathbb{R}$ with ${\alpha_0}^2 < -4$ (or, equivalently, $c \geq c_{\rm crit} - 2 = 98$), we have*

$$F(\Lambda,\alpha_0) \cong \begin{cases} M(\Lambda,\alpha_0) & \text{for } i(\Lambda+\alpha_0\rho) \in \eta D_+\,, \\ \overline{M}(\Lambda,\alpha_0) & \text{for } -i(\Lambda+\alpha_0\rho) \in \eta D_+\,, \end{cases} \tag{2.84}$$

where $D_+ = \{\lambda \in \mathfrak{h}^*_{\mathbb{R}} \mid (\lambda,\alpha) \geq 0, \forall \alpha \in \Delta_+\}$ denotes the fundamental Weyl chamber, and $\eta = \mathrm{sign}(-i\alpha_0)$.

It immediately follows from Corollary 2.28 that for almost all $\Lambda \in P$ we have an isomorphism $M(\Lambda,\alpha_0) \cong F(\Lambda,\alpha_0)$ of $\mathcal{W}_3$ modules. Note, however, that since the (generalized) eigenvalues of $\mathcal{W}_{3,0}$ are algebraic in (the components of) Λ, they are in fact equal (and have the same multiplicity) on $M(\Lambda,\alpha_0)$ and $F(\Lambda,\alpha_0)$ for *all* $\Lambda \in \mathfrak{h}^*_{\mathbb{C}}$. This in turn implies the equality of the characters and since the characters ch L are algebraically independent (see Theorem 2.2) it follows immediately from (2.12) and Lemma 2.15 that

Theorem 2.29

i. *For all $\Lambda \in \mathfrak{h}^*_{\mathbb{C}}$ and all irreducible modules L we have $(M(\Lambda,\alpha_0) : L) = (F(\Lambda,\alpha_0) : L)$.*

ii. *There is a 1–1 correspondence between primitive vectors in $M(\Lambda,\alpha_0)$ and in $F(\Lambda,\alpha_0)$.*

2.3 Verma Modules and Fock Modules at $c = 2$

2.3.1 Generalities

In this section we study in more detail the structure of Verma modules, irreducible modules and Fock spaces for central charge $c = 2$. This is the case of most interest for the rest of this paper, where we study the 4D $\mathcal{W}_3$ string – i.e., the off-critical $\mathcal{W}_3$ string with two flat embedding coordinates. These embedding coordinates correspond to the "matter" free fields in the above for $c^M = 2$, thus motivating the interest in such Fock modules. The results for the remaining modules are required to obtain a framework in which calculations for $c = 2$ Fock spaces are feasible. This becomes more clear below, and in the following section.

Remarkably, the $\widehat{\mathfrak{sl}_3}$ structure which appears for $c = 2$ allows us to derive strong results; in particular, we obtain the weights and multiplicities of primitive vectors in $c = 2$ Verma modules. Although the construction of a level-1 representation of $\widehat{\mathfrak{sl}_3}$ on $c = 2$ Fock spaces is standard, we give a brief review in Appendix B. This serves to set conventions, as well as to illustrate the concept of a Vertex Operator Algebra (VOA) associated with a given set of Fock spaces corresponding to highest weight vectors on a lattice.

A preliminary result is the character of irreducible representations at $c = 2$.

Theorem 2.30 [Bo,BoScl] *For the irreducible modules $L(\Lambda, 0)$ at $c = 2$ with $\Lambda \in P_+$, we have*

$$\begin{aligned} \mathrm{ch}_{L(\Lambda,0)}(q) &= \frac{1}{\prod_{n\geq 1}(1-q^n)^2} \sum_{w\in W} \epsilon(w)\, q^{\frac{1}{2}|w(\Lambda+\rho)-\rho|^2} \\ &= \frac{q^{\frac{1}{2}|\Lambda|^2}}{\prod_{n\geq 1}(1-q^n)^2} \prod_{\alpha\in\Delta_+} (1-q^{(\lambda+\rho,\alpha)})\,. \end{aligned} \tag{2.85}$$

Theorem 2.31 *Consider $c = 2$, i.e., $\alpha_0 = 0$, and $\Lambda \in \mathfrak{h}^*_{\mathbb{C}}$.*

i. The Fock space $F(\Lambda, 0)$ is completely reducible for all $\Lambda \in \mathfrak{h}^*_{\mathbb{C}}$.

ii. For all $\Lambda \in P$ we have

$$F(\Lambda,0) \cong \bigoplus_{\Lambda'\in P_+} m_\Lambda^{\Lambda'} L(\Lambda',0)\,, \tag{2.86}$$

where $m_\Lambda^{\Lambda'}$ is equal to the multiplicity of the weight Λ in the irreducible finite dimensional representation $\mathcal{L}(\Lambda')$ of $\mathfrak{sl}_3$ with highest weight Λ'.

iii. $(F(\Lambda,0) : L(\Lambda',0)) = m_\Lambda^{\Lambda'}$.

Proof. (i) By Theorems 2.23 and 2.24 we have a positive definite Hermitian form $(-|-)_F$, contravariant with respect to $\overline{\omega}_{\mathcal{W}}$, on the Fock space $F(\Lambda, 0)$, i.e., the $\mathcal{W}_3$ module $F(\Lambda, 0)$ is unitary with respect to $(-|-)_F$. As in, e.g., Prop. 3.1 of [KaRa] this immediately implies the complete reducibility of $F(\Lambda, 0)$.

(ii) From the Frenkel-Kac-Segal vertex operator construction it follows that $\bigoplus_{\Lambda\in P} F(\Lambda,0)$ is an $\widehat{\mathfrak{sl}_3}$ module at level 1. In fact, it is known that

$$F \equiv \bigoplus_{\Lambda\in P} F(\Lambda,0) \cong L^{\widehat{\mathfrak{sl}_3}}(\Lambda_0)\oplus L^{\widehat{\mathfrak{sl}_3}}(\Lambda_1)\oplus L^{\widehat{\mathfrak{sl}_3}}(\Lambda_2)\,, \tag{2.87}$$

where $L^{\widehat{\mathfrak{sl}_3}}(\Lambda_i)$, $i=0,1,2$, denotes the integrable $\widehat{\mathfrak{sl}_3}$ highest weight module at level-1 with highest weight Λ_i. Under the horizontal algebra $\mathfrak{sl}_3$, F decomposes as

$$F \cong \bigoplus_{\Lambda'\in P_+} \mathcal{L}(\Lambda')\otimes V(\Lambda')\,. \tag{2.88}$$

But, since $\mathcal{W}_3$ is in the commutant of $\mathfrak{sl}_3$ [BBSS], it acts on the "multiplicity spaces" $V(\Lambda')$. In fact, by comparing the characters on each side of (2.88) and using Theorem 2.30, one easily verifies that $V(\Lambda')\cong L(\Lambda',0)$ (see, e.g., [KaPe]). Now, decomposing F under $\mathfrak{h}$ immediately gives (2.86).
(iii) Follows directly from (ii). □

Remarks

i. Note that since the weight multiplicities $m_\Lambda^{\Lambda'}$ are Weyl invariant, i.e., $m_{w\Lambda}^{\Lambda'} = m_\Lambda^{\Lambda'}$, for all $w\in W$, we have an isomorphism

$$F(\Lambda,0)\cong F(w\Lambda,0)\,. \tag{2.89}$$

Similar isomorphisms do *not* hold for $\alpha_0\neq 0$.
ii. All the results given to this point directly extend to $c=\ell$ representations of $\mathcal{W}_{\ell+1}$, where $\mathfrak{sl}_3$ is replaced by $\mathfrak{sl}_{\ell+1}$.
iii. If $\Lambda\in P_+$ we have the following explicit formula – specific to $\mathfrak{sl}_3$ – for the weight multiplicities $m_\Lambda^{\Lambda'}$

$$\begin{aligned}\sum_{\beta\in Q_+} m_\Lambda^{\Lambda+\beta} e^\beta &= \frac{1}{(1-e^{\alpha_1})(1-e^{\alpha_2})(1-e^{\alpha_3})}\\ &\quad - \frac{e^{(\Lambda+\rho,\alpha_1)\alpha_2}}{(1-e^{\alpha_2})(1-e^{\alpha_3})(1-e^{\alpha_1+2\alpha_2})}\\ &\quad - \frac{e^{(\Lambda+\rho,\alpha_2)\alpha_1}}{(1-e^{\alpha_1})(1-e^{\alpha_3})(1-e^{2\alpha_1+\alpha_2})}\,.\end{aligned} \tag{2.90}$$

Part (ii) of Theorem 2.31 can also be argued more heuristically, along the lines of the decomposition theorem for Virasoro Fock modules at $c=1$. One proceeds by explicitly constructing a standard set of singular vectors in the Fock space. Then, by comparing the character of the irreducible modules built on those singular vectors with the character of the Fock space, one concludes that this set exhausts all possible singular vectors. The standard set of singular vectors is naturally determined by the following construction: Consider, for $\Lambda\in P$, the "screening operators" $Q_i\ :\ F(\Lambda+\alpha_i,0)\longrightarrow F(\Lambda,0)$ associated to the simple roots α_i, $i=1,\ldots,\ell$ of $\mathfrak{g}$, defined by

$$Q_i = \oint \frac{dz}{2\pi i} \, e^{-i\alpha_i \cdot \phi(z)} . \tag{2.91}$$

It is straightforward to check that, for each i, Q_i is a $\mathcal{W}$ homomorphism. Also, the Q_i satisfy the Serre relations of $U_q(\mathfrak{n}_-)$ for $q = -1$ [BMP1]. (With appropriate phase cocycles in the definition of the Q_i one easily modifies this to the Serre relations of $\mathfrak{n}_-$.) Clearly then, provided it is nonvanishing, the image of the highest weight vector $|\Lambda + \alpha_i\rangle \in F(\Lambda + \alpha_i, 0)$ under Q_i is a singular vector in $F(\Lambda, 0)$. More generally, the image of $|\Lambda + \beta\rangle$ under the composite operator $Q_\beta = Q_{i_1} \cdots Q_{i_n}$, $\beta = \alpha_{i_1} + \ldots + \alpha_{i_n}$, yields a singular vector in $F(\Lambda, 0)$, provided this image is nonvanishing. From the inequality

$$h(\Lambda + \beta + r\alpha_i) \geq h(\Lambda + \beta) \qquad \text{iff} \quad r \leq (\Lambda + \beta, \alpha_i) , \tag{2.92}$$

it follows trivially that

$$(Q_i)^r |\Lambda + \beta\rangle = 0 \qquad \text{if } r \geq (\Lambda + \beta + \rho, \alpha_i) . \tag{2.93}$$

Further, because of the algebra of the Q_i, we may identify Q_β with a state at weight Λ in the $\mathfrak{sl}_3$ Verma module with highest weight $\Lambda + \beta$. Then (2.93) implies that the combinations of screening operators which act nontrivially on $|\Lambda + \beta\rangle$ can be, at most, identified with the weight Λ subspace of the irreducible quotient of the Verma module $M_{\Lambda+\beta}$. In other words, the number of nonvanishing singular vectors in $F(\Lambda, 0)$ of type $Q_\beta|\Lambda + \beta\rangle$, $\beta \in Q$, $\Lambda + \beta \in P$, is at most equal to $m_\Lambda^{\Lambda+\beta}$, the multiplicity of the weight Λ in the irreducible $\mathfrak{g}$ module with highest weight $\Lambda + \beta$. The proof would now be complete if we could show that the number of nonvanishing singular vectors constructed in this way is exactly *equal* to $m_\Lambda^{\Lambda+\beta}$. For then one can easily sum the characters of the irreducible modules built on these singular vectors, using Theorem 2.30, and finds that the result is exactly equal to the character of the Fock space $F(\Lambda, 0)$. Thus it would follow that these singular vectors in fact exhaust the set of all singular vectors in $F(\Lambda, 0)$. Hence the last step for the proof along these lines involves a careful study of the integral representations of the singular vectors constructed above. We have not carried out this step. However, it seems that the proof presented earlier could be interpreted exactly as a "nonvanishing theorem" for these integrals.

This concludes our discussion of the $c = 2$ Fock spaces. We now turn our attention to $c = 2$ (i.e., $\alpha_0 = 0$) Verma modules. Unfortunately, the precise submodule structure of Verma modules is unknown. We can, however, conclude a lot from the known structure of the Fock modules. First of all

Theorem 2.32 *Let $M(\Lambda, 0) \xrightarrow{\imath'} F(\Lambda, 0) \xrightarrow{\imath''} \overline{M}(\Lambda, 0)$ be the $\mathcal{W}$ homomorphisms of Theorem 2.26. We have $\imath'(M(\Lambda, 0)) \cong L(\Lambda, 0)$ or, in other words, $\imath'(I(\Lambda, \alpha_0)) = 0$. Similarly, $\imath''(F(\Lambda, 0)) \cong L(\Lambda, 0)$.*

Proof. Follows from the complete reducibility of $F(\Lambda, 0)$ and the fact that $M(\Lambda, 0)$ is generated by $\mathcal{W}_{3,-}$. □

Furthermore, since the composition series for $c = 2$ Fock modules is now completely known, and Verma modules have the same composition factors, we have

Theorem 2.33 *Let $\Lambda, \Lambda' \in P$ and let $w' \in W$ such that $w'\Lambda' \in P_+$, then*
i. $(M(\Lambda,0):L(\Lambda',0)) = m_\Lambda^{w'\Lambda'}$.
ii. $\mathrm{Prim}(M^{(\kappa)}(\Lambda,0)) \subset \coprod_{\Lambda'\in P_+\,;\,m_\Lambda^{\Lambda'}\neq 0} M^{(\kappa)}(\Lambda,0)_{(h(\Lambda'),w(\Lambda'))}$.
iii. $\mathrm{Hom}_{\mathcal{W}_3}(M^{(\kappa')}(\Lambda',0), M^{(\kappa)}(\Lambda,0))$ *is nontrivial only if* $m_\Lambda^{w'\Lambda'} \neq 0$.

Proof. (i) Follows from Theorems 2.29 (i) and 2.31 (iii).
(ii) Follows from (i), Lemma 2.15 (or Theorem 2.29 (ii)) and the filtration (2.49).
(iii) Follows from (ii), Lemma 2.17 (recall that $\alpha_0 = 0$ for $c = 2$) and the fact that the image of of $v_{\kappa'-1} \in M^{(\kappa')}(\Lambda',0)$ is a (nontrivial) p-singular vector in $M^{(\kappa)}(\Lambda,0)$. □

Remark. Obviously, for $\Lambda' \in P_+$, $\Lambda \in P$, we have $m_\Lambda^{\Lambda'} \neq 0$ only if $\Lambda' - \Lambda \in Q_+$. For $c = 2$ it therefore makes sense to extend the action W on $\mathfrak{h}^*$ to $\widehat{W}$ by defining

$$t_\alpha \Lambda = \Lambda + \alpha, \quad \alpha \in Q\,, \Lambda \in \mathfrak{h}^*\,, \tag{2.94}$$

where we have used that $\widehat{W} \cong W \ltimes T$, i.e., every $\widehat{w} \in \widehat{W}$ can be (uniquely) decomposed as $\widehat{w} = wt_\alpha$ for some $w \in W$, $\alpha \in Q$. Using this affine Weyl group action, Theorem 2.33 (ii) can now be formulated as the statement that the weights of primitive vectors in a generalized Verma module $M^{(\kappa)}(\Lambda,0)$ are on the orbit of Λ under $\widehat{W}$.

2.3.2 Explicit Examples

Let us introduce some more notation. For any set of vectors $S = \{v_1, v_2, \ldots\} \subset M(\Lambda,\alpha_0)$ we denote by $M(S) = M(v_1, v_2, \ldots)$ the submodule of $M(\Lambda,\alpha_0)$ generated by $\{v_1, v_2, \ldots\}$. Further, in the remainder of this chapter we generically use the symbol w for a primitive vector which is not p-singular, v for a p-singular vector which is not singular and u for a singular vector. By Theorem 2.33, the weights of the primitive vectors in $M(\Lambda,0)$ are concentrated on the orbit of Λ under the coset $\widehat{W}/W$, so we find it convenient to label primitive vectors by the Dynkin labels of the corresponding weight; i.e., we use the notation $w_{s_1s_2}$ for a primitive vector of weight $(h(\Lambda), w(\Lambda))$ where $\Lambda = s_1\Lambda_1 + s_2\Lambda_2$. This notation is adopted for the u and v vectors also. Moreover, we label $c = 2$ (generalized) Verma modules by the Dynkin indices of their highest weight (inside square brackets); i.e., we use the notation $M[s_1, s_2]$ for $M(s_1\Lambda_1 + s_2\Lambda_2, 0)$ etc. ($\alpha_0 = 0$ is implicitly understood in this notation).

Let us now discuss the example of Sect. 2.2.2in more detail. Consider the Verma module $M[0,0]$. Its highest weight vector is, conforming to the conventions above, denoted by u_{00}. We have already seen that $M[0,0]$, at L_0 level $h = 1$,

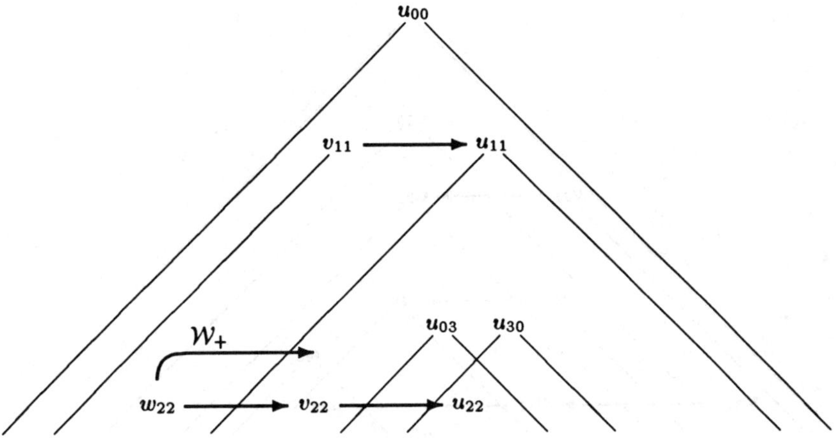

Fig. 2.1. Embedding structure for $M[0,0]$

consists of a two-dimensional Jordan block under W_0 corresponding to weight $(h = 1, w = 0)$ $(\Lambda = \Lambda_1 + \Lambda_2)$; i.e., at this weight there is a singular vector u_{11} and a p-singular vector v_{11}. Note that since $m_0^{\Lambda_1+\Lambda_2} = 2$ this is consistent with Theorem 2.33 (ii). Next, at energy level $h = 3$ we find two singular vectors u_{30} and u_{03} in accordance with $m_0^{3\Lambda_1} = m_0^{3\Lambda_2} = 1$.

At energy level $h = 4$ something interesting happens. Explicit computation shows that there are only two p-singular vectors, while on the other hand $m_0^{2\Lambda_1+2\Lambda_2} = 3$. The resolution of this paradox is that – besides the singular vector u_{22}, and p-singular vector v_{22} – there is a primitive vector w_{22} at this level. In fact, the generalized eigenspace corresponding to $\Lambda = 2\Lambda_1 + 2\Lambda_2$ (i.e., $h = 4$, $w = 0$) has dimension four and decomposes into $3 + 1$ dimensional Jordan blocks. The remaining vector, i.e., the vector in the 1-dimensional block, is in the irreducible module. As far as the content of the submodules generated by these primitive vectors is concerned, explicit computation shows that $u_{30}, u_{03}, v_{22} \in M(u_{11})$, $w_{22} \in M(v_{11})$, $u_{22} \in M(u_{30})$, $u_{22} \in M(u_{03})$, but $w_{22} \notin M(u_{11})$ and $v_{22} \notin M(u_{30}, u_{03})$. Combining the fact that $w_{22} \in M(v_{11})$ but $w_{22} \notin M(u_{11})$, with the fact that w_{22} is primitive, leads, in particular, to the conclusion that $\mathcal{W}_{3,+} \cdot w_{22} \subset M(u_{11})$; i.e., w_{22} becomes singular in the quotient module $M[0,0]/M(u_{11})$. We have checked this by explicit calculation as well.

All of this information is summarized in Fig. 2.1. The figure contains all primitive vectors up to level 6 (the level increases going down), the horizontal arrows between the primitive vectors refer to the action of $W_0 - w$. The cones built on a set of vectors S depict the module generated by S, i.e., $M(S)$.

One may deduce from the above that (a possible choice for) the Jordan-Hölder series $\mathrm{JH}_N(M[0,0])$, $N \leq 3$, is given by (see Remark (ii) after Theorem 2.4)

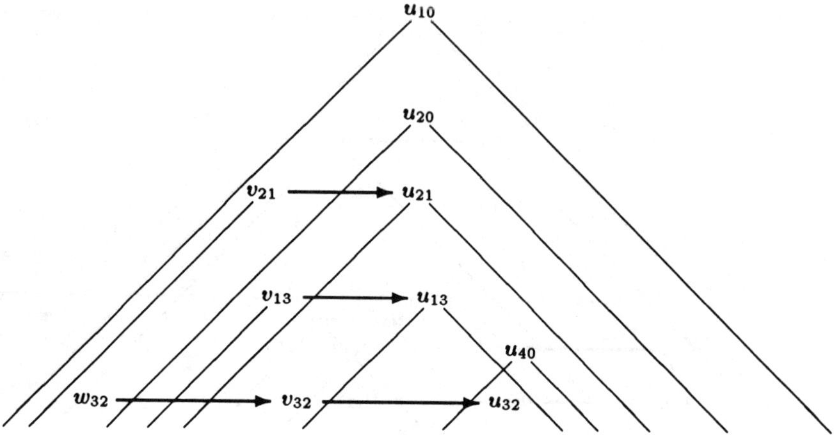

Fig. 2.2. Embedding structure for $M[1,0]$

$$M \supset M(v_{11}) \supset M(u_{11}) \supset M(u_{30},u_{03}) \supset M(u_{30}). \tag{2.95}$$

The quotients are isomorphic (up to $N=3$) with $L[0,0], L[1,1], L[1,1], L[0,3]$ and $L[3,0]$, respectively. For $N \geq 4$, though, the quotient $M(v_{11})/M(u_{11})$ is no longer irreducible due to the appearance of the primitive vector w_{22}. The following is, however, a viable Jordan-Hölder series for $N \leq 6$

$$\begin{aligned} M &\supset M(v_{11}) \supset M(w_{22},u_{11}) \supset M(u_{11}) \supset M(u_{30},u_{03},v_{22}) \\ &\supset M(u_{30},v_{22}) \supset M(v_{22}) \supset M(u_{22}). \end{aligned} \tag{2.96}$$

In Appendix A we have summarized some explicit computations regarding the submodule structure of $c=2$ Verma modules. In these tables we have labelled the Verma modules as well as irreducible modules by the the Dynkin indices of their highest weights, e.g., $M[s_1,s_2]$, as before. The triality of Λ is defined, as usual, by $(s_1+2s_2) \bmod 3$.

Tables A.1–A.4 provide a list of primitive vectors (arranged in Jordan blocks) for (generalized) Verma modules of low lying highest weights and levels.[8] A prime on a primitive vector in $M^{(2)}[s_1,s_2]$ indicates that this vector is in the kernel of the natural homomorphism $M^{(2)}[s_1,s_2] \to M[s_1-1,s_2-1]$. Tables A.5 and A.6 list the dimensions of the level h subspaces of irreducible $c=2$ modules and Tables A.7–A.9 list the dimensions for some submodules of, respectively,

[8] For $h \geq 9$, Tables A.1–A.4, do not necessarily give the entire Jordan blocks, i.e., it is possible that the Jordan blocks contain additional vectors which are not primitive. Also, there often exist additional Jordan blocks at the same weight (h,w) as the ones in the table, e.g., $M[0,0]$ has an additional 1-dimensional Jordan block at $(h,w) = (4,0)$ – the corresponding vector $\tilde{u}_{22}$ is in $L[0,0]$.

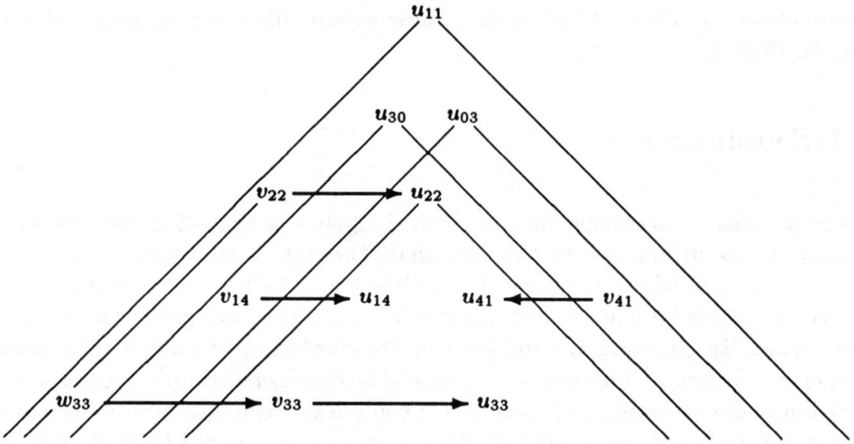

Fig. 2.3. Embedding structure for $M[1,1]$

$M[0,0], M[1,0]$ and $M[1,1]$, generated by primitive vectors. All computations were done with the help of Mathematica^{TM}, except for those in Tables A.5 and A.6 which follow from Theorem 2.30, and some cases for which the submodule is known to be isomorphic to a Verma module (see the discussion in Sect. 2.2.2).

With the help of the tables in Appendix A one can verify that, for example, the quotients in the JH series (2.95) and (2.96) are indeed irreducible up to the asserted level. Additional examples, like the one discussed above, can be worked out using the tables. For illustrational purposes we give the embedding structures of $M[1,0]$ and $M[1,1]$ in Figs. 2.2 and 2.3.

As another example of a JH series, that can be read off from the tables, we give $\mathrm{JH}_N(M[1,1])$ valid for $N \leq 8$

$$\begin{aligned} M[1,1] &\supset M(u_{30}, u_{03}, v_{22}) \supset M(u_{30}, u_{03}, w_{33}) \supset \\ &\supset M(u_{03}, w_{33}, v_{41}) \supset M(u_{22}, v_{41}, v_{14}, w_{33}) \supset \\ &\supset M(v_{41}, v_{14}, w_{33}) \supset M(v_{41}, v_{14}) \supset \ldots . \end{aligned} \tag{2.97}$$

We conclude this section with the following observation. While for the Virasoro algebra all submodules of Verma modules are generated by singular vectors (see, e.g., [FeFu]), here we have

Corollary 2.34 *Not every submodule of a $\mathcal{W}_3$ Verma module is generated by p-singular vectors.*

Proof. The submodule $M(w_{22}, u_{11})$ of $M[0,0]$ in the example above provides a counterexample. □

The Corollary above is another manifestation of the fact that the $\mathcal{W}_3$ algebra behaves, in many respects, as a rank 3 Lie algebra (in fact as $\widehat{\mathfrak{sl}_3}$), while the Virasoro algebra is a rank 2 Lie algebra whose submodule structure is considerably simpler [FeFu].

2.4 Resolutions

An important construction in homological algebra is that of a "resolution" of a module. Its utility lies in the fact that, through a resolution of a module V, many computations involving the module V can be reduced to computations involving the modules in the resolution of V – e.g., by means of spectral sequence techniques. By choosing the modules in the resolution to have certain simple properties – exactly which properties should be considered simple depends on the problem under investigation – the latter computations might become tractable. In the physical problem of the 4D $\mathcal{W}_3$ string we are required to work with free fields, since these are the embedding coordinates of the string into spacetime. Thus, in the context of this paper, the need for resolutions follows from the complicated nature of the free field realization in Theorem 2.24. Given the Fock space decomposition in Theorem 2.31 for $c = 2$, it is enough to understand resolutions of irreducible modules.

Definition 2.35 *A resolution of a $\mathcal{W}_3$ module $V \in \mathcal{O}$ is a $\mathbb{Z}$ graded complex $(\mathcal{C}, \delta)$ of $\mathcal{W}_3$ modules with a differential δ of degree 1, i.e., $\delta : \mathcal{C}^{(n)} \longrightarrow \mathcal{C}^{(n+1)}$, $\delta^2 = 0$, such that $H^n(\delta, \mathcal{C}) \cong \delta_{n,0} V$.*

As an example of a resolution, consider[9]

Theorem 2.36 *There exists a resolution $\mathcal{C}^{(n)}$ for the $c = 2$ irreducible module $L(\Lambda, 0)$, $\Lambda \in P_+$, in terms of Fock spaces. Here*

$$\mathcal{C}^{(n)} \cong \bigoplus_{\{w \in W \mid \ell(w) = n\}} F(w(\Lambda + \rho) - \rho, 0)\,. \tag{2.98}$$

Proof. Follows directly from the Fock space decomposition (2.86). □

Interestingly, the resolution is of finite length – it contains a finite number of Fock spaces, here labelled by the Weyl group of $\mathfrak{sl}_3$. This result is clearly consistent with the character formula (2.85). In fact, with the differential constructed from screening charges as discussed below the proof of Theorem 2.31,

[9] More generally, Fock space resolutions for the $\mathcal{W}_3$ irreducible modules with a completely degenerate highest weight ("minimal models") were constructed by applying the Quantum Drinfel'd-Sokolov reduction to the Fock space resolutions of admissible $\widehat{\mathfrak{sl}_3}$ modules [BMP2,FKW].

the structure of the resolutions reflect those of the (dual to the) BGG resolutions for irreducible finite dimensional $\mathfrak{sl}_3$ modules.

2.4.1 Verma Module Resolutions of $c = 2$ Irreducible $\mathcal{W}_3$ Modules

Verma modules, and also generalized Verma modules, have the "simple property" that they are, in a sense, free over $\mathcal{W}_{3,-}$. This is the main reason that Verma module resolutions (also called BGG resolutions) are important homological constructions. As we have already seen in Sect. 2.3.2, resolutions of $c = 2$ irreducible $\mathcal{W}_3$ modules in terms of Verma modules will not exist in general; e.g., the kernel of the canonical projection $M[0,0] \to L[0,0]$ is isomorphic with the image of a *generalized* Verma module, namely $M^{(2)}[1,1]$, in $M[0,0]$. However, in this section we present, for any given $c = 2$ irreducible $\mathcal{W}_3$ module $L(\Lambda,0)$, $\Lambda \in P_+$, the construction of a resolution, to be denoted by $(\mathcal{M}(\Lambda,0),\delta)$, in terms of generalized Verma modules; i.e., a resolution where each of the terms $\mathcal{M}^{(n)}(\Lambda,0)$ is the direct sum of a (finite) number of generalized Verma modules of $\mathcal{W}_3$. By construction we have $\mathcal{M}^{(n)}(\Lambda,0) = 0$ for $n > 0$. It turns out that also $\mathcal{M}^{(n)}(\Lambda,0) = 0$ for n sufficiently negative, namely $n < -4$, so that the resolutions are of "finite length." It should be remarked that the fact that such resolutions exist in the first place is rather remarkable, since, as we have seen in Corollary 2.34, not every submodule of a $\mathcal{W}_3$ Verma module is generated by p-singular vectors.

Let us now, assuming their existence, try to construct such generalized Verma module resolutions by combining the various results of the previous sections. By Theorem 2.33 (iii), nontrivial homomorphisms $M^{(\kappa')}(\Lambda',0) \to M^{(\kappa)}(\Lambda,0)$ with $\Lambda,\Lambda' \in P$ exist only if $m_\Lambda^{w'\Lambda'} \neq 0$, where $w' \in W$ is such that $w'\Lambda' \in P_+$. Using the redundancy in parametrization by Λ (Lemma 2.17 with $\alpha_0 = 0$), in determining the various terms $\mathcal{M}^{(n)}(\Lambda,0)$ in a resolution of an irreducible module $L(\Lambda,0)$, it is sufficient to only consider sums of generalized Verma modules $M^{(\kappa)}(\Lambda',0)$ with $\Lambda' \in P_+$ such that $m_\Lambda^{\Lambda'} \neq 0$.

Furthermore, since

$$w^{-1}(w(\Lambda+\rho)-\rho) = \Lambda+\rho-w^{-1}\rho\,, \tag{2.99}$$

one might think that, in analogy with Theorem 2.36, only (generalized) Verma modules with highest weights $\Lambda' = \Lambda+\rho-w^{-1}\rho$, $w \in W$ – corresponding to translations $t_{\rho-w^{-1}\rho}$ in (2.94) – will enter the resolution. This turns out to be false. In addition, as we will see later, weights corresponding to the translation t_ρ will arise.[10]

As will become clear in Sect. 3.4.1 it is useful to introduce an extension $\widetilde{W}$ of the Weyl group W of $\mathfrak{sl}_3$, $\widetilde{W} \equiv W \cup \{\sigma_1,\sigma_2\}$ and extend the length function on W to $\widetilde{W}$ by assigning $\ell(\sigma_1) = 1$ and $\ell(\sigma_2) = 2$. Similarly, we can extend the "twisted length" $\ell_w(\sigma) \equiv \ell(w^{-1}\sigma) - \ell(w^{-1})$, $w,\sigma \in W$ to $\sigma \in \widetilde{W}$ by defining the multiplications

[10]Unfortunately, we have no intrinsic understanding why exactly this particular subset of $T \equiv \{t_\alpha \mid \alpha \in Q\}$ occurs in the generalized Verma module resolutions at $c = 2$.

$$w\sigma_i = \sigma_i\,, \qquad i = 1,2\,, \quad w \in W\,. \tag{2.100}$$

Furthermore, $\widetilde{W}$ acts on $\mathfrak{h}^*$ by $\sigma_i\lambda = 0$, $i = 1,2$. Note that this action is consistent with the multiplications (2.100). Then, motivated by (2.99), we define the "circle action" of $\widetilde{W}$ on $\mathfrak{h}^*$ by

$$\sigma \circ \Lambda = \Lambda + \rho - \sigma\rho\,, \qquad \sigma \in \widetilde{W}\,. \tag{2.101}$$

To denote the weights in the resolution we will use both the notation $\sigma \circ \Lambda$ as well as their Dynkin labels. Below we provide a translation table for quick reference.

Table 2.1. The circle action of $\widetilde{W}$

σ	$\sigma \circ \Lambda$	$[\sigma \circ \Lambda]$	$\ell(\sigma)$
1	Λ	$[s_1, s_2]$	0
r_1	$\Lambda + \alpha_1$	$[s_1 + 2, s_2 - 1]$	1
r_2	$\Lambda + \alpha_2$	$[s_1 - 1, s_2 + 2]$	1
σ_1	$\Lambda + \alpha_1 + \alpha_2$	$[s_1 + 1, s_2 + 1]$	1
r_{12}	$\Lambda + 2\alpha_1 + \alpha_2$	$[s_1 + 3, s_2]$	2
r_{21}	$\Lambda + \alpha_1 + 2\alpha_2$	$[s_1, s_2 + 3]$	2
σ_2	$\Lambda + \alpha_1 + \alpha_2$	$[s_1 + 1, s_2 + 1]$	2
r_3	$\Lambda + 2\alpha_1 + 2\alpha_2$	$[s_1 + 2, s_2 + 2]$	3

As we will see, not all $\sigma \in \widetilde{W}$ enter the (generalized) Verma module resolution of $L(\Lambda, 0)$.[11] It proves useful to define a subset $W(\Lambda) \subset \widetilde{W}$ for all $\Lambda \in P_+$ as follows

$$W(\Lambda) = \begin{cases} \widetilde{W} & \text{if } \Lambda \in P_{++}\,, \\ \{1, r_i, \sigma_1, r_{12}, r_{21}, r_3\} & \text{if } (\Lambda, \alpha_i) = 0,\ \Lambda \neq 0\,, \\ \{1, \sigma_1, r_{12}, r_{21}, r_3\} & \text{if } \Lambda = 0\,. \end{cases} \tag{2.102}$$

The resolutions $(\mathcal{M}(\Lambda,0), \delta)$ of $L(\Lambda, 0)$ have the following structure: Only generalized Verma modules $M^{(\kappa)}(\sigma \circ \Lambda, 0)$ with $\sigma \in W(\Lambda)$ occur. For any given $\sigma \in W(\Lambda)$ a (generalized) Verma module with either $\kappa = 1$ or $\kappa = 2$ and highest weight $\sigma \circ \Lambda$ occurs as a direct summand of $\mathcal{M}^{(n)}(\Lambda, 0)$ for $n = -\ell(\sigma)$. If $M^{(2)}(\sigma \circ \Lambda, 0)$ occurs as a direct summand of $\mathcal{M}^{(-\ell(\sigma))}(\Lambda, 0)$, then $M(\sigma \circ \Lambda, 0)$ occurs as a direct summand of $\mathcal{M}^{(-\ell(\sigma)-1)}(\Lambda, 0)$ provided $\Lambda \neq 0$. Otherwise, i.e., if $M(\sigma \circ \Lambda, 0)$ occurs as a direct summand of $\mathcal{M}^{(-\ell(\sigma))}(\Lambda, 0)$ and/or $\Lambda = 0$, the generalized Verma modules with highest weight $\sigma \circ \Lambda$ will not occur as a summand of $\mathcal{M}^{(n)}(\Lambda, 0)$ for $n \neq -\ell(\sigma)$.

A more precise statement is contained in the following

[11] This is, of course, intimately related to the fact that $m^{\Lambda}_{\Lambda - \alpha_i} = 0$ iff $(\Lambda, \alpha_i) = 0$.

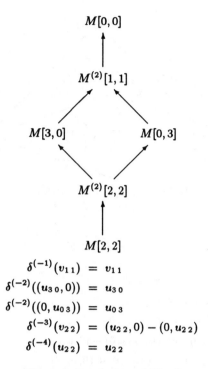

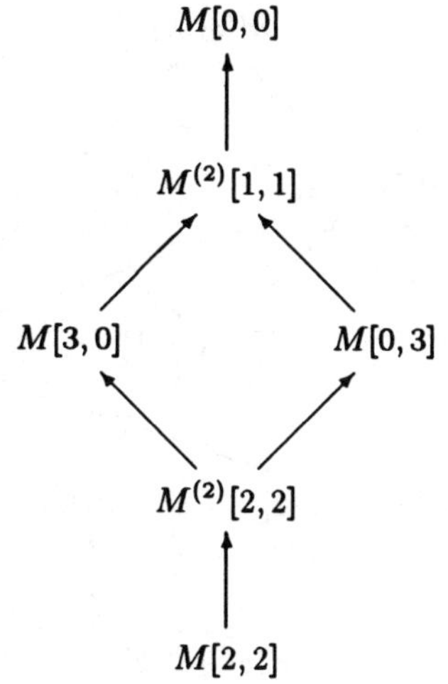

$$\begin{aligned}
\delta^{(-1)}(v_{1\,1}) &= v_{1\,1} \\
\delta^{(-2)}((u_{3\,0}, 0)) &= u_{3\,0} \\
\delta^{(-2)}((0, u_{0\,3})) &= u_{0\,3} \\
\delta^{(-3)}(v_{2\,2}) &= (u_{2\,2}, 0) - (0, u_{2\,2}) \\
\delta^{(-4)}(u_{2\,2}) &= u_{2\,2}
\end{aligned}$$

Fig. 2.4. Resolution of $L[0,0]$

Conjecture 2.37 *The resolution, $(\mathcal{M}(\Lambda, 0), \delta)$, of an irreducible $\mathcal{W}_3$ module $L(\Lambda, 0)$, $\Lambda \in P_+$, is one of three types, depending on whether $\Lambda \in P_{++}$, $\Lambda \in P_+ \backslash P_{++}$ but $\Lambda \neq 0$, or $\Lambda = 0$. The resolutions are depicted in Figs. 2.4–2.6. In these pictures, each $\mathcal{M}^{(n)}$ (n decreases going downward) is the direct sum of the generalized Verma modules on the same horizontal line, and the differentials $\delta^{(n)} : \mathcal{M}^{(n)} \to \mathcal{M}^{(n+1)}$ are given by the collection of homomorphisms represented by the arrows. The homomorphisms are fully determined by the image of the lowest vector, i.e., $v_{\kappa-1}$, in each highest weight Jordan block of the generalized Verma modules $M^{(\kappa)}(\Lambda', 0)$.*

The evidence for Conjecture 2.37. We have explicitly carried out the program of constructing and checking the resolution in four different cases; namely, for $L[0,0]$, $L[1,0]$, $L[2,0]$, and $L[1,1]$. This is done as follows: First we examine the primitive vector structure of $M[s_1, s_2]$, and of the (generalized) Verma modules with the same highest weights as the primitive vectors, and so on. This information is given, down to a finite L_0 level, in Tables A.1–A.4 of Appendix A (see also the discussion in Sect. 2.3.2). At the first step in the resolution, we choose $\mathcal{M}^{(-1)}[s_1, s_2]$ so that the image of $\mathcal{M}^{(-1)}[s_1, s_2]$ in $M[s_1, s_2]$ is precisely the maximal ideal $I[s_1, s_2]$. From the multiplicities in Tables A.1 and A.2 one

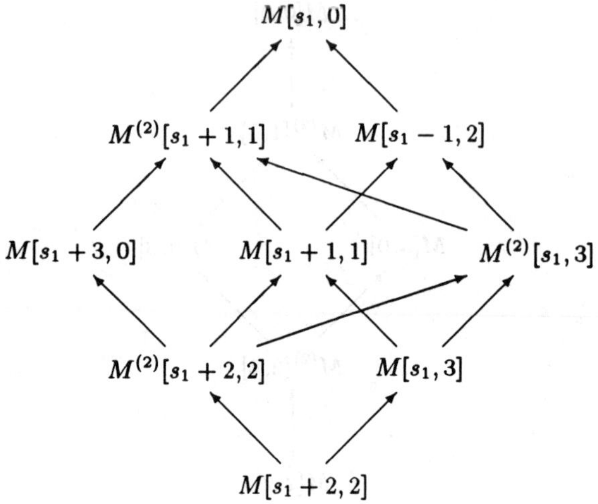

$$\begin{aligned}
\delta^{(-1)}((v_{s_1+1\,1}, 0)) &= v_{s_1+1\,1} \\
\delta^{(-1)}((0, u_{s_1-1\,2})) &= u_{s_1-1\,2} \\
\delta^{(-2)}((u_{s_1+3\,0}, 0, 0)) &= (u'_{s_1+3\,0}, 0) \\
\delta^{(-2)}((0, u_{s_1+1\,1}, 0)) &= (u_{s_1+1\,1}, 0) - (0, u_{s_1+1\,1}) \\
\delta^{(-2)}((0, 0, v_{s_1\,3})) &= -(v_{s_1\,3}, 0) + (0, v_{s_1\,3}) \\
\delta^{(-3)}((v_{s_1+2\,2}, 0)) &= \tfrac{1}{12}(u_{s_1+2\,2}, 0, 0) + (0, v_{s_1+2\,2}, 0) + (0, 0, v_{s_1+2\,2}) \\
\delta^{(-3)}((0, u_{s_1\,3})) &= (0, u_{s_1\,3}, 0) + (0, 0, u_{s_1\,3}) \\
\delta^{(-4)}(u_{s_1+2\,2}) &= -(u_{s_1+2\,2}, 0) + (0, u_{s_1+2\,2})
\end{aligned}$$

Fig. 2.5. Resolution of $L[s_1, 0]$, $s_1 > 0$

now concludes that the previously constructed homomorphism has a nontrivial kernel, i.e., that the various summands of $\mathcal{M}^{(-1)}[s_1, s_2]$ have some "overlap" in $M[s_1, s_2]$. This is taken care of by a proper choice of $\mathcal{M}^{(-2)}[s_1, s_2]$, and so on. This reasoning by itself leads to the "minimal Ansatz" for the resolutions as depicted in the figures. Secondly, we fix the normalization of all homomorphisms constituting the differential by imposing the condition that $\delta^{(n+1)}\delta^{(n)} = 0$ on the highest weight vectors. The last step, the actual verification of the resolution, now comes down to the explicit calculation of the dimensions of the images $\mathcal{I}^{(n)} \subset \mathcal{M}^{(n)}$ of the homomorphisms $\delta^{(n-1)} : \mathcal{M}^{(n-1)} \to \mathcal{M}^{(n)}$ at each step in the resolution. Then we must prove, for all $n \leq 0$, that at each L_0 level h,

$$\dim \mathcal{M}^{(n)}_{(h)} = \dim \mathcal{I}^{(n)}_{(h)} + \dim \mathcal{I}^{(n+1)}_{(h)}, \tag{2.103}$$

where we have defined, for convenience, $\mathcal{I}^{(1)} \equiv L[s_1, s_2]$.

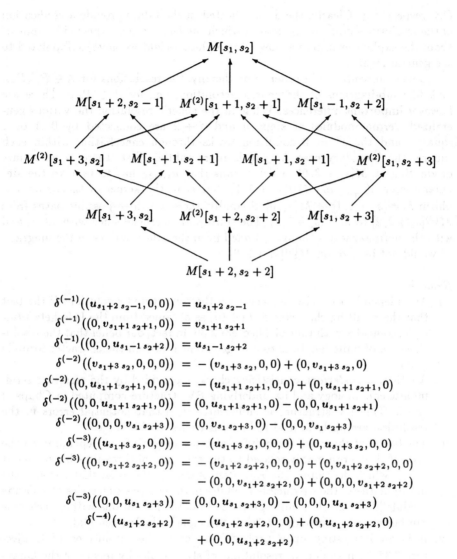

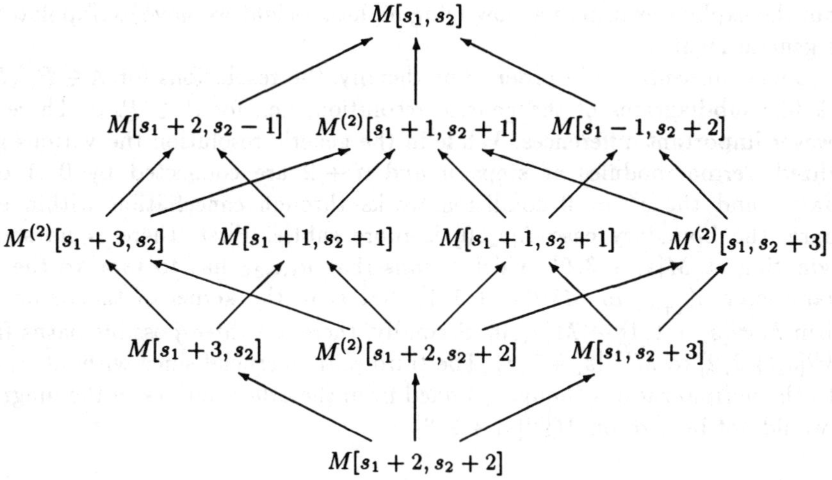

$$
\begin{aligned}
\delta^{(-1)}((u_{s_1+2\,s_2-1},0,0)) &= u_{s_1+2\,s_2-1} \\
\delta^{(-1)}((0,v_{s_1+1\,s_2+1},0)) &= v_{s_1+1\,s_2+1} \\
\delta^{(-1)}((0,0,u_{s_1-1\,s_2+2})) &= u_{s_1-1\,s_2+2} \\
\delta^{(-2)}((v_{s_1+3\,s_2},0,0,0)) &= -(v_{s_1+3\,s_2},0,0)+(0,v_{s_1+3\,s_2},0) \\
\delta^{(-2)}((0,u_{s_1+1\,s_2+1},0,0)) &= -(u_{s_1+1\,s_2+1},0,0)+(0,u_{s_1+1\,s_2+1},0) \\
\delta^{(-2)}((0,0,u_{s_1+1\,s_2+1},0)) &= (0,u_{s_1+1\,s_2+1},0)-(0,0,u_{s_1+1\,s_2+1}) \\
\delta^{(-2)}((0,0,0,v_{s_1\,s_2+3})) &= (0,v_{s_1\,s_2+3},0)-(0,0,v_{s_1\,s_2+3}) \\
\delta^{(-3)}((u_{s_1+3\,s_2},0,0)) &= -(u_{s_1+3\,s_2},0,0,0)+(0,u_{s_1+3\,s_2},0,0) \\
\delta^{(-3)}((0,v_{s_1+2\,s_2+2},0)) &= -(v_{s_1+2\,s_2+2},0,0,0)+(0,v_{s_1+2\,s_2+2},0,0) \\
&\quad -(0,0,v_{s_1+2\,s_2+2},0)+(0,0,0,v_{s_1+2\,s_2+2}) \\
\delta^{(-3)}((0,0,u_{s_1\,s_2+3})) &= (0,0,u_{s_1\,s_2+3},0)-(0,0,0,u_{s_1\,s_2+3}) \\
\delta^{(-4)}(u_{s_1+2\,s_2+2}) &= -(u_{s_1+2\,s_2+2},0,0)+(0,u_{s_1+2\,s_2+2},0) \\
&\quad +(0,0,u_{s_1+2\,s_2+2})
\end{aligned}
$$

Fig. 2.6. Resolution of $L[s_1, s_2]$, $s_1, s_2 > 0$

To compute the dimension of the image at a specific L_0 level in a given module is straightforward in principle: We calculate the action of the standard basis vector of $U(\mathcal{W}_{3,-})$ – as given in Definition 2.13 – on the p-singular vectors of interest, the level of each basis vector being chosen so that the result is an vector in the given module at L_0 level h. Then we calculate the rank of the matrix of coefficients of these vectors in the standard L_0 level h basis of the given module. The computations are done using Mathematica™. The results for $L[0,0]$, $L[1,0]$, $L[1,1]$ and $L[2,0]$ are displayed in Tables C.1, C.2, C.3 and

C.4, respectively. Clearly, the data collected in the tables provide a verification of the resolutions down to L_0 level at which the last space is expected to appear. From the explicit examples at low lying highest weight we have extrapolated to the general result.

Some comments are in order. Superficially, the resolutions for $\Lambda \in P_+ \backslash P_{++}$ look like subdiagrams of the generic resolution, i.e., for $\Lambda \in P_{++}$. There are however important differences. While in the generic resolution the various generalized Verma modules at steps n and $n+2$ are connected by 0, 1 or 2 squares, and the $\delta^2 = 0$ condition works through cancellation within each square, the boundary case $L[s_1, 0]$ is more subtle. First, there is no square originating at $M[s_1+3, 0]$, which means that $u_{s_1+3 0}$ has to map to the singular vector $u'_{s_1+3 0}$ in $M^{(2)}[s_1+1, 1]$ that is in the kernel of the homomorphism $M^{(2)}[s_1+1, 1] \rightarrow M[s_1, 0]$. Secondly, there are three possible paths from $M^{(2)}[s_1+2, 2]$ to $M^{(2)}[s_1+1, 1]$. The third path is crucial since without it, and with the normalizations (uniquely) fixed from the other squares in the diagram, δ^2 would not be zero on $M^{(2)}[s_1+2, 2]$.

Remarks

i. An independent consistency check on the conjectured resolutions is the fact that the resulting character of $L(\Lambda, 0)$, as obtained from the Lefschetz principle, coincides with that of Theorem 2.30. In fact, our belief that the resolutions are of finite length is to a large extent based on the character formula (2.85).
ii. Another, *a posteriori*, consistency check is provided by the resulting semi-infinite cohomology and its underlying BV structure computed in Chaps. 3 and 5. This BV structure is sufficiently rigid that potential errors in the resolution are likely to lead to inconsistencies at this stage.
iii. The fact that there are three different types of resolutions depending on the type of Λ, is presumably related to the existence of three possible posets at $c=2$, which determine the Kazhdan-Lusztig polynomial that encodes the multiplicities of the irreducible modules in the composition series of a Verma module [dVvD2]. It is quite probable that the resolutions within each case can be related by invoking a "shift principle" a la Jantzen [Ja].
iv. It is an interesting open problem to derive the resolutions of Conjecture 2.37 from analogous resolutions of $\widehat{\mathfrak{sl}_3}$ modules by means of the Quantum Drinfel'd-Sokolov reduction.

This concludes our discussion of the structure theory of $\mathcal{W}_3$ modules. In the next section we discuss how to apply the above results in the computation of the semi-infinite cohomology of the $\mathcal{W}_3$ algebra.

3 BRST Cohomology of the 4D $\mathcal{W}_3$ String

3.1 Complexes of Semi-infinite Cohomology of the $\mathcal{W}_3$ Algebra

The notion of semi-infinite cohomology of the $\mathcal{W}_3$ algebra with values in a positive energy module was first introduced in [TM]. In this chapter we briefly summarize an extension of this construction to tensor products of *two* positive energy modules [BLNW1].

3.1.1 The $\mathcal{W}_3$ Ghost System

The first step in the construction of a complex for the semi-infinite cohomology of the $\mathcal{W}_3$ algebra is the same as that for the case of the Virasoro or affine Lie algebras (see, e.g., [Fe,FGZ]). Corresponding to the currents $T(z)$ and $W(z)$, we introduce two anticommuting bc ghost systems $(b^{[j]}, c^{[j]})$, with $j = 2$ and $j = 3$, respectively. The nonvanishing OPEs of the ghost fields are

$$c^{[j]}(z)b^{[j']}(w) \sim \frac{\delta^{jj'}}{z-w}, \quad b^{[j]}(z)c^{[j']}(w) \sim \frac{\delta^{jj'}}{z-w}, \tag{3.1}$$

so that the mode operators, $c_n^{[j]}$ and $b_n^{[j]}$, defined by the expansions,

$$c^{[j]}(z) = \sum_{n\in\mathbb{Z}} c_n^{[j]} z^{-n+j-1}, \quad b^{[j]}(z) = \sum_{n\in\mathbb{Z}} b_n^{[j]} z^{-n-j}, \tag{3.2}$$

satisfy the anticommutation relations of a Clifford algebra:

$$[c_m^{[j]}, c_n^{[j']}]_+ = 0, \quad [b_m^{[j]}, b_n^{[j']}]_+ = 0, \quad [c_m^{[j]}, b_n^{[j']}]_+ = \delta^{jj'}\delta_{m+n,0}, \quad m,n \in \mathbb{Z}. \tag{3.3}$$

The dimensions of the fields $b^{[j]}(z)$ and $c^{[j]}(z)$ are equal to j and $-j+1$, respectively, and follow from the stress-energy tensor

$$T^{gh\,[j]}(z) = -(j-1)(\partial b^{[j]}c^{[j]})(z) - j(b^{[j]}\partial c^{[j]})(z), \quad j = 2,3. \tag{3.4}$$

Let F^{gh} denote the ghost Fock space defined as the standard positive energy module of the Clifford algebra (3.3). It is freely generated by $c_{-n}^{[j]}$, $n \geq 0$, and $b_{-n}^{[j]}$, $n > 0$, from the "physical" ghost vacuum $|0\rangle_{\mathrm{gh}}$, satisfying

$$c_n^{[j]}|0\rangle_{\mathrm{gh}} = 0, \quad n \geq 1, \qquad b_n^{[j]}|0\rangle_{\mathrm{gh}} = 0, \quad n \geq 0, \quad j = 2,3. \tag{3.5}$$

A standard basis in F^{gh} consists of the elements

$$g_{k_1\ldots k_K;\ell_1\ldots\ell_L;m_1\ldots m_M;n_1\ldots n_N} = c^{[2]}_{-k_1}\cdots b^{[2]}_{-\ell_1}\cdots c^{[3]}_{-m_1}\cdots b^{[3]}_{-n_1}\cdots|0\rangle_{\rm gh}, \qquad (3.6)$$

where $k_1 > \ldots > k_K \geq 0$, etc. Exactly as discussed in Sect. 2.2.3, there is an isomorphism between the states in (3.6) and the chiral algebra $\mathfrak{W}^{\rm gh}$ of fields obtained by a finite number of normal products of a finite number of derivatives of the basic fields $(b^{[j]}, c^{[j]})$: for any state $|\mathcal{O}\rangle \in F^{\rm gh}$, there is a corresponding field $\mathcal{O}(z) \in \mathfrak{W}^{\rm gh}$ such that $|\mathcal{O}\rangle = \lim_{z\to 0}\mathcal{O}(z)|0\rangle$, where $|0\rangle$ is the $SL(2,\mathbb{C})$ invariant vacuum (for the ghost system, $|0\rangle = b^{[2]}_{-1}b^{[3]}_{-1}b^{[3]}_{-2}|0\rangle_{\rm gh}$). Both $\mathfrak{W}^{\rm gh}$ and $F^{\rm gh}$ are graded by the ghost number gh$(\cdot)$, with the usual assignment gh$(c^{[j]}) = 1$ and gh$(b^{[j]}) = -1$, and normalized such that the ghost number of the identity operator, i.e., the $SL(2,\mathbb{C})$ vacuum, is equal to zero. In this normalization the ghost number of the physical ghost vacuum, $|0\rangle_{\rm gh}$, which corresponds to the operator $c^{[2]}\partial c^{[3]}c^{[3]}(z)$, is equal to three.

It is convenient to also define the Fock space $\overline{F}^{\rm gh}$, precisely as $F^{\rm gh}$ but now with a vacuum $|\bar{0}\rangle_{\rm gh}$, satisfying

$$c^{[j]}_n|\bar{0}\rangle_{\rm gh} = 0\,, \quad n \geq 0\,, \qquad b^{[j]}_n|\bar{0}\rangle_{\rm gh} = 0\,, \quad n \geq 1\,, \quad j = 2,3\,. \qquad (3.7)$$

We have a $\mathfrak{W}^{\rm gh}$ isomorphism $\overline{F}^{\rm gh} \cong F^{\rm gh}$ by identifying $|\bar{0}\rangle_{\rm gh} = c^{[2]}_0 c^{[3]}_0|0\rangle_{\rm gh}$. This isomorphism preserves ghost number if we assign ghost number five to $|\bar{0}\rangle_{\rm gh}$. We may now introduce the $\mathbb{C}$-linear anti-involution $\omega_{\rm gh}$ of $\mathfrak{W}^{\rm gh}$ defined by

$$\omega_{\rm gh}(c^{[j]}_n) = c^{[j]}_{-n}\,, \qquad \omega_{\rm gh}(b^{[j]}_n) = b^{[j]}_{-n}\,. \qquad (3.8)$$

Similarly to the discussion of Fock spaces in Sect. 2.2.3, we have

Theorem 3.1 *There exists a unique bilinear form $\langle -|-\rangle_{\rm gh} : \overline{F}^{\rm gh} \times F^{\rm gh} \longrightarrow \mathbb{C}$, contravariant with respect to $\omega_{\rm gh}$, such that $\langle\bar{0}|0\rangle_{\rm gh} = 1$. This form is nondegenerate on $\overline{F}^{gh,8-n} \times F^{gh,n}$.*

3.1.2 The BRST Current and the Differential

Theorem 3.2 [BLNW1] *Let V^M and V^L be two arbitrary positive energy modules of the $\mathcal{W}_3$ algebra. Consider the current*

$$\begin{aligned} J(z) &= c^{[3]}\big(\tfrac{1}{\sqrt{\beta^M}}W^M - \tfrac{i}{\sqrt{\beta^L}}W^L\big) + c^{[2]}\big(T^M + T^L + \tfrac{1}{2}T^{gh\,[2]} + T^{gh\,[3]}\big) \\ &\quad + (T^M - T^L)b^{[2]}c^{[3]}\partial c^{[3]} - \mu\, b^{[2]}\partial c^{[3]}\partial^2 c^{[3]} + \tfrac{2}{3}\mu\, b^{[2]}c^{[3]}\partial^3 c^{[3]} + \tfrac{3}{2}\partial^2 c^{[2]}\,, \end{aligned} \qquad (3.9)$$

where $\mu = (1-17\beta^M)/(10\beta^M)$ and $\beta^{M,L} = 16/(22+5c^{M,L})$ (see Sect. 2.1.2). Then the operator

$$d = \oint \frac{dz}{2\pi i}\, J(z)\,, \qquad (3.10)$$

acting on $V^M \otimes V^L \otimes F^{\rm gh}$, satisfies $d^2 = 0$ if and only if $c^M + c^L = 100$.

The current (3.9) is a natural generalization of the BRST current constructed in [TM]. In particular, the leading terms

$$J(z) = c^{[3]}\left(\frac{1}{\sqrt{\beta^M}}W^M - \frac{i}{\sqrt{\beta^L}}W^L\right) + c^{[2]}\left(T^M + T^L\right) + \dots, \qquad (3.11)$$

have the form one would expect if $\mathcal{W}_3$ were a Lie algebra acting on the tensor product of two modules. It has been shown in [BLNW1] that the completion of (3.11) by the higher order terms in (3.9) is unique, up to a total derivative, if one requires that the corresponding charge d is a differential of ghost number one, i.e., $d^2 = 0$. Thus the following definition is quite natural.

Definition 3.3 *Let V^M and V^L be positive energy modules of the $\mathcal{W}_3$ algebra with $c^M + c^L = 100$. Then the complex $(V^M \otimes V^L \otimes F^{\mathrm{gh}}, d)$ graded by the ghost number (degree), and with the differential d of ghost number one, is the complex of semi-infinite (BRST) cohomology of the $\mathcal{W}_3$ algebra with values in the tensor product $V^M \otimes V^L$. The corresponding cohomology will be denoted by $H(\mathcal{W}_3, V^M \otimes V^L)$ and called the noncritical $\mathcal{W}_3$ cohomology.*

Remarks

i. When V^L is the trivial $\mathcal{W}_3$ module, the above complex reduces to the original complex introduced in [TM]. We will call the corresponding cohomology (with values in a single $\mathcal{W}_3$ module) the critical $\mathcal{W}_3$ cohomology.
ii. Alternative derivations of the BRST current (3.9) were given in [BSS] and [dBGo2].

One should note that the existence of an extension of the complex from the critical to the noncritical case is by no means obvious, because, unlike for Lie algebras, the tensor product of two $\mathcal{W}_3$ modules does not have a natural $\mathcal{W}_3$ module structure. A more conceptual explanation of the result in Theorem 3.2 has been given in [BLNW2,dBGo2] and, more recently, [DSTS], where it is argued that noncritical complexes may be constructed from a suitable complex of semi-infinite cohomology of an affine Lie algebra, using the fact that the $\mathcal{W}_3$ algebra itself is a (Quantum Drinfel'd-Sokolov) reduction of an affine Lie algebra (see, e.g., [BeOo,FeFr2,Fi]). It seems, however, that in the cases we want to study explicitly in this book the precise relation between the two cohomologies is to a large extent conjectural – by extrapolating the results for the $\mathcal{W}_2$ (Virasoro) string (see, e.g., [ASY,AGSY,Sa]) – and thus we will not pursue this point of view further.

In the following we will also need the "operator version" of the cohomology, in which the modules are replaced by chiral algebras of operators. More precisely, let $\mathfrak{V}$, $\mathfrak{V}^M$ and $\mathfrak{V}^L$ be chiral algebras that decompose as $\mathcal{W}_3$ modules into direct sums of positive energy $\mathcal{W}_3$ modules, with the central charges $c = 100$ and $c^M + c^L = 100$, respectively. Then we have "operator valued" complexes given by $\mathfrak{C} = \mathfrak{V} \otimes \mathfrak{V}^{\mathrm{gh}}$ in the critical case and $\mathfrak{C} = \mathfrak{V}^M \otimes \mathfrak{V}^L \otimes \mathfrak{V}^{\mathrm{gh}}$ in the noncritical case. Let $\mathcal{O}(z)$ be a field in the chiral algebra $\mathfrak{C}$. The action of the differential d is given by the OPE with the BRST current $J(z)$, namely

$$(d\mathcal{O})(z) = \oint_{C_z} \frac{dw}{2\pi i} J(w)\, \mathcal{O}(z)\,, \tag{3.12}$$

where the contour C_z surrounds the point $w = z$ counterclockwise. It is straightforward to verify that $(\mathfrak{C}, d)$ is a complex. We will denote the corresponding "operator valued" cohomology by $H(\mathcal{W}_3, \mathfrak{C})$.

One can use the relation between an operator $\mathcal{O}(z)$ and the corresponding state $|\mathcal{O}\rangle$ at the level of the whole complex given by the analogue of (2.59). This allows one to pass from an "operator valued" to a "state valued" complex. In cases where there is an equivalence between the state and operator formulations – as discussed above for the ghost system or in Sect. 2.2.3 for the Fock spaces – we will switch freely between the two depending on which one is more convenient. One should remember, however, that for certain classes of modules, e.g., Verma modules, the operator valued counterpart of the complex may not exist.

A natural problem is to understand the algebraic structure on the cohomology space $H(\mathcal{W}_3, \mathfrak{C})$ that is induced from the underlying chiral algebra $\mathfrak{C}$. It turns out that if $\mathfrak{C}$ is a VOA, then $H(\mathcal{W}_3, \mathfrak{C})$ has the structure of a BV algebra. We will discuss this in detail in Chap. 5. First, however, we need to define more precisely what is the cohomology problem we want to solve.

3.2 The $\mathcal{W}_3$ Cohomology Problem for the 4D $\mathcal{W}_3$ String

The spectrum of physical states of 4D $\mathcal{W}_3$ gravity is computed as noncritical $\mathcal{W}_3$ cohomology with values in the tensor product of two Fock modules, $F(\Lambda^M, 0) \otimes F(\Lambda^L, 2i)$; i.e., the background charges of the matter and the Liouville Fock spaces are $\alpha^M = 0$ and $\alpha^L = 2i$, with the corresponding central charges $c^M = 2$ and $c^L = 98$. In principle the matter and the Liouville momenta, Λ^M and Λ^L, are arbitrary. However, for reasons that will be explained shortly, we will assume in addition that $(\Lambda^M, -i\Lambda^L)$ is restricted to lie on a lattice $L \subset \mathfrak{h}^*_{\mathbb{R}} \times \mathfrak{h}^*_{\mathbb{R}}$ characterized by the following properties:

i. $\lambda \in P$ for all $(\lambda, \mu) \in L$.
ii. $(\Lambda_i, \Lambda_i) \in L$, $i = 1, 2$.
iii. L is an integral lattice (of signature $(2,2)$), i.e.,

$$\lambda \cdot \lambda' - \mu \cdot \mu' \in \mathbb{Z}\,, \tag{3.13}$$

for all (λ, μ), $(\lambda', \mu') \in L$.

Lemma 3.4 *The maximal lattice L satisfying (i)-(iii) consists of weights (λ, μ) such that*

$$\lambda, \mu \in P\,, \quad \lambda - \mu \in Q\,. \tag{3.14}$$

Proof. Set $\lambda' = \mu' = \Lambda_i$ in (3.13). Then for all $(\lambda, \mu) \in L$ we find

$$\Lambda_i \cdot (\lambda - \mu) \in \mathbb{Z}\,, \quad i = 1, 2\,, \tag{3.15}$$

which, together with (i), implies (3.14). Conversely, given a lattice satisfying (3.14), and thus (i) and (ii), we may use the identity

$$\lambda \cdot \lambda' - \mu \cdot \mu' = (\lambda - \mu) \cdot \lambda' + \mu \cdot (\lambda' - \mu') \in \mathbb{Z}\,, \tag{3.16}$$

to deduce (iii). □

The choice of the lattice L is partly motivated by the following result.

Theorem 3.5 *Let $\mathfrak{C}$ be the chiral algebra corresponding to*

$$\mathcal{C} = \bigoplus_{(\Lambda^M, -i\Lambda^L) \in L} F(\Lambda^M, 0) \otimes F(\Lambda^L, 2i) \otimes F^{\mathrm{gh}}\,. \tag{3.17}$$

Then $(\mathfrak{C}, \mathcal{C})$ can be equipped with a structure of a VOA.

Proof. An operator $\mathcal{O}(z) \in \mathfrak{C}$ is of the form

$$\mathcal{O}(z) = P[\partial\phi^{M,i}, \partial\phi^{L,i}, c^{[j]}, b^{[j]}, \ldots]\, \mathcal{V}_{\Lambda^M, -i\Lambda^L}(z)\,, \tag{3.18}$$

where $P[\ldots]$ is a polynomial in the fields $\partial\phi^{M,i}$, $\partial\phi^{L,i}$, $i = 1, 2$, and $c^{[j]}$, $b^{[j]}$, $j = 2, 3$, and their derivatives, while $\mathcal{V}_{\Lambda^M, -i\Lambda^L}(z) = V_{\Lambda^M, \Lambda^L}(z)\, c_{\Lambda^M, \Lambda^L}$, the vertex operator corresponding to the vacuum state

$$|\Lambda^M, \Lambda^L\rangle = |\Lambda^M, 0\rangle \otimes |\Lambda^L, 2i\rangle \otimes b^{[2]}_{-1} b^{[3]}_{-1} b^{[3]}_{-2} |0\rangle_{\mathrm{gh}}\,, \tag{3.19}$$

is, the product of a phase-cocycle c_{Λ^M, Λ^L} with the normal ordered exponent

$$V_{\Lambda^M, \Lambda^L}(z) = e^{i\Lambda^M \cdot \phi^M + i\Lambda^L \cdot \phi^L}(z)\,. \tag{3.20}$$

The conformal dimension of the operator $P[\ldots]$ will be called the operator level of the operator $\mathcal{O}$.

The OPE of two operators in $\mathfrak{C}$ with the momenta $(\Lambda^M_A, \Lambda^L_A)$ and $(\Lambda^M_B, \Lambda^L_B)$, respectively, is schematically of the form

$$\mathcal{O}_{\Lambda^M_A, \Lambda^L_A}(z)\, \mathcal{O}_{\Lambda^M_B, \Lambda^L_B}(w) = \sum_{n \in \mathbb{Z}} (z - w)^{h_{AB} + n}\, \mathcal{O}^{(n)}_{\Lambda^M_A + \Lambda^M_B, \Lambda^L_A + \Lambda^L_B}\,, \tag{3.21}$$

where $h_{AB} = \Lambda^M_A \cdot \Lambda^M_B + \Lambda^L_A \cdot \Lambda^L_B$. Here the common factor $(z-w)^{h_{AB}}$ comes from the contraction of exponentials, the remaining contractions clearly only modify this by integer powers of $(z - w)$. By setting the momenta of the operators to lie on the lattice L, we find, using (iii), that all OPEs are meromorphic.

To prove that $(\mathfrak{C}, \mathcal{C})$ is in fact a VOA, we must still show that it is possible to choose the phase-cocycles, c_{Λ^M, Λ^L}, such that the analytic continuation of the right hand side in (3.21) is consistent with the graded commutativity of the OPE determined by the ghost number of operators. The existence of the required phase-cocycles is proved in the following lemma. □

Lemma 3.6 *Let $\xi : Q \to P$ be a linear map satisfying*

$$\xi(\alpha)\cdot\alpha' - \xi(\alpha')\cdot\alpha = \alpha\cdot\alpha' \mod 2\,, \tag{3.22}$$

then

$$c_{\Lambda^M,\Lambda^L} = e^{i\pi(\xi(\Lambda^M+i\Lambda^L)-i\Lambda^L)\cdot(p^M+ip^L)}\eta(\Lambda^M+i\Lambda^L)\,, \tag{3.23}$$

where $\eta : Q \to \mathbb{Z}/2\mathbb{Z}$ is defined in (B.18), are the required phase-cocycles turning $(\mathfrak{C}, \mathcal{C})$ into a VOA.

Remark. As reviewed in Appendix B, the map ξ defines, through (B.12), a phase-cocycle in the vertex operator construction of $\widehat{\mathfrak{sl}_3}$. A particular choice for ξ is given in (B.13).

Proof. Following (B.12), let us set

$$c'_{\Lambda^M,\Lambda^L} = e^{i\pi\xi^M((\Lambda^M,\Lambda^L))\cdot p^M + i\pi\xi^L((\Lambda^M,\Lambda^L))\cdot p^L}\,. \tag{3.24}$$

Then the linear map $(\xi^M, \xi^L) : L \to P \times iP$ must satisfy

$$\xi^M_A\cdot\Lambda^M_B + \xi^L_A\cdot\Lambda^L_B - \xi^M_B\cdot\Lambda^M_A + \xi^L_B\cdot\Lambda^L_A = \Lambda^M_A\cdot\Lambda^M_B + \Lambda^L_A\cdot\Lambda^L_B \mod 2\,, \tag{3.25}$$

where $\xi^M_A = \xi^M((\Lambda^M_A, \Lambda^L_A))$, etc. This may be rewritten as a mod 2 equation

$$\begin{aligned}\xi^M_A\cdot(\Lambda^M_B+i\Lambda^L_B) - (\Lambda^M_A+i\Lambda^L_A)\cdot\xi^M_B - i(\xi^M_A+i\xi^L_A)\cdot\Lambda^L_B + i\Lambda^L_A\cdot(\xi^M_B+i\xi^L_B)\\ = (\Lambda^M_A+i\Lambda^L_A)\cdot(\Lambda^M_B+i\Lambda^L_B) - i\Lambda^L_A\cdot(\Lambda^M_B+i\Lambda^L_B) - i(\Lambda^M_A+i\Lambda^L_A)\cdot\Lambda^L_B\,,\end{aligned} \tag{3.26}$$

which is solved by $\xi^L = i\xi^M$ and $\xi^M((\Lambda^M, \Lambda^L)) = \xi(\Lambda^M + i\Lambda^L) - i\Lambda^L$, as one immediately verifies by using (3.14) and (3.22). Then, according to (B.15), the required phase-cocycles are given by $c_{\Lambda^M,\Lambda^L} = c'_{\Lambda^M,\Lambda^L}\eta(\Lambda^M + i\Lambda^L)$, where $\eta : Q \to \mathbb{Z}/2\mathbb{Z}$ is defined in (B.18). □

Let us comment on the conditions (i)-(iii) on the lattice L. One expects that the most interesting subsector of the cohomology should arise for maximally degenerate matter Fock modules of the $\mathcal{W}_3$ algebra. For, if the Fock module is degenerate just along one root direction then the calculation will essentially reduce to the Virasoro case, and if it is irreducible we will obtain at most the vacuum state as nontrivial cohomology – a result that follows from reduction theorems discussed in the next section. As discussed in Sect. 2.3, at $c = 2$ the maximally degenerate Fock modules have integral weights, which explains (i). Condition (iii), via Theorem 3.5, allows us to study the cohomology as a BV algebra and thus is equally natural. The remaining condition can be justified only *a posteriori*, as by explicit cohomology computation we will find that the ground ring of the theory, i.e., the ghost number zero subalgebra of the full cohomology, has generators with weights $(\Lambda^M, -i\Lambda^L) = (\Lambda_j, \Lambda_j)$, $j = 1, 2$, as required by (ii). However, we should stress that the cohomology problem is well defined for all weights, and that at this point our choice merely selects what should be the most interesting subsector both from the mathematical and the physical point of view.

To summarize, let us formulate the main mathematical problem in the quantization of the 4D $\mathcal{W}_3$ string.

Problem *For the VOA, $(\mathfrak{C}, \mathcal{C})$, given in (3.17), compute the semi-infinite cohomology $H(\mathcal{W}_3, \mathfrak{C})$ and determine explicitly its BV algebra structure.*

In the following sections we present a (partially conjectural) solution to this problem. Given the length of the analysis and its reliance on technical results, it may be useful at this point to outline the main steps.

The problem clearly splits into two parts: a computation of the cohomology, $H(\mathcal{W}_3, \mathfrak{C})$, and a study of its global structure. The two steps are of course related, as the BV algebra structure of $H(\mathcal{W}_3, \mathfrak{C})$ provides quite a lot of information on the cohomology itself.

An explicit computation of the cohomology requires a rather detailed understanding of the action of the $\mathcal{W}_3$ algebra on the complex. In this respect the results of Sect. 2.4.1 are adequate for the subcomplex in which the shifted Liouville momentum $-i\Lambda^L + 2\rho$ is in the fundamental Weyl chamber. Let us denote this subcomplex by $\mathfrak{C}_1$. More generally, we denote the subcomplex with $-i\Lambda^L + 2\rho \in w^{-1} P_+$ by $\mathfrak{C}_w$. (Note that $\mathfrak{C}_1$ is not closed under OPEs!) A series of technical results in Sect. 3.3 then allows a straightforward computation of $H(\mathcal{W}_3, \mathfrak{C}_1)$ in Sect. 3.4. The general form of the result suggests an extension to an arbitrary Weyl chamber. This is discussed in Sect. 3.5. Then the complete BV algebra is studied in the last part of the book.

3.3 Preliminary Results

3.3.1 A Comment on the Relative Cohomology

We begin with some results on the general structure of the $\mathcal{W}_3$ cohomology, in both the critical and the noncritical cases. We will use V to denote either a single positive energy $\mathcal{W}_3$ module or a tensor product of two such modules.

Consider the operators

$$L_n^{\text{tot}} = [d, b_n^{[2]}]_+ , \quad W_n^{\text{tot}} = [d, b_n^{[3]}]_+ , \quad n \in \mathbb{Z} . \tag{3.27}$$

Then the $L_n^{\text{tot}} = L_n + L_n^{[2]} + L_n^{[3]}$ define a positive energy representation of the "total" Virasoro algebra on $V \otimes F^{\text{gh}}$, with vanishing central charge and diagonalizable L_0^{tot}. The eigenspaces of L_0^{tot} yield a decomposition of the complex into finite dimensional subcomplexes. By the usual argument (see, e.g., [FGZ]), nontrivial cohomology can arise only in the subcomplex annihilated by L_0^{tot}. However, the operators L_n^{tot} and W_n^{tot} do not generate a "total" $\mathcal{W}_3$ algebra,[1]

[1] It has been shown in [BLNW2] that $T^{\text{tot}}(z)$, $W^{\text{tot}}(z)$, together with $J(z)$, $b^{[2]}(z)$ and $b^{[3]}(z)$, form a subset of generators of the topological $N = 2$ $\mathcal{W}_3$ superalgebra.

as would have been the case if $\mathcal{W}_3$ were a Lie algebra. Moreover, following the discussion in Chap. 2, $W_0^{\rm tot}$ is in general nondiagonalizable on the complex. As a consequence, nontrivial cohomology states need not be annihilated by $W_0^{\rm tot}$.

Lemma 3.7 *Nontrivial cohomology may arise only in the subcomplex whose elements $|\Phi\rangle$ satisfy*

$$L_0^{\rm tot}|\Phi\rangle = 0\,, \tag{3.28}$$

and

$$(W_0^{\rm tot})^N|\Phi\rangle = 0\,, \tag{3.29}$$

for some $N > 0$.

Proof. The first condition (3.28) follows by diagonalizing $L_0^{\rm tot}$ on the complex and is the same as in the case of the Virasoro algebra [FGZ]. Since the subcomplex corresponding to Ker $L_0^{\rm tot}$ is finite dimensional, it can be decomposed into a direct sum of generalized eigenspaces of $W_0^{\rm tot}$ that are preserved by d, since $[d, W_0^{\rm tot}]_- = 0$. Thus we may assume that

$$(W_0^{\rm tot} - w)^N|\Phi\rangle = 0\,, \tag{3.30}$$

for some $w \in \mathbb{C}$ and $N > 0$. Together with (3.27), this implies

$$w^N|\Phi\rangle = db_0^{[3]}\Big(\sum_{n=1}^{N}(-1)^{n+1}\binom{N}{n}w^{N-n}(W_0^{\rm tot})^{n-1}\Big)|\Phi\rangle\,, \tag{3.31}$$

provided $d|\Phi\rangle = 0$. Thus $|\Phi\rangle$ is a trivial cohomology state whenever $w \neq 0$. □

One can define the complex of relative $\mathcal{W}_3$ cohomology with respect to the "Cartan subalgebra" $\mathcal{W}_{3,0}$ as the intersection

$$\text{Ker } W_0^{\rm tot} \cap \text{Ker } L_0^{\rm tot} \cap \text{Ker } b_0^{[2]} \cap \text{Ker } b_0^{[3]} \subset V \otimes F^{\rm gh}\,, \tag{3.32}$$

with the differential d, which clearly preserves this subspace. The corresponding cohomology will be called relative. However, unlike in cases where the Cartan algebra acts semi-simply on the complex, this relative cohomology is not only difficult to compute (e.g., in most nontrivial examples considered below it is practically impossible to determine the relative subcomplex explicitly) but also cumbersome to relate to the full cohomology.

It turns out, however, that the description of $H(\mathcal{W}_3, \mathfrak{C})$ may nevertheless be simplified, as the explicit results below suggest that the $\mathcal{W}_3$ cohomology carries a (noncanonical) quartet structure, first recognized in the critical case in [PSSW]. The lowest ghost number members of the quartets have been called "prime states," and, for enumeration purposes, they play the analogous role to relative cohomology states.

3.3.2 Reduction Theorems

Theorem 3.8 [BMP5] *Let $M^{(\kappa)}(\Lambda^M, \alpha_0^M)$ be an arbitrary generalized Verma module and $\overline{M}(\Lambda^L, \alpha_0^L)$ a contragredient Verma module with $c^M + c^L = 100$. The cohomology $H(\mathcal{W}_3, M^{(\kappa)}(\Lambda^M, \alpha_0^M) \otimes \overline{M}(\Lambda^L, \alpha_0^L))$ is nonvanishing if and only if*

$$-i(\Lambda^L + \alpha_0^L \rho) = w(\Lambda^M + \alpha_0^M \rho)\,, \tag{3.33}$$

for some $w \in W$, in which case it is spanned by the states

$$v_0\,, \quad c_0^{[2]} v_0\,, \quad c_0^{[3]} v_{\kappa-1}\,, \quad c_0^{[3]} c_0^{[2]} v_{\kappa-1}\,,$$

where $v_i = v_i^M \otimes \overline{v}^L \otimes |0\rangle_{\mathrm{gh}}$, $i = 0, \ldots, \kappa - 1$ (see Definition 2.21).

Remark. Theorem 3.8 is a generalization of a similiar result for the semi-infinite cohomology of the Virasoro and affine Lie algebras [Fe,FGZ] (see also [BMP4]).

Proof. Consider linear functions ν, called the ν degree, on $M^{(\kappa)}(\Lambda^M, \alpha_0^M)$, $\overline{M}(\Lambda^L, \alpha_0^L)$, F^{gh} with values in $\mathbb{C}[\varepsilon^{-1}, \varepsilon]$, which map standard basis elements (2.47), (2.20) and (3.6) into powers of an indeterminate ε,

$$\begin{aligned}
\nu\,(e^{(i)}_{m_1 \ldots m_M; n_1 \ldots n_N}) &= \varepsilon^{-M-2N}\,, \quad i = 0, \ldots, \kappa - 1\,, \\
\nu(\overline{e}_{m_1 \ldots m_M; n_1 \ldots n_N}) &= \varepsilon^{M+2N}\,, \\
\nu(g_{k_1 \ldots k_K; \ell_1 \ldots \ell_L; m_1 \ldots m_M; n_1 \ldots n_N}) &= \varepsilon^{L-K+2N-2M}\,.
\end{aligned} \tag{3.34}$$

Extend ν as a multiplicative map to the tensor product $\mathcal{C} = M^{(\kappa)}(\Lambda^M, \alpha_0^M) \otimes \overline{M}(\Lambda^M, \alpha_0^M) \otimes F^{\mathrm{gh}}$. Let $\mathcal{C}_{(m)}$ denote the linear span of elements of ν degree ε^m, i.e., $\mathcal{C}_{(m)} = \nu^{-1}(\varepsilon^m)$.[2] Then an operator A on $\mathcal{C}$ has the ν degree equal n if $A\,\mathcal{C}_{(m)} \subset \mathcal{C}_{(m+n)}$. In such case we will simply say that A acts like ε^n. For an arbitrary A, let A_{ε^n} be its component of degree n. The action of the generators of the $\mathcal{W}_3$ algebra on the basis vectors (2.47) and those of the corresponding contragredient basis (2.20) can be studied explicitly using the commutation relations (2.1). It is then straightforward to determine the degrees present in the decomposition of each generator acting on $\mathcal{C}$. In schematic notation, we find

$$\begin{aligned}
L^M_{-n} &\sim \frac{1}{\varepsilon} + 1 + \varepsilon + \cdots\,, \quad n > 0\,, & L^L_{-n} &\sim 1 + \varepsilon + \varepsilon^2 + \cdots\,, \quad n \geq 0\,, \\
L^M_{n} &\sim 1 + \varepsilon + \varepsilon^2 + \cdots\,, \quad n \geq 0 & L^L_{n} &\sim \frac{1}{\varepsilon} + 1 + \varepsilon + \cdots\,, \quad n > 0\,, \\
W^M_{-n} &\sim \frac{1}{\varepsilon^2} + \frac{1}{\varepsilon} + 1 + \cdots\,, \quad n > 0\,, & W^L_{-n} &\sim \frac{1}{\varepsilon} + 1 + \varepsilon + \cdots\,, \quad n \geq 0\,, \\
W^M_{n} &\sim \frac{1}{\varepsilon} + 1 + \varepsilon + \cdots\,, \quad n \geq 0\,, & W^L_{n} &\sim \frac{1}{\varepsilon^2} + \frac{1}{\varepsilon} + 1 + \cdots\,, n > 0\,.
\end{aligned} \tag{3.35}$$

---

[2] This decomposition generalizes the filtration of Verma modules introduced in the proof of Theorem 2.12.

In particular, the lowest degree components of generators L_{-n}^M, L_n^L, W_{-n}^M, and W_n^L, $n > 0$, map a given basis vector onto another one, which is obtained simply by increasing the power of the corresponding generator in (2.47) or (2.20). Thus those operators commute. The ν degrees of the ghost and antighost mode operators are

$$c_n^{[2]} \sim \varepsilon, \quad b_n^{[2]} \sim \frac{1}{\varepsilon}, \quad c_n^{[3]} \sim \varepsilon^2, \quad b_n^{[3]} \sim \frac{1}{\varepsilon^2}, \quad n \in \mathbb{Z}. \tag{3.36}$$

By expanding the differential (3.10) in terms mode operators, and then using (3.35) and (3.36), we find that d is a sum of operators with nonnegative degrees. Thus, we have a spectral sequence $(\mathcal{E}_r, d_r)$, $r \geq 0$, induced from the filtration defined by the ν degree.[3] Recall that the first term in this sequence is given by $\mathcal{E}_0 = \mathcal{C}$, while the differential is $d_0 = (d)_{\varepsilon^0}$. Then $\mathcal{E}_{r+1} = H(d_r, \mathcal{E}_r)$ and d_{r+1} is induced from $\sum_{i=0}^{r+1}(d)_{\varepsilon^i}$. In the present case

$$d_0 = \sum_{n>0} \left(c_n^{[2]}(L_{-n}^M)_{1/\varepsilon} + c_{-n}^{[2]}(L_n^L)_{1/\varepsilon} + c_n^{[3]}(W_{-n}^M)_{1/\varepsilon^2} + c_{-n}^{[3]}(W_n^L)_{1/\varepsilon^2} \right), \tag{3.37}$$

i.e., $(\mathcal{E}_0, d_0)$ is the Koszul complex of the abelian algebra generated by the leading terms of the $\mathcal{W}_3$ algebra generators. By the standard argument (see, e.g., [Kn]) there is a contracting homotopy for the differential (3.37), and its cohomology is therefore concentrated on the states of the form

$$v_0, \quad c_0^{[2]} v_0, \quad c_0^{[3]} v_0, \quad c_0^{[2]} c_0^{[3]} v_0,$$

$$\vdots \tag{3.38}$$

$$v_{\kappa-1}, \quad c_0^{[2]} v_{\kappa-1}, \quad c_0^{[3]} v_{\kappa-1}, \quad c_0^{[2]} c_0^{[3]} v_{\kappa-1}.$$

Those states span $\mathcal{E}_1$, on which the differential d_1 is explicitly given by

$$d_1 = c_0^{[2]}(L_0^M + L_0^L) + c_0^{[3]}\left(\frac{1}{\sqrt{\beta^M}} W_0^M - \frac{i}{\sqrt{\beta^L}} W_0^L \right). \tag{3.39}$$

By evaluating this operator on (3.38) we find that its cohomology is nonvanishing if and only if

$$h(\Lambda^M, \alpha_0^M) + h(\Lambda^L, \alpha_0^L) = 0 \quad \text{and} \quad w(\Lambda^M, \alpha_0^M) - iw(\Lambda^L, \alpha_0^L) = 0. \tag{3.40}$$

It follows from Lemma 2.17 that the most general solution to those conditions is given by the weights Λ^M and Λ^L satisfying $-i(\Lambda^L + \alpha_0^M \rho) = w(\Lambda^M + \alpha_0^L \rho)$ for some $w \in W$. The nonvanishing cohomology, i.e., the $\mathcal{E}_2$ term, is then spanned by the states

$$v_0, \quad c_0^{[2]} v_0, \quad c_0^{[3]} v_{\kappa-1}, \quad c_0^{[2]} c_0^{[3]} v_{\kappa-1}. \tag{3.41}$$

Since those states are annihilated by d, we also have $d_2 = d_3 = \ldots = 0$, so that the spectral sequence collapses at this term, $\mathcal{E}_2 = \mathcal{E}_3 = \ldots = \mathcal{E}_\infty$, and (3.41) yields the cohomology $H(\mathcal{W}_3, M^{(\kappa)}(\Lambda^M, \alpha_0^M) \otimes \overline{M}(\Lambda^L, \alpha_0^L))$. $\qquad \square$

---

[3] See, e.g., [BMP4] for a more extensive discussion of this spectral sequence and its applications in the context of cohomology of the Virasoro and affine Lie algebras.

The spectral sequence argument in the proof above relies on the existence of a filtration with respect to which the degrees of all generators were bounded from below by some power of ε. Replacing one of the modules by an arbitrary module with a suitable filtration gives the following vanishing theorem.

Theorem 3.9 *Let F be a $\mathcal{W}_3$ module and $\nu : F \to \mathbb{C}[\lambda^{-1},\lambda]$ a degree on F, such that the ν degrees of all $\mathcal{W}_3$ generators are bounded from below, i.e.,*

$$L_n = \sum_{k\geq k_0} (L_n)_{\lambda^k}\,, \quad W_n = \sum_{k\geq k_0} (W_n)_{\lambda^k}\,, \quad n\in\mathbb{Z}\,, \tag{3.42}$$

for some $k_0 \in \mathbb{Z}$. Then

$$H^n(\mathcal{W}_3, M^{(\kappa)}(\Lambda^M,\alpha_0^M)\otimes F) = 0\,, \quad \text{for} \quad n\leq 2\,, \tag{3.43}$$

and

$$H^n(\mathcal{W}_3, F\otimes \overline{M}(\Lambda^L,\alpha_0^L)) = 0\,, \quad \text{for} \quad n\geq 6\,. \tag{3.44}$$

Proof. In the first case consider the ν degree on $M^{(\kappa)}(\Lambda^M,\alpha_0^M)\otimes F^{\rm gh}$ defined as in the proof of Theorem 3.8, but with $\varepsilon = \lambda^{|k_0|+1}$. Then the ν degree extended to $M^{(\kappa)}(\Lambda^M,\alpha_0^M)_N\otimes F\otimes F^{\rm gh}$ yields a spectral sequence $(\mathcal{E}_r, d_r)$ with

$$d_0 = \sum_{m>0} c_m^{[2]}(L_{-m}^M)_{1/\varepsilon} + c_m^{[3]}(W_{-m}^M)_{1/\varepsilon^2}\,. \tag{3.45}$$

As before the cohomology of d_0 simply picks up the highest weight vectors v_i^M in the Verma module, i.e.,

$$\mathcal{E}_1 \cong \bigoplus_{i=0}^{\kappa-1} \mathbb{C}v_i^M \otimes F\otimes F_>^{\rm gh}\,, \tag{3.46}$$

where $F_>^{\rm gh}$ is generated by $c_{-m}^{[2]}$ and $c_{-m}^{[3]}$, $m\geq 0$, acting on $|0\rangle_{\rm gh}$. Thus all states in $\mathcal{E}_1$ have ghost numbers $3+n$, with $n\geq 0$, which implies (3.43).

The second part of the theorem is proved similarly, except that $\mathcal{E}_1 \cong \mathbb{C}\overline{v}^L \otimes F\otimes F_<^{\rm gh}$, where $F_<^{\rm gh}$ is generated by $c_0^{[2]}$, $c_0^{[3]}$, and $b_{-n}^{[2]}$, $b_{-n}^{[3]}$, $n>0$, from $|0\rangle_{\rm gh}$. □

Lemma 3.10 *Define a ν degree on a Fock space, $F(\Lambda,\alpha)$, by setting*

$$\nu(f_{m_1\ldots m_M;n_1\ldots n_N}) = \lambda^{M+N}\,. \tag{3.47}$$

Then the action of $\mathcal{W}_3$ on $F(\Lambda,\alpha)$ is bounded as in (3.42) with $k_0 = -3$.

Proof. Note that α_{-n}^i, α_n^i, and α_0^i, $i = 1,2$, $n>0$, act as λ, $1/\lambda$ and 1, respectively. The lemma follows by examining the explicit formula (2.71) for the generators. □

3.3.3 The $\mathfrak{sl}_3 \oplus (\mathfrak{u}_1)^2$ Symmetry of $H(\mathcal{W}_3, \mathfrak{C})$

The vertex operator realization of $\widehat{\mathfrak{sl}_3}$, reviewed in Appendix B, can be extended to act on the complex $\mathfrak{C}$ by the currents

$$H^i(z) = i\partial\phi^i(z)\,, \quad i = 1,2\,, \qquad E^\alpha(z) = \mathcal{V}_{\alpha,0}(z)\,, \quad \alpha \in \Delta\,. \tag{3.48}$$

From the explicit form of (3.20) and (3.23) we find that this realization only acts on the matter degrees of freedom, but for the phase factor which depends on the Liouville momentum. The corresponding $\mathfrak{sl}_3$ generators commute with d, and thus their action descends to the cohomology.

Additional symmetry operators that commute with d are the Liouville momenta, $-ip^{L,i}$, with the corresponding currents $\partial\phi^{L,i}(z)$, $i = 1,2$. They obviously commute with the $\mathfrak{sl}_3$ algebra as well. The resulting $\mathfrak{sl}_3 \oplus (\mathfrak{u}_1)^2$ symmetry of $H(\mathcal{W}_3, \mathfrak{C})$ will greatly simplify the following dicussion.

The levels, h, of operators in $\mathfrak{C}$ at a given ghost number are bounded from below, as the only operators with nonpositive dimension are $\partial^n c^{[2]}$, $n = 0, 1$, and $\partial^n c^{[3]}$, $n = 0, 1, 2$. If in addition we require that a given operator be annihilated by $L_0^{\rm tot}$, we have

$$h = \tfrac{1}{2}|-i\Lambda^L + 2\rho|^2 - \tfrac{1}{2}|\Lambda^M|^2 - 4\,, \quad (\Lambda^M, -i\Lambda^L) \in L\,. \tag{3.49}$$

Thus for a fixed Liouville momentum, Λ^L, and a ghost number, n, but arbitrary matter weight, Λ^M, there is a finite dimensional subspace of operators in $\mathfrak{C}$ whose level satisfies (3.49). This subspace is clearly closed under the action of $\mathfrak{sl}_3 \oplus (\mathfrak{u}_1)^2$, which immediately yields the following result.

Theorem 3.11 *The cohomology $H(\mathcal{W}_3, \mathfrak{C})$ decomposes into a direct sum of finite dimensional irreducible modules of $\mathfrak{sl}_3 \oplus (\mathfrak{u}_1)^2$.*

3.3.4 A Bilinear Form on $\mathfrak{C}$ and $H(\mathcal{W}_3, \mathfrak{C})$

By combining the $\mathbb{C}$-linear anti-involutions on $\mathcal{A}^M$, $\mathcal{A}^L$ and $\mathfrak{W}^{\rm gh}$ we obtain a $\mathbb{C}$-linear anti-involution $\omega = \omega_{\mathcal{A}^M} \otimes \omega_{\mathcal{A}^L} \otimes \omega_{\rm gh}$ on $\mathfrak{C}$. A straightforward calculation show that the differential d behaves naturally under this anti-involution, namely

$$\omega(d) \;=\; d\,. \tag{3.50}$$

Similarly, let $C(\Lambda^M, \Lambda^L)$ denote the complex $F(\Lambda^M, 0) \otimes F(\Lambda^L, 2i) \otimes F^{\rm gh}$, then by combining (3.50) with the results of Theorems 2.22 and 3.1, we immediately have the following results

Theorem 3.12

i. There exists a unique bilinear form

$$\langle - | - \rangle_C \;:\; C(\Lambda^M, w_0 \cdot \Lambda^L) \times C(\Lambda^M, \Lambda^L) \longrightarrow \mathbb{C}\,,$$

contravariant with respect to ω, *and such that* $\langle \Lambda^M, w_0 \cdot \Lambda^L | \Lambda^M, \Lambda^L \rangle_C = 1$. *This form is nondegenerate on* $C^{8-n}(\Lambda^M, w_0 \cdot \Lambda^L) \times C^n(\Lambda^M, \Lambda^L)$.

ii. The differential d *is symmetric with respect to the form* $\langle - | - \rangle_C$.

iii. The form (i) induces a nondegenerate bilinear form on $H^{8-n}(\mathcal{W}_3, F(\Lambda^M, 0) \otimes F(w_0 \cdot \Lambda^L, 2i)) \times H^n(\mathcal{W}_3, F(\Lambda^M, 0) \otimes F(\Lambda^L, 2i))$.

Corollary 3.13 *There is an isomorphism*

$$H^n(\mathcal{W}_3, F(\Lambda^M, 0) \otimes F(\Lambda^L, 2i)) \cong H^{8-n}(\mathcal{W}_3, F(\Lambda^M, 0) \otimes F(w_0 \cdot \Lambda^L, 2i)), \quad (3.51)$$

for all $(\Lambda^M, -i\Lambda^L) \in L$ *and* $n \in \mathbb{Z}$, *which extends to an isomorphism of* $H(\mathcal{W}_3, \mathfrak{C})$ *as an* $\mathfrak{sl}_3 \oplus (\mathfrak{u}_1)^2$ *module.*

Remark. We will refer to (3.51) as the "duality" of the cohomology.

3.4 The Cohomology in the "Fundamental Weyl Chamber"

In this section we determine $H(\mathcal{W}_3, \mathfrak{C})$ in the fundamental Weyl chamber, i.e., for the Liouville weights satisfying $-i\Lambda^L + 2\rho \in P_+$. This computation relies on several results derived earlier: the isomorphism $F(\Lambda^L, 2i) \cong \overline{M}(\Lambda^L, 2i)$ that holds for $-i\Lambda^L + 2\rho$ in the fundamental Weyl chamber (Corollary 2.28); the reduction theorem for the $\mathcal{W}_3$ cohomology with values in a tensor product of a (generalized) Verma and a contragredient Verma modules (Theorem 3.8); and explicit resolutions of the irreducible $\mathcal{W}_3$ modules (Conjecture 2.37) together with the decomposition of Fock modules at $c = 2$ (Theorem 2.31).

The cohomology $H(\mathcal{W}_3, F(\Lambda^M, 0) \otimes F(\Lambda^L, 2i))$ is then obtained as follows: First using the decomposition theorem for the matter Fock space (Theorem 2.31), and the isomorphism in the Liouville sector, it is sufficient to compute the cohomology $H(\mathcal{W}_3, L(\Lambda, 0) \otimes \overline{M}(\Lambda^L, 2i))$, where $\Lambda = \Lambda^M + \beta$, $\beta \in Q_+$. The latter cohomology can be studied through a spectral sequence associated with the resolution of the irreducible module $L(\Lambda, 0)$ in terms of generalized Verma modules obtained in Sect. 2.4.1. Using the reduction theorem it is then easy to show that this spectral sequence collapses no later than at the second term, and to compute its limit explicitly. The main result for the cohomology is given in Theorems 3.17 and 3.19, and in Appendix E.

3.4.1 $H(\mathcal{W}_3, L(\Lambda, 0) \otimes F(\Lambda^L, 2i))$ with $-i\Lambda^L + 2\rho \in P_+$

In Sect. 2.4.1 we have argued that for a given irreducible $\mathcal{W}_3$ module $L(\Lambda, 0)$, $\Lambda \in P_+$, there exists a resolution $(\mathcal{M}, \delta)$ of $L(\Lambda, 0)$ in terms of $c = 2$ (generalized) Verma modules of highest weight $(h(\sigma \circ \Lambda), w(\sigma \circ \Lambda))$, where $\sigma \circ \Lambda \equiv \Lambda + \rho - \sigma\rho$ and σ runs over the set $W(\Lambda) \subset \widetilde{W}$ given in (2.102) (see also Table 2.1). Replacing $L(\Lambda, 0)$ with this resolution allows us to calculate $H(\mathcal{W}_3, L(\Lambda, 0) \otimes F(\Lambda^L, 2i))$

via relatively standard techniques applied to the resulting double complex. A cursory inspection of the resolutions displayed in Figs. 2.4-2.6 shows that there are only a few ways in which (generalized) Verma modules with the same highest weights arise; in particular, they are distinguished by how they are joined by the arrows representing the nontrivial homomorphisms comprising the differential δ. As this structure is important in the calculation, we will first discuss these different possibilities explicitly.

The first case, Case I, is that a given space is isolated, i.e., it is not joined by arrows to a space with the same highest weights. This occurs in all the resolutions of $L(\Lambda,0)$, $\Lambda \in P_+$, but there are actually two subcases: in Case Ia the isolated space is a Verma module, $M(\sigma\circ\Lambda,0)$, which appears for $\sigma\in W(\Lambda)\cap W$ at step $-\ell(\sigma)$; in Case Ib it is a generalized Verma module $M^{(2)}(\sigma\circ\Lambda,0)$, which only appears for $\sigma=\sigma_1$ at step $-\ell(\sigma_1)=-1$ in the resolution of $L(0,0)$. The next case, Case II, has exactly two spaces with the same highest weights joined by an arrow. This only occurs in the following way,

$$M(\sigma\circ\Lambda,0) \longrightarrow M^{(2)}(\sigma\circ\Lambda,0)\,, \tag{3.52}$$

and is present in all resolutions. For $\sigma\in W(\Lambda)\cap W$ the "top space," $M^{(2)}(\sigma\circ\Lambda,0)$, appears at step $-\ell(\sigma)$, while for $\sigma=\sigma_1$ it appears (at step -1) only in the resolutions of $L(\Lambda,0)$ for $\Lambda\in P_+\backslash P_{++}$. The last case, Case III, has exactly three spaces with the same highest weights joined by an arrow. This only occurs as

$$M(\sigma\circ\Lambda,0)\oplus M(\sigma\circ\Lambda,0) \longrightarrow M^{(2)}(\sigma\circ\Lambda,0)\,, \tag{3.53}$$

for $\sigma=\sigma_i$ in the resolution of $L(\Lambda,0)$ for $\Lambda\in P_{++}$ (the top space occurring at step -1).

We may now state the result.

Theorem 3.14 *Let* $-i\Lambda^L+2\rho\in P_+$*. Then*
i. $H(\mathcal{W}_3, L(\Lambda,0)\otimes F(\Lambda^L,2i))\neq 0$ *if and only if*

$$-i\Lambda^L+2\rho=\Lambda+\rho-\sigma\rho\;=\;\sigma\circ\Lambda\,, \tag{3.54}$$

for some $\sigma\in W(\Lambda)$.
ii. For a given Λ, Λ^L *and* σ *satisfying (3.54),*

$$\dim H^m(\mathcal{W}_3, L(\Lambda,0)\otimes F(\Lambda^L,2i)) \;=\; d(m,3-\ell(\sigma))\,, \tag{3.55}$$

where

$$d(m,n)=\delta_{m,n}+2\,\delta_{m,n+1}+\delta_{m,n+2}\,, \tag{3.56}$$

i.e., each $\sigma\in W(\Lambda)$ *gives rise to an independent "quartet" of cohomology states at ghost numbers* n, $n+1$, $n+1$ *and* $n+2$*, respectively, where* $n=3-\ell(\sigma)$.

$$
\begin{array}{lccccc}
 & 0 & & c_0^{[2]}c_0^{[3]}v_0 & & 0 \\
E_1' : & 0 & \longrightarrow & c_0^{[2]}v_0\,,\quad c_0^{[3]}v_0 & \longrightarrow & 0\,, \\
 & 0 & & v_0 & & 0
\end{array}
$$

$$E_1' = E_2' = \ldots = E_\infty' .$$

$$
\begin{array}{lccccc}
 & 0 & & c_0^{[2]}c_0^{[3]}v_1 & & 0 \\
E_1' : & 0 & \longrightarrow & c_0^{[2]}v_0\,,\quad c_0^{[3]}v_1 & \longrightarrow & 0\,, \\
 & 0 & & v_0 & & 0
\end{array}
$$

$$E_1' = E_2' = \ldots = E_\infty' .$$

Fig. 3.1. Cases Ia and Ib

Proof. Consider the double complex $(\mathcal{M} \otimes F(\Lambda^L, 2i) \otimes F^{\mathrm{gh}}, d, \delta)$, obtained by "replacing" the irreducible module $L(\Lambda, 0)$ with the corresponding resolution.[4] Since $H^n(\delta, \mathcal{M}) \cong \delta^{n,0}\, L(\Lambda, 0)$, the first spectral sequence associated with this double complex (see, e.g., [BoTu]) collapses at the first term to yield

$$
\begin{aligned}
E_\infty^{p,q} &\cong H^p(\mathcal{W}_3, H^q(\delta, \mathcal{M}) \otimes F(\Lambda^L, 2i)) \\
&\cong \delta^{q,0} H^p(\mathcal{W}_3, L(\Lambda, 0) \otimes F(\Lambda^L, 2i))\,.
\end{aligned}
\tag{3.57}
$$

The E_2' term of the second spectral sequence is given by

$$E'^{p,q}_2 \cong H^q(\delta, H^p(\mathcal{W}_3, \mathcal{M} \otimes F(\Lambda^L, 2i)))\,, \tag{3.58}$$

and can be computed explicitly using the isomorphism $F(\Lambda^L, 2i) \cong \overline{M}(\Lambda^L, 2i)$, the reduction theorem of Sect. 3.3.2, and the explicit form of the resolutions (see Theorem 2.37). Since $H(\mathcal{W}_3, M^{(\kappa)}(\sigma \circ \Lambda, 0) \otimes \overline{M}(\Lambda^L, 2i))$ vanishes unless Λ^L satisfies (3.54), the first part of the theorem follows immediately. Moreover, the reduction theorem implies that this cohomology, when nonvanishing, arises only at the highest weight of the given $M(\sigma \circ \Lambda, 0)$. Thus, depending on Λ and σ, we find (E_r', δ_r), $r \geq 2$, to be given by one of the three cases (which correspond precisely to those introduced above) shown in Figs. 3.1-3.3.

[4] Once more the technique employed here is quite standard, and the reader can consult [BMP4] for an elementary exposition in a similar context of the semi-infinite cohomology of the Virasoro algebra.

$$
\begin{array}{lccccccc}
 & 0 & & c_0^{[2]}c_0^{[3]}v_0 & & c_0^{[2]}c_0^{[3]}v_1' & & 0 \\
E_1' : & 0 & \longrightarrow & c_0^{[2]}v_0\,,\ \ c_0^{[3]}v_0 & \xrightarrow{\delta_1} & c_0^{[2]}v_0'\,,\ \ c_0^{[3]}v_1' & \longrightarrow & 0\,, \\
 & 0 & & v_0 & & v_0' & & 0
\end{array}
$$

$$
\begin{array}{lccccccc}
 & 0 & & c_0^{[2]}c_0^{[3]}v_0 & & c_0^{[2]}c_0^{[3]}v_1' & & 0 \\
E_2' : & 0 & \longrightarrow & c_0^{[3]}v_0 & \xrightarrow{\delta_2} & c_0^{[3]}v_1' & \longrightarrow & 0\,, \\
 & 0 & & 0 & & 0 & & 0
\end{array}
$$

$$E_2' = E_3' = \ldots = E_\infty' .$$

Fig. 3.2. Case II

The notation used in the figures is the same as in Theorem 3.8, except that we have denoted the highest weight states from different spaces by primes, and at the third term of the spectral resolution in Case III we have introduced $v_0^{(\pm)} \equiv v_0 \pm v_0'$. Each diagram represents a double-graded complex $E_r'^{p,q}$, with the ghost number, p, increasing in the vertical direction, starting with $p = 3$, which is the ghost number of the state v_0. The horizontal grading, q, is induced from the resolutions. In particular, each quartet of states in E_1' arises at the position, q, of a Verma module $M(\sigma \circ \Lambda, 0)$ or a generalized Verma module $M^{(2)}(\sigma \circ \Lambda, 0)$ in the resolution. The differential $\delta_1 : E_1'^{p,q} \to E_1'^{p,q+1}$ is obtained from the differential δ in the resolutions; i.e., it maps, up to a sign, a given state in the quartet onto the identical one in the quartet at the next step (if such is present, it maps to zero otherwise). For example, in Case II we find $\delta_1(v_0) = v_0'$, $\delta_1(c_0^{[2]}v_0) = c_0^{[2]}v_0'$, but $\delta_1(c_0^{[3]}v_0) = 0$, etc. The resulting E_2' terms are spanned by the elements listed in the diagrams. Since $\delta_2 : E_1'^{p,q} \to E_1'^{p-1,q+2}$, we find that in all cases δ_2 vanishes identically, and thus the sequence collapses. The second part of the theorem then follows by comparing the limits of the two spectral sequences using

$$\bigoplus_{p+q=n} E_\infty^{p,q} \cong \bigoplus_{p+q=n} E_\infty'^{p,q}\,, \tag{3.59}$$

i.e., the so-called "zig-zag procedure." □

$$
\begin{array}{lccccccc}
 & 0 & & c_0^{[2]}c_0^{[3]}v_0 & c_0^{[2]}c_0^{[3]}v_0' & & c_0^{[2]}c_0^{[3]}v_1'' & 0 \\
E_1' : & 0 & \rightarrow & c_0^{[2]}v_0\,,\quad c_0^{[3]}v_0\,, & c_0^{[2]}v_0'\,,\quad c_0^{[3]}v_0' & \xrightarrow{\delta_1} & c_0^{[2]}v_0''\,,\quad c_0^{[3]}v_1'' & \rightarrow\ 0\,, \\
 & 0 & & v_0 & v_0' & & v_0'' & 0
\end{array}
$$

$$
\begin{array}{lccccccc}
 & 0 & & c_0^{[2]}c_0^{[3]}v_0^{(-)}\,, & c_0^{[2]}c_0^{[3]}v_0^{(+)} & & c_0^{[2]}c_0^{[3]}v_1'' & 0 \\
E_2' : & 0 & \rightarrow & c_0^{[2]}v_0^{(-)}\,,\quad c_0^{[3]}v_0^{(-)}\,, & c_0^{[3]}v_0^{(+)} & \xrightarrow{\delta_2} & c_0^{[3]}v_1'' & \rightarrow\ 0\,, \\
 & 0 & & v_0^{(-)} & & & 0 & 0
\end{array}
$$

$$E_2' = E_3' = \ldots = E_\infty' .$$

Fig. 3.3. Case III

In view of Theorem 3.14, it is rather natural to seek an explicit description of $H(\mathcal{W}_3, L(\Lambda, 0) \otimes F(\Lambda^L, 2i))$ in terms of quartets. A quartet with states of ghost number n, $n+1$, $n+1$ and $n+2$ is parametrized by its lowest lying member, which will be called a "prime state," following the terminology introduced in [PSSW] for similar states in the critical $\mathcal{W}_3$ cohomology. We should stress, however, that the decomposition of the cohomology into quartets is at the level of vector spaces only, and that we have no intrinsic characterization of prime states as specific cohomology classes. Let $H_{\rm pr}(\mathcal{W}_3, L(\Lambda, 0) \otimes F(\Lambda^L, 2i))$ denote the space of prime states. Then part (ii) of Theorem 3.14 can be restated simply as follows

Theorem 3.15 *Consider Λ, Λ^L and $\sigma \in W(\Lambda)$ as in (3.54). Then*

$$H_{\rm pr}^n(\mathcal{W}_3, L(\Lambda, 0) \otimes F(\Lambda^L, 2i)) \cong \begin{cases} \mathbb{C} & \text{if } n = 3 - \ell(\sigma)\,, \\ 0 & \text{otherwise}\,, \end{cases} \tag{3.60}$$

and there is a (noncanonical) isomorphism (of vector spaces)

$$H^n \cong H_{\rm pr}^n \oplus H_{\rm pr}^{n-1} \oplus H_{\rm pr}^{n-1} \oplus H_{\rm pr}^{n-2} . \tag{3.61}$$

We would like to conclude with a comment on a possible role of the relative cohomology. One may be tempted to conjecture, by extrapolating the known result for the Virasoro algebra [BMP3,FGZ,LiZu1], that the full cohomology is (noncanonically) isomorphic to the direct sum of relative cohomologies "shifted" by the ghosts' zero modes; i.e., schematically, $H \cong H_{\rm rel} \oplus c_0^{[2]} H_{\rm rel} \oplus c_0^{[3]} H_{\rm rel} \oplus c_0^{[2]} c_0^{[3]} H_{\rm rel}$. If this was the case, it would be natural to identify prime states with the relative cohomology states. It would also explain the quartet structure of the cohomology. Unfortunately, as discussed earlier, the relative cohomology, as well as its relation to the full cohomology, is difficult to analyze, and we cannot give any general arguments that would support such a conjecture. However, one finds, at least in cases we have studied explicitly, that representatives of prime states in cohomology can be chosen such that they are annihilated by $b_0^{[2]}$ and $b_0^{[3]}$ (and thus by $L_0^{\rm tot}$ and $W_0^{\rm tot}$).

3.4.2 $H(\mathcal{W}_3, F(\Lambda^M,0) \otimes F(\Lambda^L,2i))$ with $-i\Lambda^L + 2\rho \in P_+$

The result for the cohomology $H(\mathcal{W}_3, F(\Lambda^M,0) \otimes F(\Lambda^L,2i))$, with $-i\Lambda^L + 2\rho \in P_+$, follows immediately by applying Theorem 3.14 to the decomposition of $F(\Lambda^M,0)$ into irreducible modules $L(\Lambda,0)$ given in Theorem 2.31. The nontrivial contributions to the cohomology for a given Liouville momentum come from $L(\Lambda,0)$ in the decomposition such that Λ satisfies (3.54). Thus it is clearly convenient to put such weights together.

Definition 3.16 *For $\Lambda' + 2\rho \in P_+$, define $R(\Lambda')$ as the set of all $\Lambda \in P_+$ such that $\Lambda' + 2\rho = \sigma \circ \Lambda$, for some $\sigma \in W(\Lambda)$.*

We have then proven the following result.

Theorem 3.17 *Let $-i\Lambda^L + 2\rho \in P_+$. Then*

$$\dim H^n_{\rm pr}(\mathcal{W}_3, F(\Lambda^M,0) \otimes F(\Lambda^L,2i)) = \sum_{\Lambda \in R(-i\Lambda^L)} \sum_{\sigma \in W(\Lambda)} \delta_{n,3-\ell(\sigma)}\, m^{\Lambda}_{\Lambda^M}\,, \tag{3.62}$$

In particular, $H(\mathcal{W}_3, F(\Lambda^M,0) \otimes F(\Lambda^L,2i)) \neq 0$ if and only if Λ^L satisfies (3.54) for some $\Lambda \in P_+$ and $\sigma \in W(\Lambda)$ and $m^{\Lambda}_{\Lambda^M} \neq 0$.

The appearance of the multiplicities $m^{\Lambda}_{\Lambda^M}$ in (3.62) is well understood in the light of Theorem 3.11. Since all states in $F(\Lambda^M,0) \otimes F(\Lambda^L,2i)$ have the same weight, $(\Lambda^M, -i\Lambda^L)$, with respect to $\mathfrak{sl}_3 \oplus (\mathfrak{u}_1)^2$, the multiplicities simply reflect the decomposition of $H(\mathcal{W}_3, \mathfrak{C})$ into finite dimensional $\mathfrak{sl}_3 \oplus (\mathfrak{u}_1)^2$ modules.

We may make this structure even more manifest as follows. Fix a pair (Λ, σ), $\Lambda \in P_+$ and $\sigma \in W(\Lambda)$, and then determine the Liouville weight Λ^L via (3.54). Now consider all matter Fock spaces, $F(\Lambda^M,0)$, that give rise to nonvanishing cohomology through the irreducible module $L(\Lambda,0)$ in their decomposition. From (3.62) we see that they fill up precisely one $\mathfrak{sl}_3 \oplus (\mathfrak{u}_1)^2$ module $\mathcal{L}(\Lambda) \otimes \mathbb{C}_{-i\Lambda^L}$ in the "prime cohomology" – or, more rigorously, a quartet of such modules in the full cohomology.

Lemma 3.18 *For $-i\Lambda^L + 2\rho$ in the fundamental Weyl chamber, the decomposition of $H(\mathcal{W}_3, \mathfrak{C})$ into quartets of $\mathfrak{sl}_3 \oplus (\mathfrak{u}_1)^2$ irreducible modules is in one to* *one correspondence with the space of pairs (Λ, σ), where $\Lambda \in P_+$ and $\sigma \in W(\Lambda)$. Moreover, the space of such pairs is a sum of disjoint cones in the $(\Lambda^M, -i\Lambda^L)$ weight space that are isomorphic with P_+.*

Proof. The first part of the lemma is just a summary of the previous discussion, so let us proceed to the cone decomposition. From the definition of $W(\Lambda)$ in (2.102) it is clear that if (Λ, σ) is a pair then so also is $(\Lambda + \lambda, \sigma)$, for all $\lambda \in P_+$. Moreover, the set of weights that give rise to a given σ is determined by inequalities, each of the form $(\Lambda, \alpha) \geq 0$, from which the cone structure follows. □

This cone-like structure for the decomposition of the cohomology into irreducible modules of $\mathfrak{sl}_3 \oplus (\mathfrak{u}_1)^2$, and its correspondence to an extension of the Weyl group, will play a key role in our extension of the result to other Weyl chambers. Let us therefore take a closer look at how this correspondence arises in the fundamental chamber. The "tips" of the cones in Lemma 3.18 can be determined explicitly by examining the sets $W(\Lambda)$. Let $\mathcal{S}^n$ be the set of cone tips at ghost number n. The result for $\mathcal{S}^n$ is given in Table 3.1. Moreover, we notice that the "shift," $(-i\Lambda^L + 2\rho) - \Lambda^M$, is constant throughout each cone and equal to $\rho - \sigma\rho$, $\sigma \in \widetilde{W}$. Thus the set of cones, as parametrized by the shifts, say, are in correspondence with the extension of the Weyl group, $\widetilde{W}$.

To conclude this section, we summarize the result for the cohomology in the fundamental Weyl chamber.

Theorem 3.19 *The cohomology $H(\mathcal{W}_3, \mathfrak{C}_1)$ is isomorphic as an $\mathfrak{sl}_3 \oplus (\mathfrak{u}_1)^2$ module to the direct sum of quartets of irreducible modules parametrized by disjoint cones $\{(\Lambda, \Lambda') + (\lambda, \lambda) \,|\, \lambda \in P_+\}$; i.e.,*

$$H^n_{\text{pr}}(\mathcal{W}_3, \mathfrak{C}_1) \cong \bigoplus_{(\Lambda,\Lambda') \in \mathcal{S}^n} \bigoplus_{\lambda \in P_+} \mathcal{L}(\Lambda + \lambda) \otimes \mathbb{C}_{\Lambda' + \lambda}\,, \tag{3.63}$$

where the sets $\mathcal{S}^n$ (tips of the cones) are given in Table 3.1.

Table 3.1. The sets $\mathcal{S}^n$

n	$\mathcal{S}^n$
0	$(0, 0)$
1	$(0, -2\Lambda_1 + \Lambda_2)$, $(\Lambda_1 + \Lambda_2, 0)$, $(0, \Lambda_1 - 2\Lambda_2)$
2	$(\Lambda_1, -2\Lambda_1)$, $(0, -\Lambda_1 - \Lambda_2)$, $(\Lambda_2, -2\Lambda_2)$
3	$(0, -2\Lambda_1 - 2\Lambda_2)$

Remarks

i. The nonvanishing cohomology in the fundamental Weyl chamber arises in ghost numbers $0, \ldots, 5$.

ii. The pattern of the cohomology cones in Table 3.1 can be conveniently represented as a plot on the lattice of shifted Liouville momenta, $-i\Lambda^L + 2\rho$, see Appendix E.
iii. The precise form of the resolutions, $(\mathcal{M}, \delta)$, required an explicit computation of the embedding patterns of (generalized) Verma modules. An independent partial confirmation of those results is provided by a computation of cohomology spaces for low lying (shifted) Liouville weights, i.e., an explicit verification of (3.63). This is summarized in Appendix D.
iv. Given the isomorphism

$$H^n(\mathcal{W}_3, F(\Lambda^M, 0) \otimes F(\Lambda^L, 2i)) \cong H^{8-n}(\mathcal{W}_3, F(\Lambda^M, 0) \otimes F(w_0 \cdot \Lambda^L, 2i)), \tag{3.64}$$

proved in Sect. 3.3.4, Theorem 3.19 also gives a complete result for the cohomology $H(\mathcal{W}_3, \mathfrak{C}_{w_0})$, i.e., for $-i\Lambda^L + 2\rho \in P_-$. At the level of prime cohomology states (3.64) reads

$$H^n_{\rm pr}(\mathcal{W}_3, F(\Lambda^M, 0) \otimes F(w_0 \cdot \Lambda^L, 2i)) \cong H^{6-n}_{\rm pr}(\mathcal{W}_3, F(\Lambda^M, 0) \otimes F(w_0 \cdot \Lambda^L, 2i)). \tag{3.65}$$

The reflection by the Weyl group element accompanied by a shift in the ghost number in (3.65) suggests a generalization of Theorem 3.19 to the other Weyl chambers which we will discuss in the following section.

3.5 The Conjecture for $H(\mathcal{W}_3, \mathfrak{C})$

3.5.1 Introduction

In this section we derive a conjecture for $H(\mathcal{W}_3, \mathfrak{C})$ by assuming that there is a "symmetry" with respect to the action of the Weyl group on the (shifted) Liouville momentum. In other words, if we define $\mathfrak{C}_w$, $w \in W$, to be the subcomplex of $\mathfrak{C}$ with $-i\Lambda^L + 2\rho \in w^{-1}P_+$, then all cohomologies $H(\mathcal{W}_3, \mathfrak{C}_w)$ should be related in some sense. For the case $w = w_0$, we saw at the end of the last section that this relation is determined by duality. Moreover, we learned there that, loosely speaking, each Weyl group reflection of the Liouville momentum should be accompanied by a shift in the ghost number of the cohomology. (This is also suggested by examining an analogous problem in the cohomology of Lie algebras, as well as the so-called generic regime of the $\mathcal{W}_3$ cohomology (see, e.g., [BLNW2,BMP7]).) Our aim is then to correlate the Weyl reflection to other chambers with the correspondence between the set of cones and the extended Weyl group discussed earlier, incorporating the ghost number shift appropriately. One can clearly expect that there might be additional subtleties if $-i\Lambda^L + 2\rho$ lies close to the boundary of a Weyl chamber. Thus we develop an ansatz for the cohomology with $-i\Lambda^L + 2\rho$ lying sufficiently deep inside a Weyl chamber (referred to as the "bulk region") and then use the results of explicit cohomology computations to extend it to a complete conjecture.

3.5.2 A Vanishing Theorem

Let us begin with the observation that $H^n(\mathcal{W}_3, \mathfrak{C})$ may be nonzero only in a finite range of ghost numbers. The restriction is given by the following vanishing theorem.

Theorem 3.20 *The cohomology $H^n(\mathcal{W}_3, \mathfrak{C})$ is nonvanishing at most in ghost numbers $n = 0, \ldots, 8$.*

Proof. Consider $H^n(\mathcal{W}_3, F(\Lambda^M, 0) \otimes F(\Lambda^L, 2i))$. Let $L(\Lambda, 0)$ be an irreducible module in the decomposition of $F(\Lambda^M, 0)$. Then for all Verma modules $M^{(\kappa)}(\sigma \circ \Lambda, 0)$, $\kappa = 1$ or 2, $\sigma \in W(\Lambda)$, in the resolution of $L(\Lambda, 0)$, the cohomology $H^n(\mathcal{W}_3, M^{(\kappa)}(\sigma \circ \Lambda, 0) \otimes F(\Lambda^L, 2i))$ vanishes for $n \leq 2$ (see Theorem 3.9). A straightforward repetition of the double complex argument in the proof of Theorem 3.14, would then give $3 - 4 = -1$ as the lower bound for the ghost number of the cohomology. The ghost number -1 cohomology could only arise from the Verma module, $M(w_0 \circ \Lambda, 0)$, at level -4 in the resolution; or, using the language of the double complex to be more precise, from the ghost number 3 state in $E'^{3,-4}_1 \cong H^3(\mathcal{W}_3, M(w_0 \circ \Lambda, 0) \otimes F(\Lambda^L, 2i))$. To demonstrate that the lower bound is correct as stated in the theorem we must therefore show that $\delta'_1 : E'^{3,-4}_1 \to E'^{3,-3}_1$ is an embedding.

To see this, consider the factor in δ'_1 that arises from the embedding $M(w_0 \circ \Lambda, 0) \to M^{(2)}(w_0 \circ \Lambda, 0)$. From the isomorphism $M^{(2)}(w_0 \circ \Lambda, 0)/M(w_0 \circ \Lambda, 0) \cong M(w_0 \circ \Lambda, 0)$ we have a short exact sequence (see (2.50))

$$0 \longrightarrow M(w_0 \circ \Lambda, 0) \longrightarrow M^{(2)}(w_0 \circ \Lambda, 0) \longrightarrow M(w_0 \circ \Lambda, 0) \longrightarrow 0 . \tag{3.66}$$

By applying $H^3(\mathcal{W}_3, - \otimes F(\Lambda^L, 2i))$ to (3.66), we obtain a long exact sequence, from which the required embedding is proved using $H^2(\mathcal{W}_3, M(w_0 \circ \Lambda, 0) \otimes F(\Lambda^L, 2i)) = 0$. The upper bound on the ghost number follows from (3.64). □

Note that for $-i\Lambda^L + 2\rho \in P_+$ the ghost number of the nonvanishing cohomology is between 0 and 5, which saturates the lower bound imposed by Theorem 3.20. As discussed in Remark (iv) of the previous section, if the Liouville weight is reflected by w_0, i.e., $-i\Lambda^L + 2\rho \in w_0 P_+$, there is a corresponding shift in the ghost number, which now ranges between 3 and 8, thus saturating the upper bound of the allowed values. We expect that the cohomology in the intermediate Weyl chambers interpolates between those two extreme cases. If we require consistency with duality (see Theorem 3.13), and symmetry with respect to interchange of the fundamental weights Λ_1 and Λ_2, there is just one natural possibility left.

Conjecture 3.21 *The cohomology $H^n(\mathcal{W}_3, \mathfrak{C}_w)$, $w \in W$, is nonvanishing at most in ghost numbers $n = \ell(w), \ldots, \ell(w) + 5$.*

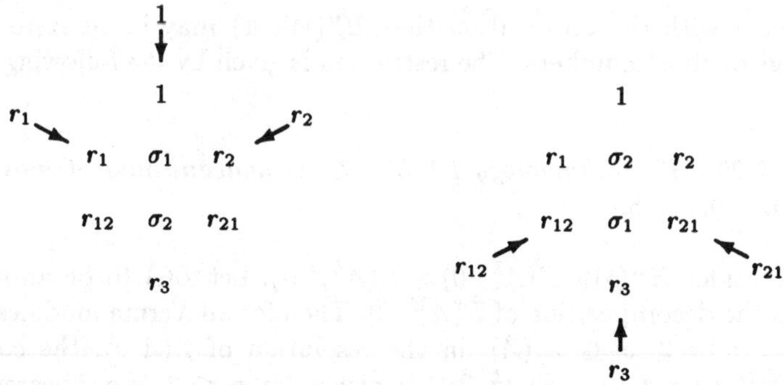

Fig. 3.4. Each twisted length $\ell_w(\sigma)$ increases in the direction of the corresponding arrow "$w \rightarrow$" and is constant along the transverse directions, as illustrated in Fig. 3.5.

3.5.3 $H(\mathcal{W}_3, F(\Lambda^M, 0) \otimes F(\Lambda^L, 2i))$

For weights $-i\Lambda^L + 2\rho \in P_+$, Theorem 3.19 states that the $\mathfrak{sl}_3 \oplus (\mathfrak{u}_1)^2$ content of prime cohomology is a direct sum of the eight cones in Table 3.1. We have seen that these cones, parametrized by $\sigma \in \widetilde{W}$, arise at ghost number $3 - \ell(\sigma)$: their tips are given in the table; their cone shift $(-i\Lambda^L + 2\rho) - \Lambda^M$, which is constant throughout the cone, is easily calculated to be $\rho - \sigma\rho$. Conversely, notice that if we know that a cone with a shift $\delta\lambda$ arises in cohomology, its tip is determined by the "lowest" weight $\Lambda \in P_+$ for which $(\Lambda^M, \Lambda^L) = (\Lambda, i(\Lambda + \delta\lambda - 2\rho))$ gives rise to a nontrivial cohomology state. Thus, given the cone structure and the set of all possible cone shifts, we could determine the complete cohomology by simply finding all such lowest weights Λ – a finite computation that can be carried out.

We will, indeed, assume that the cone structure generalizes to the other Weyl chambers. Hence we must formulate an ansatz for the cone shifts. The case $w = w_0$ suggests that the cones which arise for $-i\Lambda^L + 2\rho \in w^{-1}P_+$, $w \in W$, should be related by a Weyl reflection to the cones in the fundamental Weyl chamber. More precisely, let us introduce the notion of a w-twisted cone.

Definition 3.22 *We define a w-twisted cone as a set of weights $\{(\Lambda, \Lambda') + (\lambda, w^{-1}\lambda) \,|\, \lambda \in P_+\}$, where the tip (Λ, Λ') has $\Lambda \in P_+$. The shift characterizing this cone is given by $w(\Lambda' + 2\rho) - \Lambda$.*

Clearly the shift does not change when the cone is reflected from one Weyl chamber to another. The natural extension of our previous results is the following conjecture for the decomposition of the cohomology in the bulk region.

Conjecture 3.23 *For $-i\Lambda^L + 2\rho$ sufficiently inside the Weyl chamber $w^{-1}P_+$ the cohomology $H(\mathcal{W}_3, \mathfrak{C}_w)$ is a direct sum of w-twisted cones with the shifts $\rho - \sigma\rho$, $\sigma \in \widetilde{W}$.*

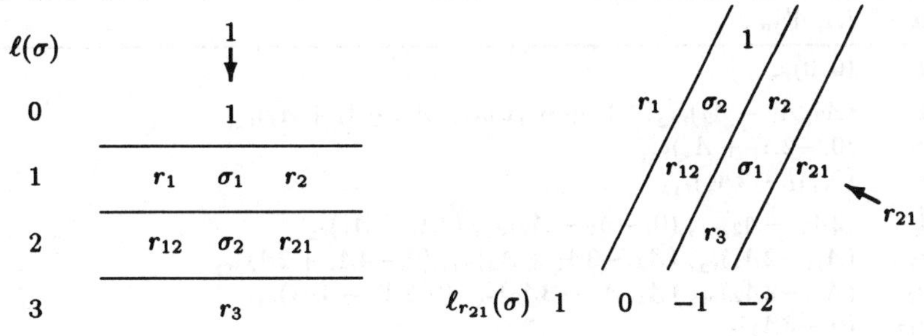

Fig. 3.5. Examples of twisted lengths for $w = 1$ and $w = r_{21}$.

It remains to find a proper ansatz for a shift in the ghost number corresponding to a given reflection w. Given $w \in W$ there is a natural generalization of the length ℓ, called the twisted length [BMP1,FeFr1], which is defined by $\ell_w(w') = \ell(w^{-1}w') - \ell(w^{-1})$, $w' \in W$. This twisted length may again be extended to $\widetilde{W}$ by using $w^{-1}\sigma_i = \sigma_i$, and a simple algorithm for computing it is given in Figs. 3.4 and 3.5.

Once again, a natural generalization of our previous results follows.

Conjecture 3.24 *The ghost number of the prime cohomology state in a w-twisted cone with the shift $\rho - \sigma\rho$, is equal to $3 - \ell_w(\sigma)$, $w \in W$, $\sigma \in \widetilde{W}$.*

Remarks

i. By construction, Conjectures 3.23 and 3.24 correctly reproduce the cones and their ghost numbers in the $w = 1$ and $w = w_0$ Weyl chambers. In the other chambers they yield the correct range of ghost numbers, in particular, those suggested by Conjecture 3.21.

ii. In the context of Lie algebras or affine Lie algebras, the twisted length functions arise naturally in the resolutions of highest weight irreducible modules in terms of twisted Verma or Wakimoto modules [BMP4]. If analogous twisted resolutions for positive energy $\mathcal{W}_3$ modules exist, one would expect to be able to prove, following the steps in Sect. 3.4, that the structure of the full cohomology is as conjectured above.

iii. One can also arrive at Conjectures 3.23 and 3.24 under seemingly weaker assumptions, by studying the BV algebra structure of $H(\mathcal{W}_3, \mathbb{C})$. This is discussed in Sect. 5.4.

In the following we assume the validity of Conjectures 3.23 and 3.24, and proceed to study their consequences. As discussed above, to determine the full cohomology we need now only calculate the cone tips. We have carried out an exhaustive computation of the dimensions of the cohomologies for low lying weights, the results are summarized in Appendix D.

Table 3.2. The sets $\mathcal{S}^n_w$ – the weights $(\Lambda,\Lambda')_\sigma$ satisfy $\Lambda'+2\rho = w^{-1}(\Lambda+\rho-\sigma\rho)$

n	w	$(\Lambda,\Lambda')_\sigma$
0	1	$(0,0)_{r_3}$
1	1	$(\Lambda_2,\Lambda_1-\Lambda_2)_{r_{12}}$, $(\Lambda_1+\Lambda_2,0)_{\sigma_2}$, $(\Lambda_1,-\Lambda_1+\Lambda_2)_{r_{21}}$
	r_1	$(0,-2\Lambda_1+\Lambda_2)_{r_{21}}$
	r_2	$(0,\Lambda_1-2\Lambda_2)_{r_{12}}$
2	1	$(2\Lambda_2,-\Lambda_2)_{r_1}$, $(0,-\Lambda_1-\Lambda_2)_{\sigma_1}$, $(2\Lambda_1,-\Lambda_1)_{r_2}$
	r_1	$(\Lambda_1,-2\Lambda_1)_{r_2}$, $(\Lambda_2,-3\Lambda_1+\Lambda_2)_{\sigma_2}$, $(0,-4\Lambda_1+2\Lambda_2)_{r_3}$
	r_2	$(\Lambda_2,-2\Lambda_2)_{r_1}$, $(\Lambda_1,\Lambda_1-3\Lambda_2)_{\sigma_2}$, $(0,2\Lambda_1-4\Lambda_2)_{r_3}$
	r_{12}	$(0,-3\Lambda_2)_{r_2}$
	r_{21}	$(0,-3\Lambda_1)_{r_1}$
3	1	$(\Lambda_1+\Lambda_2,-\Lambda_1-\Lambda_2)_1$
	r_1	$(\Lambda_2,-2\Lambda_1-\Lambda_2)_1$, $(\Lambda_1,-4\Lambda_1+\Lambda_2)_{\sigma_1}$, $(\Lambda_2,-5\Lambda_1+2\Lambda_2)_{r_{12}}$
	r_2	$(\Lambda_1,-\Lambda_1-2\Lambda_2)_1$, $(\Lambda_2,\Lambda_1-4\Lambda_2)_{\sigma_1}$, $(\Lambda_1,2\Lambda_1-5\Lambda_2)_{r_{21}}$
	r_{12}	$(\Lambda_2,-\Lambda_1-3\Lambda_2)_1$, $(0,\Lambda_1-5\Lambda_2)_{r_{21}}$, $(\Lambda_2,-5\Lambda_2)_{\sigma_2}$
	r_{21}	$(\Lambda_1,-3\Lambda_1-\Lambda_2)_1$, $(0,-5\Lambda_1+\Lambda_2)_{r_{12}}$, $(\Lambda_1,-5\Lambda_1)_{\sigma_2}$
	r_3	$(0,-2\Lambda_1-2\Lambda_2)_1$
4	r_1	$(0,-4\Lambda_1-\Lambda_2)_{r_1}$
	r_2	$(0,-\Lambda_1-4\Lambda_2)_{r_2}$
	r_{12}	$(\Lambda_2,-2\Lambda_1-4\Lambda_2)_{r_1}$, $(\Lambda_1,-\Lambda_1-5\Lambda_2)_{\sigma_1}$, $(0,-6\Lambda_2)_{r_3}$
	r_{21}	$(\Lambda_1,-4\Lambda_1-2\Lambda_2)_{r_2}$, $(\Lambda_2,-5\Lambda_1-\Lambda_2)_{\sigma_1}$, $(0,-6\Lambda_1)_{r_3}$
	r_3	$(0,-3\Lambda_1-3\Lambda_2)_{\sigma_2}$, $(2\Lambda_1,-4\Lambda_1-3\Lambda_2)_{r_2}$, $(2\Lambda_2,-3\Lambda_1-4\Lambda_2)_{r_1}$
5	r_{12}	$(0,-2\Lambda_1-5\Lambda_2)_{r_{12}}$
	r_{21}	$(0,-5\Lambda_1-2\Lambda_2)_{r_{21}}$
	r_3	$(\Lambda_1,-5\Lambda_1-3\Lambda_2)_{r_{21}}$, $(\Lambda_1+\Lambda_2,-4\Lambda_1-4\Lambda_2)_{\sigma_1}$, $(\Lambda_2,-3\Lambda_1-5\Lambda_2)_{r_{12}}$
6	r_3	$(0,-4\Lambda_1-4\Lambda_2)_{r_3}$

From this we determine the cone tips to be as listed in Table 3.2, where $\mathcal{S}^n_w$ denotes the set of w-twisted cone tips at ghost number n. Finally, then, we have

Theorem 3.25 *The cohomology $H(\mathcal{W}_3,\mathfrak{C})$ is isomorphic, as an $\mathfrak{sl}_3\oplus(\mathfrak{u}_1)^2$ module, to the direct sum of quartets of irreducible $\mathfrak{sl}_3\oplus(\mathfrak{u}_1)^2$ modules with the highest weights in a set of disjoint cones $\{(\Lambda,\Lambda')+(\lambda,w^{-1}\lambda)\,|\,\lambda\in P_+,\ (\Lambda,\Lambda')\in\mathcal{S}_w\}$; i.e.,*

$$H^n_{\rm pr}(\mathcal{W}_3,\mathfrak{C}) \cong \bigoplus_{w\in W}\ \bigoplus_{(\Lambda,\Lambda')\in\mathcal{S}^n_w}\ \bigoplus_{\lambda\in P_+} \mathcal{L}(\Lambda+\lambda)\otimes\mathbb{C}_{\Lambda'+w^{-1}\lambda}\,, \tag{3.67}$$

where the sets $\mathcal{S}^n_w$ (tips of the cones) are given in Table 3.2.

Remark. Since a given cone and its Weyl reflection may overlap, the theorem requires an explicit decomposition of $H(\mathcal{W}_3,\mathfrak{C})$ into disjoint cones in the overlap region. In all cases we deal with the ambiguity by including the complete common region in only one of the cones. In particular, this explains why some of the tips in the fundamental Weyl chamber given in Table 3.1 are shifted with respect to those in Table 3.2 with $w = 1$.

Example. To illustrate this ambiguity, let us consider as an example all the cones in $H^1(\mathcal{W}_3, \mathbb{C})$ characterized by the shift $3\Lambda_1$. We have already found such a cone in the untwisted sector, namely $\{(0, \Lambda_1 - 2\Lambda_2) + (\lambda, \lambda) \mid \lambda \in P_+\}$, which appears in Theorem 3.19. Examining the tables in Appendix D, we conclude from the appearance of the quartet with weights $(\Lambda_2, 2\Lambda_1 - 3\Lambda_2)$ that there is an r_2-twisted cone with the same shift, $3\Lambda_1$. The lowest weight state, i.e., the tip of this r_2-twisted cone, would also be $(0, \Lambda_1 - 2\Lambda_2)$ – in fact the common boundary of the two cones is one dimensional, at $(n\Lambda_1, (n+1)\Lambda_1 - 2\Lambda_2)$, $n > 0$.

By retracing the steps which give Theorem 3.17 from Theorem 3.15, we may derive from Theorem 3.25 the result for $H(\mathcal{W}_3, L(\Lambda, 0) \otimes F(\Lambda^L, 2i))$ when Λ^L is arbitrary.

Corollary 3.26 *Let $\Lambda \in P_+$.*

i. The cohomology $H^n(\mathcal{W}_3, L(\Lambda, 0) \otimes F(\Lambda^L, 2i))$ is nontrivial only if there exist $w \in W$, $\sigma \in \widetilde{W}$ such that

$$-i\Lambda^L + 2\rho \;=\; w^{-1}(\Lambda + \rho - \sigma\rho)\,. \tag{3.68}$$

ii. For w, σ, Λ and Λ^L as in (3.68), the cohomology $H^n_{\rm pr}(\mathcal{W}_3, L(\Lambda, 0) \otimes F(\Lambda^L, 2i))$ is one-dimensional if

$$n = 3 - \ell_w(\sigma) = 3 + \ell(w^{-1}) - \ell(w^{-1}\sigma)\,, \tag{3.69}$$

and

$$\begin{array}{rcr} \sigma \in W\,, & \Lambda \in P_+\,, & w \in W\,,\\ \sigma \in \{\sigma_1, \sigma_2\}\,, & \Lambda \in P_{++}\,, & w \in W\,,\\ \sigma = \sigma_1\,, & (\Lambda, \alpha_i) = 0\,, \Lambda \neq 0\,, & w \in <r_i>\backslash W\,,\\ \sigma = \sigma_2\,, & (\Lambda, \alpha_i) = 0\,, \Lambda \neq 0\,, & w \in r_i(<r_i>\backslash W)\,, \end{array}$$

and vanishes otherwise.

In the case that certain weights $(\Lambda, -i\Lambda^L)$ and certain ghost number n satisfy (i) and (ii) for more than one choice of (w, σ), the above should be understood in the sense that the corresponding cohomology is nevertheless one-dimensional.

4 Batalin-Vilkovisky Algebras

In this chapter we collect some general results on BV algebras, and study a class of examples that will be important for describing explicitly the BV algebra structure of $H(\mathcal{W}_3, \mathfrak{C})$. Most of this chapter can be read independently from the rest of the book.

The notion of BV algebras first appeared in the work of the mathematician J. Koszul [Ko], where they were called (exact) coboundary G algebras (see also [K-S]). Independently, and at roughly the same time, the physicists Batalin and Vilkovisky [BaVi] constructed a particular example of a BV operator and applied it to the quantization of gauge theories (see also [Wi1]). Recently, Lian and Zuckerman [LiZu2], Schwarz and Penkava [PeSc] and Getzler [Gt] – building upon earlier work of Witten and Zwiebach [Wi2,WiZw] – recognized that BV algebras provide a framework for describing operator algebras in a large class of topological field theories: in particular, in two-dimensional string theory.

4.1 G Algebras and BV Algebras

4.1.1 Definitions

Definition 4.1 [Gs] *A G algebra (or Gerstenhaber algebra)* $(\mathfrak{A}, \cdot, [-,-])$ *is a* $\mathbb{Z}$ *graded, supercommutative, associative algebra under the "dot" product,* $\cdot$*, and a* $\mathbb{Z}$ *graded Lie superalgebra under the bracket,* $[-,-]$ *(of degree* -1*), such that the (odd) bracket acts as a superderivation of the algebra, i.e.,* $\mathfrak{A} = \bigoplus_{n\in\mathbb{Z}} \mathfrak{A}^n$,

$$\begin{aligned} \cdot \quad &: \quad \mathfrak{A}^m \times \mathfrak{A}^n \longrightarrow \mathfrak{A}^{m+n}\,, \\ [-,-] \quad &: \quad \mathfrak{A}^m \times \mathfrak{A}^n \longrightarrow \mathfrak{A}^{m+n-1}\,, \end{aligned}$$

and for any homogeneous $a, b, c \in \mathfrak{A}$ *(we define* $|a| = n$ *for* $a \in \mathfrak{A}^n$*)*

i. $a \cdot b = (-1)^{|a||b|} b \cdot a,$
ii. $(a \cdot b) \cdot c = a \cdot (b \cdot c),$
iii. $[a, b] = -(-1)^{(|a|-1)(|b|-1)}[b, a],$
iv. $(-1)^{(|a|-1)(|c|-1)}[a, [b, c]] + \text{cyclic} = 0,$
v. $[a, b \cdot c] = [a, b] \cdot c + (-1)^{(|a|-1)|b|} b \cdot [a, c].$

Let $\mathfrak{A}$ be a $\mathbb{Z}$ graded supercommutative algebra. We recall that a first order superderivation of $\mathfrak{A}$ of degree $|K|$ is a map $K : \mathfrak{A}^n \to \mathfrak{A}^{n+|K|}$ such that for all $a, b \in \mathfrak{A}$,

$$K(a \cdot b) = (Ka) \cdot b + (-1)^{|K||a|} a \cdot (Kb) . \tag{4.1}$$

We will refer to (4.1) as the (super) Leibniz rule. Now, for all $a \in \mathfrak{A}$, we define the left multiplication $l_a : \mathfrak{A} \to \mathfrak{A}$ by

$$l_a(b) = a \cdot b . \tag{4.2}$$

Then (4.1) is equivalent to

$$[K, l_a]_\pm - l_{Ka} = 0 , \tag{4.3}$$

with an obvious definition of the (graded) bracket. By induction, an n-th order superderivation of degree $|K|$ is a map $K : \mathfrak{A}^n \to \mathfrak{A}^{n+|K|}$ such that $[K, l_a]_\pm - l_{Ka}$ is an $(n-1)$-th order derivation for all $a \in \mathfrak{A}$. For example, a second order derivation of degree $|K|$ satisfies

$$\begin{aligned} K(a \cdot b \cdot c) &= K(a \cdot b) \cdot c + (-1)^{|K||a|} a \cdot K(b \cdot c) + (-1)^{(|K|+|a|)|b|} b \cdot K(a \cdot c) \\ &\quad - (Ka) \cdot b \cdot c - (-1)^{|K||a|} a \cdot (Kb) \cdot c - (-1)^{|K|(|a|+|b|)} a \cdot b \cdot (Kc) , \end{aligned} \tag{4.4}$$

for all $a, b, c \in \mathfrak{A}$.

Definition 4.2 [Ko,LiZu2] *A Batalin-Vilkovisky (BV) algebra* $(\mathfrak{A}, \cdot, \Delta)$ *is a* $\mathbb{Z}$ *graded, supercommutative, associative algebra with a second order derivation* Δ *(BV operator) of degree* -1 *satisfying* $\Delta^2 = 0$.

There is a close relation between the two classes of algebras; indeed, the following lemma shows that any BV algebra has the canonical structure of a G algebra.

Lemma 4.3 [LiZu2,PeSc,Wi1] *For any BV algebra* $(\mathfrak{A}, \cdot, \Delta)$, *the bracket*

$$[a, b] = (-1)^{|a|} \Big(\Delta(a \cdot b) - (\Delta a) \cdot b - (-1)^{|a|} a \cdot (\Delta b) \Big) , \qquad a, b \in \mathfrak{A} , \tag{4.5}$$

introduces on $\mathfrak{A}$ *the structure of a G algebra. Moreover, the BV operator acts as a superderivation of the bracket, i.e.,*

$$\Delta[a, b] = [\Delta a, b] + (-1)^{|a|-1} [a, \Delta b] . \tag{4.6}$$

By combining Definition 4.2 with Lemma 4.3, we obtain

Theorem 4.4 [Gt,PeSc] *Let* $(\mathfrak{A}, \cdot, [-,-])$ *be a G algebra and* Δ *an operator of degree* -1 *satisfying* $\Delta^2 = 0$ *and (4.5). Then* $(\mathfrak{A}, \cdot, \Delta)$ *is a BV algebra.*

Proof. One must only show that Δ satisfies identity (4.4), which follows directly by evaluating the left hand side in (4.4) as, e.g., $\Delta(a \cdot (b \cdot c))$ using (4.5), and then identity (v) for the bracket. □

It is clear from the definitions above that for any G algebra the subspace $\mathfrak{A}^0$ is an Abelian algebra with respect to the dot product. Similarly, $\mathfrak{A}^1$ is a Lie algebra with respect to the bracket. Moreover, by Definition 4.1, the map $\mathfrak{A}^1 \to \mathrm{Map}(\mathfrak{A}, \mathfrak{A})$ defined by

$$a \mapsto K_a \quad \text{with} \quad K_a(b) = [a, b], \quad a \in \mathfrak{A}^1, b \in \mathfrak{A}, \tag{4.7}$$

satisfies

$$\begin{aligned} K_a(b \cdot c) &= K_a(b) \cdot c + b \cdot K_a(c), \\ [K_a, K_b]_- &= K_{[a,b]}, \end{aligned} \tag{4.8}$$

i.e., $a \to K_a$ is a Lie algebra homomorphism from $\mathfrak{A}^1$ into the Lie algebra $\mathcal{D}(\mathfrak{A})$ of derivations of the "dot algebra" $\mathfrak{A}$.

Using (4.6) we also prove

Lemma 4.5 *Let $\mathfrak{A}$ be a BV algebra and consider $a \in \mathfrak{A}^1$ which satisfies $[\Delta(a), b] = 0$ for all $b \in \mathfrak{A}$, then $K_a \Delta = \Delta K_a$.*

To orient the reader, let us just note here a standard class of examples of G algebras: the algebra of polyvector fields on a given manifold, with the operations of wedge product and Schouten bracket (the bracket induced from the Lie bracket on vector fields), is a G algebra. We next detail an abstraction of this example, before returning to the simplest such: polyvectors on $\mathbb{C}^N$.

4.1.2 The G Algebra of Polyderivations of an Abelian Algebra

We now summarize the canonical construction of the polyderivations of an arbitrary Abelian algebra. This is a standard example of a G algebra. For a complete discussion see, e.g., [FrNi,Ko,Kr,K-S,Sc].

First recall that if $(\mathcal{R}, \cdot)$ is an Abelian algebra, and M an $\mathcal{R}$ module, then the space of derivations of $\mathcal{R}$ with values in M, $\mathcal{D}(\mathcal{R}, M)$ (or simply $\mathcal{D}(\mathcal{R})$ if $M = \mathcal{R}$), consists of those $a \in \mathrm{Hom}(\mathcal{R}, M)$ that satisfy the Leibniz rule, i.e.,

$$a(x \cdot y) = x \cdot a(y) + y \cdot a(x), \tag{4.9}$$

for any $x, y \in \mathcal{R}$.

Definition 4.6 *The polyderivations $\mathcal{P}^n(\mathcal{R}, M)$ of degree*[1] *n are defined as follows:*

i. $\mathcal{P}^0(\mathcal{R}, M) \cong M$,
ii. $\mathcal{P}^1(\mathcal{R}, M) \cong \mathcal{D}(\mathcal{R}, M)$,
iii. $\mathcal{P}^n(\mathcal{R}, M)$, $n \geq 2$, *is the space of those* $a \in \mathcal{D}(\mathcal{R}, \mathcal{P}^{n-1})$ *that satisfy*

$$a(x, y) = -a(y, x), \qquad x, y \in \mathcal{R}, \tag{4.10}$$

where $a(x, y)$ denotes the element $a(x)(y) \in \mathcal{P}^{n-2}(\mathcal{R}, M)$.

[1] Note, that it is more conventional to use the term "order" in this context.

Clearly, one may simply consider the n-th degree polyderivations $\mathcal{P}^n(\mathcal{R}, M)$ as a subspace in $\mathrm{Hom}(\mathcal{R}^{\otimes n}, M)$. Then we have

Lemma 4.7 *The polyderivations $\mathcal{P}^n(\mathcal{R}, M)$ of degree n consist of those $a \in \mathrm{Hom}(\mathcal{R}^{\otimes n}, M)$ for which $a(x_1, \ldots, x_n) = a(x_1)(x_2)\ldots(x_n)$ is completely antisymmetric and satisfies the Leibniz rule (4.9) in all of the arguments $x_1, \ldots, x_n$.*

In the case that $M = \mathcal{R}$, the space of polyderivations $\mathcal{P}(\mathcal{R}, \mathcal{R})$ – in the following denoted simply by $\mathcal{P}(\mathcal{R})$ – is itself a $\mathbb{Z}$ graded algebra, $\mathcal{P}(\mathcal{R}) = \bigoplus_{n \geq 0} \mathcal{P}^n(\mathcal{R})$, with the product "$\cdot$" defined by induction[2] using

$$(a \cdot b)(x) \;=\; a \cdot b(x) + (-1)^{|b|} a(x) \cdot b\,. \tag{4.11}$$

More explicitly,

$$\begin{aligned}(a \cdot b)&(x_1, \ldots, x_{m+n}) \\ &= (-1)^{mn} \sum_{\sigma \in S_{m+n}} \mathrm{sgn}(\sigma)\, a(x_{\sigma(1)}, \ldots, x_{\sigma(m)}) \cdot b(x_{\sigma(m+1)}, \ldots, x_{\sigma(m+n)})\,.\end{aligned} \tag{4.12}$$

Definition 4.8 *The Schouten bracket on $\mathcal{P}(\mathcal{R})$ is the unique bilinear map $[-,-]_S : \mathcal{P}(\mathcal{R}) \times \mathcal{P}(\mathcal{R}) \to \mathcal{P}(\mathcal{R})$ satisfying*

$$[a, b]_S \;=\; \begin{cases} 0, & \textit{for } |a| = 0\,, |b| = 0, \\ a(b), & \textit{for } |a| = 1\,, |b| = 0, \\ -b(a), & \textit{for } |a| = 0\,, |b| = 1, \end{cases} \tag{4.13}$$

and

$$[a, b]_S(x) \;=\; [a, b(x)]_S + (-1)^{|b|-1}[a(x), b]_S\,, \quad x \in \mathcal{R}\,, \tag{4.14}$$

for all other $|a|, |b|$.

In particular, it is straightforward to check that if $|a| > 0$ and $x \in \mathcal{R}$,

$$[a, x]_S \;=\; a(x). \tag{4.15}$$

Theorem 4.9 *The algebra $(\mathcal{P}(\mathcal{R}), \cdot, [-.-]_S)$ of polyderivations of an Abelian algebra $\mathcal{R}$, with the dot product (4.11) and the Schouten bracket (4.13), is a G algebra.*

---

[2] For simplicity we do not distinguish the degree 0 space in this induction (here or later), and so we have extended the notation via $a(x) = 0$ for $a \in \mathcal{P}^0(\mathcal{R}, M)$.

4.1.3 The BV Algebra of Polyvectors on a Free Algebra, $\mathcal{C}_N$

To illustrate these ideas, we discuss in some detail a simple example of a BV algebra. It can be considered as a model for the more complicated examples in the following sections.

Let $\mathcal{C}_N \cong \mathbb{C}[x^1, \ldots, x^N]$ be a free Abelian algebra. It is straightforward to verify that the G algebra of polyderivations $\mathcal{P}(\mathcal{C}_N)$, constructed above, is nothing but the algebra of polynomial polyvector fields on $\mathbb{C}^N$, i.e.,

$$\mathcal{P}(\mathcal{C}_N) \cong \bigoplus_{n=0}^{N} \bigwedge^n \mathcal{D}(\mathcal{C}_N) \,. \tag{4.16}$$

More explicitly, it is a free $\mathbb{Z}_2$ graded algebra with the even generators $x^1, \ldots, x^N \in \mathcal{C}_N$ and the odd generators $x_1^*, \ldots, x_N^* \in \mathcal{D}(\mathcal{C}_N)$, where

$$x_i^*(x^j) \;=\; \delta_i^j \,, \quad i,j = 1, \cdots, N \,. \tag{4.17}$$

Let $\Phi = \Phi^{i_1 \ldots i_m} x_{i_1}^* \cdots x_{i_m}^*$. Then

$$\Phi^{i_1 \ldots i_m} \;=\; (-1)^{m(m-1)/2} \frac{1}{m!} \Phi(x^{i_1}, \ldots, x^{i_m}) \,. \tag{4.18}$$

Given $\Psi = \Psi^{j_1 \ldots j_n} x_{j_1}^* \cdots x_{j_n}^*$, the product $\Phi \cdot \Psi$ is just the wedge product in the exterior algebra (4.16),

$$\Phi \cdot \Psi \;=\; (\Phi^{i_1 \ldots i_m} \Psi^{j_1 \ldots j_n}) x_{i_1}^* \ldots x_{i_m}^* x_{j_1}^* \ldots x_{j_n}^* \,, \tag{4.19}$$

while the Schouten bracket is the extension of the usual bracket on vector fields,

$$\begin{aligned} [\Phi, \Psi]_S \;=\; & \sum_{k=1}^{m} (-1)^{m+k} \Phi^{i_1 \ldots i_m} x_{i_k}^* (\Psi^{j_1 \ldots j_n}) x_{i_1}^* \ldots \widehat{x}_{i_k}^* \ldots x_{i_m}^* x_{j_1}^* \ldots x_{j_n}^* \\ & - (-1)^{(m-1)(n-1)} \sum_{k=1}^{n} (-1)^{n+k} \Psi^{j_1 \ldots j_n} x_{j_k}^* (\Phi^{i_1 \ldots i_m}) \\ & \qquad \times x_{j_1}^* \ldots \widehat{x}_{j_k}^* \ldots x_{j_n}^* x_{i_1}^* \ldots x_{i_m}^* \,. \end{aligned} \tag{4.20}$$

Let $\imath(x)$, $x \in \mathcal{C}_N$, be the usual evaluation operator, $\imath(x) : \mathcal{P}^n(\mathcal{C}_N) \to \mathcal{P}^{n-1}(\mathcal{C}_N)$, defined by

$$\imath(x)\Phi(x_1, \ldots, x_{n-1}) \;=\; \Phi(x, x_1, \ldots, x_{n-1}) \,, \quad x_1, \ldots, x_{n-1} \in \mathcal{C}_N \,. \tag{4.21}$$

By the definition of the bracket we also have $\imath(x) = [-, x\,]_S$. Note that these two forms of the evaluation operator are always well-defined for any polyderivation algebra $\mathcal{P}(\mathcal{R})$. In addition, for $x = x^i$, we obtain a representation specific to $\mathcal{P}(\mathcal{C}_N)$ – by expansion in the dual basis we find $\imath(x^i) = \frac{\partial}{\partial x_i^*}$, i.e., it acts like the left derivative with respect to the Grassmann variable x_i^*.

Consider an operator

$$\Delta \;=\; -\frac{\partial}{\partial x^i} \frac{\partial}{\partial x_i^*} \,. \tag{4.22}$$

This operator is a second order derivation on $\mathcal{P}(\mathcal{C}_N)$. By direct calculation, we also verify that it turns $\mathcal{P}(\mathcal{C}_N)$ into a BV algebra.

Theorem 4.10 [Ko,Wil] *The polyvector algebra* $(\mathcal{P}(\mathcal{C}_N), \cdot, \Delta)$ *is a BV algebra. The bracket induced by* Δ *is equal to the Schouten bracket (4.20).*

Finally, let us note that there is a canonical polyvector of maximal degree, the volume form,

$$\Omega = \frac{1}{N!}\epsilon^{i_1\ldots i_N} x^*_{i_1}\cdots x^*_{i_N}, \tag{4.23}$$

and that $(\mathcal{P}(\mathcal{C}_N), \cdot, \Delta)$, as a BV algebra, is generated by $x^1, \ldots, x^N$ and Ω.

4.1.4 Algebra of Polyderivations Associated with a BV Algebra

For any BV algebra $(\mathfrak{A}, \cdot, \Delta)$, the subspace $\mathfrak{A}^0$ is an Abelian algebra with respect to the dot product. Thus, as we have just seen, there is a naturally associated G algebra: namely, $\mathcal{P}(\mathfrak{A}^0)$, the G algebra of polyderivations with the Schouten bracket $[-,-]_S$. We now study the relation between $\mathcal{P}(\mathfrak{A}^0)$ and the G algebra of $\mathfrak{A}$ with the bracket $[-,-]$ induced by Δ.

Assume that $\mathfrak{A}$ has components of only nonnegative degree, i.e., $\mathfrak{A} = \bigoplus_{n\geq 0}\mathfrak{A}^n$. Then there is a natural map, $\pi : \mathfrak{A} \to \mathcal{P}(\mathfrak{A}^0)$, which is defined by induction on the degree. For $n = 0$, π is the identity map, i.e., $\pi(x) = x$ for $x \in \mathfrak{A}^0$. It is then extended to $n > 0$ using the condition

$$\pi(a)(x) = \pi([a,x]), \tag{4.24}$$

for any $a \in \mathfrak{A}^n$ and $x \in \mathfrak{A}^0$. It is easy to verify, using the properties of the bracket, that $\pi(a)$ is indeed a polyderivation of degree n for all $a \in \mathfrak{A}^n$.

Theorem 4.11 [LiZu2] *Suppose* $\mathfrak{A}^n = 0$ *for* $n < 0$, *then the map* $\pi : \mathfrak{A} \to \mathcal{P}(\mathfrak{A}^0)$ *is a G algebra homomorphism between* $(\mathfrak{A}, \cdot, [-,-])$ *and* $(\mathcal{P}(\mathfrak{A}^0), \cdot, [-,-]_S)$.

Proof. One must show that π is a homomorphism with respect to the dot product and the bracket. Both follow easily by induction on the sum of degrees of a and b in (4.9) and (4.13)-(4.14). Indeed, in the case of the product we have $\pi(a\cdot b) = \pi(a)\cdot\pi(b)$ for $|a| = |b| = 0$. For $|a| + |b| > 0$, we find, using (4.24), properties of the bracket, and the induction hypothesis, that

$$\begin{aligned} \pi(a\cdot b)(x) &= \pi([a\cdot b, x]) \\ &= \pi(a\cdot[b,x]) + (-1)^{|b|}\pi([a,x]\cdot b) \\ &= \pi(a)\cdot\pi(b)(x) + (-1)^{|b|}\pi(a)(x)\cdot\pi(b) \\ &= (\pi(a)\cdot\pi(b))(x). \end{aligned} \tag{4.25}$$

In the case of the bracket, we first note that (4.24) is equivalent to

$$[\pi(a), \pi(b)]_S = \pi([a,b]) \quad \text{for} \quad |a| + |b| = 1. \tag{4.26}$$

Then the general step of the induction is shown similarly as above,

$$\begin{aligned}\pi([a,b])(x) &= \pi([[a,b],x]) \\ &= \pi([a,[b,x]]) + (-1)^{|b|-1}\pi([a,x],b]) \\ &= [\pi(a),\pi(b)(x)]_S + (-1)^{|b|-1}[\pi(a)(x),\pi(b)]_S \\ &= [\pi(a),\pi(b)]_S(x)\,,\end{aligned} \tag{4.27}$$

for all other $|a|, |b|$. □

In the following we will consider BV algebras for which π is an epimorphism, i.e., $\pi(\mathfrak{A}) = \mathcal{P}(\mathfrak{A}^0)$, and, in addition, $\mathcal{P}(\mathfrak{A}^0)$ is itself a BV algebra with a BV operator Δ_S. The following theorem then provides a convenient criterium to determine whether π is in fact a homomorphism of BV algebras.

Theorem 4.12 *Assume that the G algebra $\mathcal{P}(\mathfrak{A}^0)$ admits a BV algebra structure $(\mathcal{P}(\mathfrak{A}^0),\, \cdot\,, \Delta_S)$, and*

$$\pi\Delta(a) = \Delta_S\pi(a) \tag{4.28}$$

holds for all $a \in \mathfrak{A}^1$. Then π is a homomorphism of BV algebras.

Proof. Recall that $\mathfrak{A}^n = 0$ for $n < 0$. Then (4.28) is obviously satisfied for any $a \in \mathfrak{A}^0$, as both sides must vanish. The general case, with $a \in \mathfrak{A}^n$, is proved by induction. Indeed, for any $x \in \mathfrak{A}^0$,

$$\begin{aligned}\pi(\Delta a)(x) &= \pi([\Delta a, x]) \\ &= \pi(\Delta[a,x]) \\ &= \Delta_S(\pi([a,x])) \\ &= \Delta_S([\pi(a),x]_S) \\ &= [\Delta_S\pi(a),x]_S \\ &= \Delta_S\pi(a)(x)\,,\end{aligned} \tag{4.29}$$

where we have used that Δ and Δ_S act as derivations on $[-,-]$ and $[-,-]_S$, respectively. □

4.2 The BV Algebra of Polyderivations of the Ground Ring Algebra $\mathcal{R}_N$

In this section we construct explicitly the BV algebra of polyderivations of an Abelian algebra which is not free, but whose generators satisfy a single quadratic relation.

4.2.1 The "Ground Ring" Algebra $\mathcal{R}_N$

Consider the Abelian algebra $\mathcal{R}_N = \mathcal{C}_{2N}/\mathcal{I}$, where $\mathcal{I}$ is the ideal generated by the vanishing relation

$$h_{ij}\, x^i \cdot x^j = 0\,, \tag{4.30}$$

where the metric is

$$(h_{ij}) = \begin{pmatrix} 0 & 1 \\ 1 & 0 \end{pmatrix}, \quad i,j = 1,\ldots,2N\,. \tag{4.31}$$

In the following this metric will be used to raise and lower indices, e.g., we will write $x_i = h_{ij}x^j$. Denote by p the projection $p : \mathcal{C}_{2N} \to \mathcal{R}_N$. If no confusion can arise, we will write x^i both for a generator x^i in $\mathcal{C}_{2N}$ as well as its image $p(x^i)$ in $\mathcal{R}_N$, and omit the dot in the products.

For $N = 3$ the algebra $\mathcal{R}_3$ is isomorphic with the ground ring algebra in Sect. 5.2.1. We will therefore refer to $\mathcal{R}_N$ as a ground ring.

The free algebra $\mathcal{C}_{2N}$ carries a natural action of the Lie algebra of $\mathfrak{so}_{2N}$ realized by the first order derivations

$$\Lambda_{ij} = x_i x_j^* - x_j x_i^*\,, \quad i,j = 1,\ldots,2N\,. \tag{4.32}$$

Clearly, $\Lambda_{ij}(\mathcal{I}) \subset \mathcal{I}$, so the action of $\mathfrak{so}_{2N}$ descends to the ground ring $\mathcal{R}_N$, with the generators x^i transforming in the vector ($2N$-dimensional) representation. Let us define a $\mathbb{Z}$ grading of $\mathcal{R}_N$, the so-called $\mathcal{R}$ degree, by declaring an element of $\mathcal{R}$ degree m to be the sum of products of precisely m generators x^i, and let $\mathcal{R}_N^m$ denote the subspace of $\mathcal{R}_N$ of elements of $\mathcal{R}$ degree m.

Since the dot product in $\mathcal{R}_N$ is commutative, and the constraint (4.30) merely amounts to subtracting the trace, we also find the following result for the structure of the entire ground ring.

Theorem 4.13 *Each $\mathcal{R}_N^m$ has a basis consisting of elements of the form*

$$P^{i_1\ldots i_m} = x^{i_1}\cdots x^{i_m}\,, \quad m \geq 0\,. \tag{4.33}$$

In other words $\mathcal{R}_N^m$ is an irreducible finite-dimensional module of $\mathfrak{so}_{2N}$, isomorphic with the space of completely symmetric traceless $\mathfrak{so}_{2N}$ tensors of rank m. Thus, the ground ring $\mathcal{R}_N$ decomposes as a direct sum of irreducible finite-dimensional modules of $\mathfrak{so}_{2N}$ as follows

$$\mathcal{R}_N = \bigoplus_{m=0}^{\infty} \mathcal{R}_N^m\,, \tag{4.34}$$

where each $\mathcal{R}_N^m$ arises precisely once.

4.2.2 A "Hidden Symmetry" of $\mathcal{R}_N$

The ground ring $\mathcal{R}_N$ acts on itself by left multiplication. Let us denote by x^i both the generator and the corresponding multiplication operator acting on the ground ring. The natural problem is then to determine the Lie algebra of transformations of $\mathcal{R}_N$ which includes the multiplication operators together with the $\mathfrak{so}_{2N}$ symmetry generators Λ_{ij}.

Theorem 4.14 *The ground ring $\mathcal{R}_N$ is an irreducible module of $\mathfrak{so}_{2N+2}$. The explicit realization of the $\mathfrak{so}_{2N+2}$ generators is given by the following differential operators on $\mathcal{R}_N$:*

$$\begin{aligned} M_i &= x_i\,, \\ \Lambda_{ij} &= x_i\frac{\partial}{\partial x^j} - x_j\frac{\partial}{\partial x^i}\,, \\ U_i &= (N-1)\frac{\partial}{\partial x^i} + x^j\frac{\partial}{\partial x^j}\frac{\partial}{\partial x^i} - \tfrac{1}{2}x_i\frac{\partial}{\partial x^j}\frac{\partial}{\partial x_j}\,, \\ U &= x^i\frac{\partial}{\partial x^i} + (N-1)\,. \end{aligned} \tag{4.35}$$

Proof. One verifies by straightforward algebra that the operators in (4.35) preserve the constraint (4.30), and thus are well defined on $\mathcal{R}_N$. Similarly, one finds that they satisfy the commutation relations of the $\mathfrak{so}_{2N+2}$ algebra, e.g., $[U_i, M_j] = -\Lambda_{ij} + h_{ij}\,U$. □

Theorem 4.14 was first proved in [BiFl] for $N=3$, and then generalized in [GeZl].

4.2.3 Polyderivations of $\mathcal{R}_N$

A polyderivation $\Phi \in \mathcal{P}^n(\mathcal{R}_N)$ is completely determined by its value on the ground ring generators, $\mathcal{R}_N^1$; i.e., we have a natural injection from $\Phi \in \mathcal{P}^n(\mathcal{R}_N)$ into the space of multilinear alternating maps $\mathrm{Hom}(\bigwedge^n \mathcal{R}_N^1, \mathcal{R}_N)$. The problem of determining all polyderivations of $\mathcal{R}_N$ thus amounts to identifying which elements Φ in $\mathrm{Hom}(\bigwedge^n \mathcal{R}_N^1, \mathcal{R}_N)$ are in the image of this injection. This is resolved by the following criterium.

Theorem 4.15 *An endomorphism $\Phi \in \mathrm{Hom}(\bigwedge^n \mathcal{R}_N^1, \mathcal{R}_N)$ determines a polyderivation of $\mathcal{R}_N$ iff*

$$\mathfrak{d}\Phi = 0\,, \tag{4.36}$$

where $\mathfrak{d} : \mathrm{Hom}(\bigwedge^n \mathcal{R}_N^1, \mathcal{R}_N) \to \mathrm{Hom}(\bigwedge^{n-1} \mathcal{R}_N^1, \mathcal{R}_N)$ is the operator $\mathfrak{d} = x_i\imath(x^i)$.

We may also express (4.36) more explicitly by expanding Φ in the dual basis,

$$\Phi = \Phi^{i_1\ldots i_n} x^*_{i_1}\ldots x^*_{i_n}\,, \quad \Phi^{i_1\ldots i_n} \in \mathcal{R}_N\,. \tag{4.37}$$

Then we have

Theorem 4.15' *An endomorphism Φ is a polyderivation iff the coefficients of its expansion (4.37) satisfy*

$$x_i \cdot \Phi^{i i_1 \dots i_{n-1}} = 0, \quad i_1, \dots, i_{n-1} = 1, \dots, 2N. \tag{4.38}$$

Proof. If Φ is a polyderivation, the Leibniz rule yields

$$\Phi(x_i \cdot x^i, x^{i_1}, \dots, x^{i_{n-1}}) = 2x_i \cdot \Phi(x^i, x^{i_1}, \dots, x^{i_{n-1}}) = 0. \tag{4.39}$$

To extend an endomorphism $\Phi \in \mathrm{Hom}(\bigwedge^n \mathcal{R}_N^1, \mathcal{R}_N)$ to the ground ring we may assume that it acts as a derivation on the products of generators in $\mathcal{R}_N$. The conditions (4.36), or equivalently (4.38), guarantee then that if we evaluate Φ on arbitrary elements of the ground ring, the final result does not depend on the particular representation of those elements as linear combinations of products of the generators. Also, essentially by construction, the resulting Φ is a polyderivation, see Lemma 4.7. □

The second part of the proof may also be rephrased as follows: First lift Φ to a homomorphism $\widetilde{\Phi}$ in the "covering algebra" $\mathcal{C}_{2N}$, by choosing arbitrary elements $\widetilde{\Phi}^{i_1 \dots i_n}$ that project onto $\Phi^{i_1 \dots i_n}$ and set $\widetilde{\Phi} = \widetilde{\Phi}^{i_1 \dots i_n} x^*_{i_1} \dots x^*_{i_n}$. Obviously $\widetilde{\Phi}$ is a polyderivation of $\mathcal{C}_{2N}$. The condition (4.38) allows us to first project $\widetilde{\Phi}$ to a polyderivation of $\mathcal{R}_N$, and then to show that the resulting polyderivation does not depend on the choice of lift $\Phi \to \widetilde{\Phi}$.

Note that if we consider $\mathrm{Hom}(\bigwedge^* \mathcal{R}_N^1, \mathcal{R}_N)$ as a graded commutative algebra with the product induced from the exterior product in $\bigwedge^* \mathcal{R}_N^1$ and the product in $\mathcal{R}_N$, the subspace $\ker \mathfrak{d}$ is an ideal, which shows that the identification of polyderivations $\mathcal{P}(\mathcal{R}_N)$ with $\ker \mathfrak{d}$ is in fact an isomorphism of algebras.

Now we would like to solve the constraint (4.36) and determine explicitly all the polyderivations $\mathcal{P}(\mathcal{R}_N)$. Since $x^i \mathcal{R}_N^m \subset \mathcal{R}_N^{m+1}$, it is enough to consider endomorphisms taking values in a subspace of fixed $\mathcal{R}$ degree. Given the basis (4.33) in $\mathcal{R}_N$, we may choose the endomorphisms

$$P_{i_1 \dots i_m} x^*_{j_1} \dots x^*_{j_n}, \quad i_1, \dots, j_n = 1, \dots, 2N, \tag{4.40}$$

as the basis in $\mathcal{E}_m^n = \mathrm{Hom}(\bigwedge^n \mathcal{R}_N^1, \mathcal{R}_N^m)$. The action of $\mathfrak{so}_{2N}$ on the ground ring in (4.32) extends to $\mathcal{E}_m^n$, which, as an $\mathfrak{so}_{2N}$ module, is then isomorphic with the tensor product of two $\mathfrak{so}_{2N}$ modules; the first one corresponds to completely symmetric traceless tensors of rank m, and the second one to antisymmetric tensors of rank n. Since the operator $\mathfrak{d} : \mathcal{E}_m^n \to \mathcal{E}_{m+1}^{n-1}$ commutes with the action of $\mathfrak{so}_{2N}$, it can only map between the same irreducible modules in the decomposition of $\mathcal{E}_m^n$ and $\mathcal{E}_{m+1}^{n-1}$.

Recall that tensor representations of $\mathfrak{so}_{2N}$ can be conveniently enumerated using Young tableaux. We will only need a small class of "hook like" tableaux, with m boxes in the first row and one box in the subsequent n rows. Let us denote the corresponding representation of $\mathfrak{so}_{2N}$ by $[m;n]$, $m,n \geq 0$. In particular, $[0;0]$ is the identity representation, $[1;0] \simeq [0;1]$ the vector representation, $[m;0]$ corresponds to the completely symmetric traceless tensors of rank m, while $[1;n]$ to the completely antisymmetric tensors of rank $n+1$. Since an antisymmetric tensor of rank n is equivalent (dual) to a tensor of rank $2N-n$, we also have $[1;n] \simeq [1;2N-n-2]$. The final subtlety is that the representation $[m;N-1]$ is a direct sum of two irreducible ones.

The product of a traceless symmetric tensor of rank m and an antisymmetric tensor of rank n decomposes as

$$[m;0] \otimes [1;n-1] = [m+1;n-1] \oplus [m;n] \oplus [m-1;n-1] \oplus [m;n-2], \quad m,n \geq 2. \tag{4.41}$$

This can be derived in two steps. In the first we use the usual rule for multiplying Young tableaux of $\mathfrak{gl}_{2N}$, and obtain the first two terms on the right hand side but with no traces subtracted. In the second step we subtract the traces (in fact just a single trace) from the first and the second term, which yields the third and the fourth term, respectively. The decomposition (4.41) is valid for generic values of $m \geq 2$ and $2N > n \geq 2$. The following are the special cases where some terms on the right hand side in the decomposition above are not present:

$$\begin{array}{lrcl}
i. & [0;0] \otimes [1,n-1] & = & [1,n-1], \\
ii. & [1;0] \otimes [1,n-1] & = & [2;n-1] \oplus [1;n] \oplus [1;n-2], \\
iii. & [m;0] \otimes [1;0] & = & [m+1;0] \oplus [m;1] \oplus [m-1,0], \\
iv. & [m;0] \otimes [1;2N-2] & = & [m+1;2N-2] \oplus [m;2N-1] \oplus [m;2N-3], \\
v. & [m;0] \otimes [1;2N-1] & = & [m+1;2N-1].
\end{array} \tag{4.42}$$

Now, let us go back to (4.36). To illustrate the method, we first consider some of the exceptional cases. It is clear from (4.38) that there can be no polyderivation with $m=0$, so the simplest nontrivial case is that of $m=1$ and $n=1$. The decomposition of $\mathcal{E}_1^1$ with respect to the $\mathfrak{so}_{2N}$ action yields a direct sum of three modules (see (4.42) (iii)), spanned by $S_{i,j}$, $P_{i,j}$ and C, $i,j = 1\ldots,2N$, respectively, where[3]

$$S_{i,j} = x_{(i}x^*_{j)}, \quad P_{i,j} = x_{[i}x^*_{j]}, \quad C = x^i x^*_i. \tag{4.43}$$

We find

$$\mathfrak{d}S_{i,j} = x_{(i}x_{j)}, \quad \mathfrak{d}P_{i,j} = 0, \quad \mathfrak{d}C = 0, \tag{4.44}$$

which shows that the space of polyderivations $\mathcal{P}_1^1(\mathcal{R}_N)$ is spanned by $P_{i,j}$ and C. Note that $2P_{i,j} = \Lambda_{ij}$, and thus we have rederived the $\mathfrak{so}_{2N}$ symmetry generators.

---

[3] Here and in the following $(\cdots)$ and $[\cdots]$ denote the symmetrization and the antisymmetrization, respectively, both normalized with strength one; i.e., for a completely symmetric tensor $s_{(i_1\ldots i_n)} = s_{i_1\ldots i_n}$, and for a completely antisymmetric tensor $a_{[i_1\ldots i_n]} = a_{i_1\ldots i_n}$.

For $\eth : \mathcal{E}^n_m \longrightarrow \mathcal{E}^{n-1}_{m+1}$ with $m = 1$ and $n \geq 2$, the following decompositions are relevant

$$\begin{aligned} \mathcal{E}^n_1 &: \quad [2; n-1] \oplus [1; n] \oplus [1, n-2]\,, \\ \mathcal{E}^{n-1}_2 &: \quad [3; n-2] \oplus [2; n-1] \oplus [2, n-3] \oplus [1; n-2]\,. \end{aligned} \tag{4.45}$$

By comparing the two decompositions, we conclude that the $[1; n]$ submodule of $\mathcal{E}^1_1$ must lie in $\ker \eth$. Indeed, $[1; n]$ is spanned by endomorphisms of the form

$$P_{i_1, j_1 \ldots j_n} = x_{[i_1} x^*_{j_1} \ldots x^*_{j_n]}\,, \tag{4.46}$$

for which

$$\eth P_{i_1, j_1 \ldots j_n} = \sum_{k=1}^{n} (-1)^{k+1} x_{[i_1} x^*_{j_1} \ldots \widehat{x^*_{j_k}} \ldots x^*_{j_n]} = 0\,. \tag{4.47}$$

We also have

$$\eth\, x_{i_1} x^*_{j_1} \ldots x^*_{j_n} = n\, x_{i_1} x_{[j_1} x^*_{j_2} \ldots x^*_{j_n]}\,. \tag{4.48}$$

By decomposing both sides into traceless and trace components, we see that $\eth$ has a nontrivial image in both the $[2; n-1]$ and $[1; n-2]$ submodules of $\mathcal{E}^{n-1}_2$, and thus (4.46) exhaust all polyderivations in this case.

The case $m \geq 2$ and $n = 1$ is similar in that there is only one trace in the decomposition of the tensor product. However, since

$$\begin{aligned} \mathcal{E}^1_m &: \quad [m+1; 0] \oplus [m; 1] \oplus [m-1; 0]\,, \\ \mathcal{E}^0_{m+1} &: \quad [m+1, 0]\,, \end{aligned} \tag{4.49}$$

we find two $\mathfrak{so}_{2N}$ representations in the decomposition of $\mathcal{P}^1_m$. The corresponding basis is given by

$$P_{i_1 \ldots i_m, j} = x_{i_1} \ldots x_{i_{m-1}} x_{[i_m} x^*_{j]} + \cdots\,, \tag{4.50}$$

and

$$C_{i_1 \ldots i_{m-1}} = x_{i_1} \ldots x_{i_{m-1}} C\,, \tag{4.51}$$

where "$\cdots$" indicate explicit subtraction of trace terms in $i_1, \ldots, i_m, j$.

In the generic case, for $m \geq 2$ and $2N-2 \geq n \geq 2$, we have

$$\begin{aligned} \mathcal{E}^n_m &: \quad [m+1; n-1] \oplus [m; n] \oplus [m-1; n-1] \oplus [m, n-2]\,, \\ \mathcal{E}^{n-1}_{m+1} &: \quad [m+2; n-2] \oplus [m+1; n-1] \oplus [m, n-2] \oplus [m+1; n-3]\,. \end{aligned} \tag{4.52}$$

The modules $[m; n]$ and $[m-1; n-1]$ lie in $\ker \eth$, with a convenient basis given by

$$P_{i_1 \ldots i_m, j_1 \ldots j_n} = x_{i_1} \ldots x_{i_{m-1}} x_{[i_m} x^*_{j_1} \ldots x^*_{j_n]} + \cdots\,, \tag{4.53}$$

and

$$C_{i_1 \ldots i_{m-1}, j_1 \ldots j_{n-1}} = C\, P_{i_1 \ldots i_{m-1}, j_1 \ldots j_{n-1}}\,, \tag{4.54}$$

where all basis elements (4.53) and (4.54) are traceless in $i_1, \ldots, j_n$. Since

$$\partial x_{i_1} \ldots x_{i_m} x^*_{j_1} \ldots x^*_{j_n} = n\, x_{i_1} \ldots x_{i_m} x_{[j_1} x^*_{j_2} \ldots x^*_{j_n]}\,, \tag{4.55}$$

we verify that (4.54) gives all polyderivations $\mathcal{P}^n_m(\mathcal{R}_N)$. The explicit form of the trace terms that must be subtracted on the right hand side in (4.50) and (4.53) are given in Appendix F. An equivalent, but more concise, expression will be given in the next section.

Although $[1; 2N-2] \simeq [1,0]$, the $n = 2N-1$ case is quite different than that with $n = 1$. For $m = 1$ we find one solution, see (4.46),

$$P_{i_1,j_1\ldots j_{2N-1}} = x_{[i_1} x^*_{j_1} \ldots x^*_{j_{2N-1}]} = \epsilon_{i_1 j_1 \ldots j_{2N-1}} X\,. \tag{4.56}$$

We will refer to X as the "volume element" of $\mathcal{R}_N$. Explicitly,

$$X = \frac{1}{(2N)!} \epsilon^{i_1 i_2 \ldots i_{2N}}\, x_{i_1} x^*_{i_2} \ldots x^*_{i_{2N}}\,. \tag{4.57}$$

For $m \geq 2$ we have

$$\begin{aligned} \mathcal{E}^{2N-1}_m &: [m+1; 2N-2] \oplus [m; 2N-1] \oplus [m, 2N-3]\,, \\ \mathcal{E}^{2N-2}_{m+1} &: [m+2; 2N-3] \oplus [m+1; 2N-2] \oplus [m, 2N-3] \oplus [m+1; 2N-4]\,. \end{aligned} \tag{4.58}$$

This leaves just one solution spanned by the elements, see (4.53),

$$x_{i_1} \ldots x_{i_{m-1}} x_{[i_m} x^*_{j_1} \ldots x^*_{j_{2N-1}]}\,. \tag{4.59}$$

Using standard identities for $\mathfrak{so}_{2N}$ tensors, we find

$$x_i x_{[j_1} x^*_{j_2} \ldots x^*_{j_{2N}]} = -(2N-1)\, C\, h_{i[j_1} P_{j_2, j_3 \ldots j_{2N}]}\,, \tag{4.60}$$

which shows that only trace components are present in (4.59). Thus the basis in $\mathcal{P}^{2N-1}_m$, $m \geq 2$, consists of elements $C_{i_1\ldots i_{m-1}, j_1 \ldots j_{2N-2}}$ defined as in (4.54).

Finally, there is no solution for $n = 2N$, which shows that the maximal degree of a polyderivation of $\mathcal{R}_N$ is equal to $2N-1$.

Let us extend the notation for the polyderivations in (4.53) and (4.54) and set $P_{i_1\ldots i_m, j_1 \ldots j_n}$ equal to $P_{i_1 \ldots i_m}$ for $n = 0$, and to 1 for $m = n = 0$. Similarly, we set $C_{i_1 \ldots i_m, j_1 \ldots j_n}$ equal to $C_{i_1 \ldots i_m}$ for $n = 0$, and to C for $m = n = 0$. We may now summarize the complete classification of the polyderivations $\mathcal{P}(\mathcal{R}_N)$.

Theorem 4.16

i. The space of polyderivations $\mathcal{P}(\mathcal{R}_N)$ is doubly graded,

$$\mathcal{P}(\mathcal{R}_N) = \bigoplus_{n=0}^{2N-1} \bigoplus_{m=0}^{\infty} \mathcal{P}^n_m(\mathcal{R}_N)\,, \tag{4.61}$$

by the degree n of the derivation, $2N-1 \geq n \geq 0$, and the $\mathcal{R}$ degree m of the coefficients in the ground ring, $m \geq 0$. Depending on m and n, each of the subspaces $\mathcal{P}^n_m(\mathcal{R}_N)$ is a direct sum of finite-dimensional irreducible modules of $\mathfrak{so}_{2N}$ which are listed in Table 4.1.

Table 4.1. The decomposition of $\mathcal{P}^n_m(\mathcal{R}_N)$ into $\mathfrak{so}_{2N}$ modules

$m\backslash n$	$n=0$	$n=1$	$2N-2\geq n\geq 2$	$n=2N-1$
$m=0$	$[0;0]$			
$m=1$	$[1;0]$	$[1;1]\oplus[0;0]$	$[1;n]$	$[1;2N-1]$
$m\geq 2$	$[m;0]$	$[m;1]\oplus[m-1;0]$	$[m;n]\oplus[m-1;n-1]$	$[m-1,2N-2]$

ii. In $\mathcal{P}^n_m(\mathcal{R}_N)$, $m,n\geq 0$, the $[m;n]$ submodule is spanned by the polyderivations $P_{i_1\dots i_m,j_1\dots j_n}$, and the $[m-1;n-1]$ submodule by the polyderivations $C_{i_1\dots i_{m-1},j_1\dots j_{n-1}}$. In cases where a given submodule does not arise in the decomposition of $\mathcal{P}^n_m(\mathcal{R}_N)$, the corresponding polyderivations $P_{i_1\dots i_m,j_1\dots j_n}$ and/or $C_{i_1\dots i_{m-1},j_1\dots j_{n-1}}$ vanish.

From the formulae for the basis in $\mathcal{P}^n_m(\mathcal{R}_N)$ we see that at each degree n there are polyderivations of $\mathcal{R}$ degree $m=1$, which cannot be obtained as products of polyderivations of lower degrees. The question of how to describe explicitly $\mathcal{P}(\mathcal{R}_N)$ in terms of generators and relations is then answered by the following theorem.

Theorem 4.17 *The graded, graded commutative algebra $(\mathcal{P}(\mathcal{R}_N),\cdot)$ is generated, as a dot algebra, by 1, the ground ring generators x^i, degree one derivation C, and degree $n-1$ polyderivations $P_{i_1,i_2\dots i_n}$, $n=2,\dots,2N$, satisfying the relations:*

$$x_i x^i = 0\,, \tag{4.62}$$

$$x_{[i}P_{i_1,i_2\dots i_n]} = 0\,, \tag{4.63}$$

$$x^i P_{i,j_1\dots j_n} = -\tfrac{n}{n+1}CP_{j_1,j_2\dots j_n}\,, \tag{4.64}$$

$$P_{i_1,i_2\dots i_m}P_{j_1,j_2\dots j_n} = (-1)^{m-1}\tfrac{m+n-1}{n}x_{[i_1}P_{i_2,i_3\dots i_m]j_1\dots j_n}\,, \tag{4.65}$$

$$CP_{i_1,i_2\dots i_{2N}} = 0\,. \tag{4.66}$$

Proof. The identities (4.62)-(4.66) are satisfied in $\mathcal{P}(\mathcal{R}_N)$. This is easily verified using the explicit form of those polyderivations in (4.43) and (4.46). On the other hand, if we consider the algebra generated by x^i, C, and $P_{i_1,i_2\dots i_n}$, subject to relations (4.62)-(4.65), all elements in this algebra are linear combinations of the products

$$x_{i_1}\dots x_{i_m}\,,\quad x_{i_1}\dots x_{i_m}C\,,\quad x_{i_1}\dots x_{i_m}P_{j_1,j_2\dots j_n}\,,\quad x_{i_1}\dots x_{i_m}CP_{j_1,j_2\dots j_n}\,. \tag{4.67}$$

There is a natural action of the $\mathfrak{so}_{2N}$ algebra on this space, with respect to which the elements in (4.67) transform as $[m;0]$, $[m;0]$, $[m;0]\otimes[1;n-1]$ and $[m;0]\otimes[1;n-1]$, respectively. Condition (4.63) sets to zero the $[m;n]$ and $[m-1;n-1]$ components in those tensor products, while (4.64) relates the trace component in the third product in (4.67) to the single nonvanishing component of the fourth term in (4.67). This shows that the elements of this space are in one to one correspondence with the elements of $\mathcal{P}(\mathcal{R}_N)$, and, in fact, establishes the required algebra isomorphism. □

4.2.4 The G Algebra Structure of $\mathcal{P}(\mathcal{R}_N)$

The computation of the Schouten bracket, as defined in section 4.1.2, only involves evaluation of polyderivations on elements of the algebra. Thus we may use similar arguments to those which led to Theorem 4.15 to derive an explicit formula for the Schouten bracket of two polyderivations.

Theorem 4.18 *Let $\Phi = \Phi^{i_1\ldots i_m}x^*_{i_1}\cdots x^*_{i_m}$ and $\Psi = \Psi^{j_1\ldots j_n}x^*_{j_1}\cdots x^*_{j_n}$, $\Phi^{i_1\ldots i_m}$, $\Psi^{j_1\ldots j_n} \in \mathcal{R}_N$, be two polyderivations. Then the Schouten bracket $[\Phi,\Psi]_S$ can be computed explicitly as in (4.20), where we assume that $x^*_{i_1},\ldots,x^*_{j_n}$ act as derivations on the products of ground ring generators.*

The following observation is a simple consequence of the above result.

Theorem 4.19 *The Schouten bracket on $\mathcal{P}(\mathcal{R}_N)$ is homogenous in both the degree and the $\mathcal{R}$ degree, i.e.,*

$$[-,-]_S : \mathcal{P}^{n_1}_{m_1} \times \mathcal{P}^{n_2}_{m_2} \longrightarrow \mathcal{P}^{n_1+n_2-1}_{m_1+m_2-1}\,. \qquad (4.68)$$

We now explicitly calculate some fundamental brackets between certain elements of the algebra, which will be required in the next section where we determine the BV operator underlying the Schouten bracket. All of these results are obtained using Theorem 4.18 and the explicit form of the polyderivations. First we have

$$[\Lambda_{ij}, x_k]_S = h_{ik}x_j - h_{jk}x_i\,, \qquad (4.69)$$

which represents the $\mathfrak{so}_{2N}$ transformation of the ground ring generator. More generally,

$$[P_{i_1,i_2\ldots i_m}, x_i]_S = (-1)^{m-1}(n-1)h_{i[i_1}P_{i_2,i_3\ldots i_m]}\,, \qquad (4.70)$$

as well as

$$[P_{i_1,i_2\ldots i_m}, P^{j_1,j_2\ldots j_n}]_S = (-1)^{m-1}(m+n-2)\delta_{[i_1}{}^{[j_1}P_{i_2,i_3\ldots i_m]}{}^{j_2\ldots j_n]}\,. \qquad (4.71)$$

Lemma 4.20 *For any $\Phi \in \mathcal{P}^n_m(\mathcal{R}_N)$,*

$$[C,\Phi]_S = (m-n)\Phi\,. \qquad (4.72)$$

Using the Schouten bracket we can also write down explicitly the decomposition of a product of two basis elements in $\mathcal{P}(\mathcal{R}_N)$ into its traceless and trace components.

Theorem 4.21 *For any $m, m' \geq 0$ and $n, n' \geq 1$,*

$$\begin{aligned} & P^{i_1 \dots i_m i_{m+1}, i_{m+2} \dots i_{m+n}} P_{j_1 \dots j_{m'} j_{m'+1}, j_{m'+2} \dots j_{m'+n'}} = \\ & \quad \tfrac{n+n'-1}{n'} P^{i_1 \dots i_m}{}_{j_1 \dots j_{m'}}{}^{[i_{m+1}}{}_{j_{m'+1},}{}^{i_{m+2} \dots i_{m+n}]}{}_{j_{m'+2} \dots j_{m'+n'}} \\ & \qquad + (-1)^n \tfrac{1}{2N+m+m'-n-n'+2} \\ & \qquad\quad \times C \, [P^{i_1 \dots i_m i_{m+1}, i_{m+2} \dots i_{m+n}}, P_{j_1 \dots j_{m'} j_{m'+1}, j_{m'+2} \dots j_{m'+n'}}]_S \, . \end{aligned} \tag{4.73}$$

Also, the bracket on the right hand side lies in the subspace spanned by the P-type basis elements in $\mathcal{P}^{n+n'-1}_{m+m'-1}$.

Proof. The first term on the right hand side is determined so that the leading terms on both sides agree, see (4.53). Then the second term on the right hand side must account for all the traces in the product, which indeed is the case. This fact, as well as the second part of the theorem, are shown by a straightforward calculation which has been outlined in Appendix F.2. □

4.2.5 The BV Algebra Structure of $\mathcal{P}(\mathcal{R}_N)$

We will now construct a BV operator Δ_S on $\mathcal{P}(\mathcal{R}_N)$, whose bracket (4.6) coincides with the Schouten bracket. Since the latter operation is both $\mathfrak{so}_{2N}$ invariant as well as homogenous with respect to both the degree and the $\mathcal{R}$ degree, we will seek a BV operator which satisfies similar restrictions.

Theorem 4.22 *There exists at most one BV operator Δ on $\mathcal{P}(\mathcal{R}_N)$ that is $\mathfrak{so}_{2N}$ invariant, homogenous of degree minus one, i.e.,*

$$\Delta : \mathcal{P}^n_m(\mathcal{R}_N) \longrightarrow \mathcal{P}^{n-1}_{m-1}(\mathcal{R}_N) ,$$

and whose bracket $[-,-]$ coincides with the Schouten bracket $[-,-]_S$.

Proof. Let Δ_1 and Δ_2 be two such BV operators. Then their difference $D = \Delta_1 - \Delta_2$ is an $\mathfrak{so}_{2N}$ invariant first order derivation on $\mathcal{P}(\mathcal{R}_N)$, and $D : \mathcal{P}^n_m(\mathcal{R}_N) \to \mathcal{P}^{n-1}_{m-1}(\mathcal{R}_N)$. By examining the $\mathfrak{so}_{2N}$ decomposition of $\mathcal{P}(\mathcal{R}_N)$ given in Theorem 4.16 we conclude that for any BV operator Δ satisfying the assumptions above

$$\Delta P_{i_1, i_2 \dots i_m} = 0 , \quad m \geq 1 . \tag{4.74}$$

Thus $DP_{i_1,i_2 \dots i_m} = 0$, and by Theorem 4.17 and D being a derivation it follows that D is completely determined by its action on C.

Once more, since $\Delta x^i = 0$ and $\Delta \Lambda_{ij} = 0$, we find, using (4.69),

$$\Delta(x^i \Lambda_{ij}) = [x^i, \Lambda_{ij}]_S = (2N-1) x_j . \tag{4.75}$$

However,

$$x^i \Lambda_{ij} = x^i (x_i x_j^* - x_j x_i^*) = -x_j C , \tag{4.76}$$

so we also have

$$\Delta(x^i A_{ij}) = -\Delta(x_j C) = -[x_j, C] - x_j \Delta C . \tag{4.77}$$

Comparing (4.75) with (4.77), and using (4.72), we determine that

$$\Delta C = -2(N-1)1 , \quad \Delta(C x_i) = -(2N-1)x_i . \tag{4.78}$$

This shows that $DC = 0$, and concludes the proof of the theorem. □

Lemma 4.23 *Let Δ be a BV operator as in Theorem 4.22, and $\Phi \in \mathcal{P}_m^n(\mathcal{R}_N)$ satisfies $\Delta\Phi = 0$. Then*

$$\Delta(C\Phi) = -(2N+m-n-2)\Phi . \tag{4.79}$$

Proof. Using (4.5), (4.72) and (4.78), we obtain

$$\Delta(C\Phi) = -[C, \Phi] + \Delta(C)\Phi = -(m-n)\Phi - (2N-2)\Phi . \tag{4.80}$$

The main result of this section is the following explicit construction of Δ_S.

Theorem 4.24 *There exists a unique BV operator Δ_S on $\mathcal{P}(\mathcal{R}_N)$ that is $\mathfrak{so}_{2N}$ invariant, homogenous of degree minus one, and whose bracket $[-,-]$ coincides with the Schouten bracket $[-,-]_S$. It is explicitly given by*

$$\begin{aligned} \Delta_S P_{i_1 \dots i_m, j_1 \dots j_n} &= 0 , \\ \Delta_S C_{i_1 \dots i_m, j_1 \dots j_n} &= -(2N+m-n-2) P_{i_1 \dots i_m, j_1 \dots j_n} . \end{aligned} \tag{4.81}$$

Proof. First we want to argue that a BV operator Δ_S satisfying the assumptions of the theorem must be of the form (4.81). Similarly as in the proof of Theorem 4.22, the $\mathfrak{so}_{2N}$ invariance restricts Δ_S to

$$\Delta_S P_{i_1 \dots i_m, j_1 \dots j_n} = 0 , \qquad \Delta_S C_{i_1 \dots i_m, j_1 \dots j_n} = \lambda(m,n) P_{i_1 \dots i_m, j_1 \dots j_n} , \tag{4.82}$$

where $\lambda(m,n)$ are some arbitrary numbers to be determined. However, since $C_{i_1 \dots i_m, j_1 \dots j_n} = C\, P_{i_1 \dots i_m, j_1 \dots j_n}$ and $\Delta_S\, P_{i_1 \dots i_m, j_1 \dots j_n} = 0$, the second part of (4.81) follows then from Lemma 4.23.

Clearly $\Delta_S^2 = 0$, so to complete the proof we must show that the bracket of Δ_S coincides with the Schouten bracket, as the second order derivation property of Δ_S will then follow from Theorem 4.4. The equality between the bracket of Δ_S and the Schouten bracket is demonstrated by explicit computation. There are three cases: the bracket of two P's, of a P and a C, and of two C's. For the first we may simply use the general formula for the product of two P's given in Theorem 4.21. Indeed, by acting with Δ_S on both sides of (4.73) we find

$$\begin{aligned} &\Delta_S(P_{i_1 \dots i_m i_{m+1}, i_{m+2} \dots i_{m+n}} P_{j_1 \dots j_{m'} j_{m'+1}, j_{m'+2} \dots j_{m'+n'}}) \\ &= (-1)^{n-1} [P_{i_1 \dots i_m i_{m+1}, i_{m+2} \dots i_{m+n}}, P_{j_1 \dots j_{m'} j_{m'+1}, j_{m'+2} \dots j_{m'+n'}}]_S . \end{aligned} \tag{4.83}$$

The remaining two cases are proved in Appendix F.2. □

We can now characterize $\mathcal{P}(\mathcal{R}_N)$ as a BV algebra in terms of generators and relations. In comparison with Theorem 4.17, the main simplification is that all generators $P_{i_1,i_2\dots i_m}$ with $2N-1 \geq m \geq 2$ are obtained from the volume element X, see (4.56), and the ground ring generators x^i. Indeed, we may first rewrite (4.70) as

$$\begin{aligned} P_{i_1,i_2\dots i_m} &= \tfrac{m+1}{m\,(2N-m)}[x^i, P_{i,i_1\dots i_m}]_S \\ &= \tfrac{m+1}{m\,(2N-m)}\Delta_S(x^i P_{i,i_1\dots i_m})\,, \end{aligned} \tag{4.84}$$

where the last line follows from the relation between the bracket and the BV operator as well as

$$\Delta_S x^i = 0\,, \quad \Delta_S P_{i,i_1\dots i_m} = 0\,. \tag{4.85}$$

By iterating (4.84) we obtain

$$\begin{aligned} P_{i_1,i_2\dots i_{2N-k}} &= (-1)^{k(k+1)/2}\tfrac{2N}{(2N-k)\,k!} \\ &\times \epsilon_{i_1\dots i_{2N-k}j_1\dots j_k}\Delta_S(x^{j_1}\Delta_S(\dots\Delta_S(x^{j_k}X)\dots))\,, \end{aligned} \tag{4.86}$$

where $2N-1 \geq k \geq 0$. In particular, for $k = 2N-1$, we find

$$x_i = (-1)^{N(2N-1)}\tfrac{2N}{(2N-1)!}\epsilon_{i\,j_1\dots j_{2N-1}}\Delta_S(x^{j_1}\Delta_S(\dots\Delta_S(x^{j_{2N-1}}X)\dots))\,. \tag{4.87}$$

Since $\Delta_S X = 0$, we may also rewrite (4.86) in terms of multiple brackets.

Theorem 4.25 *The BV algebra $(\mathcal{P}(\mathcal{R}_N), \cdot, \Delta_S)$ is generated by 1, the ground ring generators x^i, degree one derivation C, and the volume element X of degree $2N-1$. The BV operator and the 'dot' product are completely determined using*

$$\Delta_S x^i = 0\,, \quad \Delta_S C = -2(N-1)\,1\,, \quad \Delta_S X = 0\,, \tag{4.88}$$

$$\Delta_S(x^i x^j) = 0\,, \quad \Delta_S(Cx^i) = -(2N-1)x^i\,, \tag{4.89}$$

together with (4.87) and the relations (4.62)-(4.66) expressed in terms of the right hand side in (4.86).

Proof. The proof is similar to that of Theorem 4.17. We will just outline the main steps, and leave the details for the reader. In the first step we show that the BV algebra generated by 1, x^i, C, and X, satisfying all the relations above, is spanned by the elements of the form

$$x_{i_1}\dots x_{i_m}\Delta_S(x_{j_1}\dots\Delta_S(x_{j_n}X)\dots)\,, \quad x_{i_1}\dots x_{i_m}C\Delta_S(x_{j_1}\dots\Delta_S(x_{j_n}X)\dots)\,, \tag{4.90}$$

where, in obvious notation, we set $m, n \geq 0$. Relations (4.62)-(4.66) determine then the structure of the 'dot' product between those elements. The next step is to show that the BV operator Δ_S is completely determined using (4.88) and (4.89) together with the defining relation (4.4). The only nontrivial computation is to derive the second equation in (4.81), which, in the notation of Theorem 4.25, reads

$$\Delta_S(C\Delta_S(x^{i_1}\dots\Delta_S(x^{i_n}X)\dots)) = -n\Delta_S(x^{i_1}\dots\Delta_S(x^{i_n}X)\,. \tag{4.91}$$

For $n = 1$, we find using (4.4) and (4.88), (4.89), and (4.66),

$$\begin{aligned} \Delta_S(Cx^iX) &= \Delta_S(Cx^i)X - C\Delta_S(x^iX) - (\Delta_S C)x^iX \\ &= -x^iX - C\Delta_S(x^iX) . \end{aligned} \tag{4.92}$$

Since $\Delta_S^2 = 0$, acting with Δ_S on both sides of this equation yields (4.91) for $n = 1$. The general step of the induction is then proved similarly. □

Since $\Delta_S : \mathcal{P}^n(\mathcal{R}_N) \to \mathcal{P}^{n-1}(\mathcal{R}_N)$ satisfies $\Delta_S^2 = 0$, it is natural to consider the homology of the complex $(\mathcal{P}(\mathcal{R}_N), \Delta_S)$. This homology is easily computed using Theorems 4.16 and 4.24.

Theorem 4.26 *The homology of the BV operator Δ_S on $\mathcal{P}(\mathcal{R}_N)$ is spanned by the volume element X.*

As we will see later in this book, it is interesting to construct extensions of the BV algebra $(\mathcal{P}(\mathcal{R}_N), \cdot, \Delta)$ in which the homology of Δ is trivial. In particular the BV algebra of the semi-infinite cohomology of the $\mathcal{W}_3$ algebra is an extension of this type.

4.2.6 "Chiral" Subalgebras of $\mathcal{P}(\mathcal{R}_N)$

There is a natural complex structure on $\mathcal{P}(\mathcal{R}_N)$ induced from the decomposition of the ground ring generators into the "holomorphic" generators x_σ and the "antiholomorphic" generators $x_{\dot\sigma}$, such that $(x_i) = (x_\sigma, x_{\dot\sigma})$, $\sigma, \dot\sigma = 1, \ldots, N$. With respect to this decomposition the only nonvanishing components of the metric (4.31) are the (1,1) components, $h_{\sigma\dot\sigma} = \delta_{\sigma\dot\sigma}$, and the $\mathfrak{so}_{2N}$ symmetry is broken to the $\mathfrak{sl}_N$ subalgebra generated by the derivations

$$D_{\sigma\dot\sigma} = x_\sigma x^*_{\dot\sigma} - x_{\dot\sigma} x^*_\sigma - \tfrac{1}{N} h_{\sigma\dot\sigma}(x^{\dot\rho} x^*_{\dot\rho} - x^\rho x^*_\rho) , \quad \sigma, \dot\sigma = 1, \ldots, N . \tag{4.93}$$

Let us denote by $\mathcal{P}_+(\mathcal{R}_N)$ the BV subalgebra of $\mathcal{P}(\mathcal{R}_N)$ generated by the holomorphic elements x_σ, and $P_{\sigma_1,\sigma_2\ldots\sigma_n}$, $\sigma, \sigma_1, \ldots \sigma_n = 1, \ldots, N$. Similarly, let $\mathcal{P}_-(\mathcal{R}_N)$ be the BV subalgebra generated by the anti-homolomorphic elements. We will refer to $\mathcal{P}_+(\mathcal{R}_N)$ and $\mathcal{P}_-(\mathcal{R}_N)$ as the chiral subalgebras of $\mathcal{P}(\mathcal{R}_N)$.

Theorem 4.27 *The chiral subalgebra $\mathcal{P}_+(\mathcal{R}_N)$ (resp. $\mathcal{P}_-(\mathcal{R}_N)$) is spanned by the elements $P_{\sigma_1\ldots\sigma_m,\rho_1\ldots\rho_n}$ (resp. $P_{\dot\sigma_1\ldots\dot\sigma_m,\dot\rho_1\ldots\dot\rho_n}$). The BV operator Δ_S restricted to $\mathcal{P}_+(\mathcal{R}_N)$ (respectively $\mathcal{P}_-(\mathcal{R}_N)$) vanishes.*

Proof. The first part of the theorem follows from Theorems 4.16, 4.17 and 4.21. In particular, since $x^*_{\dot\sigma}(x_\rho) = 0$, (4.73) implies that

$$x_{\sigma_1} \ldots x_{\sigma_{m-1}} P_{\sigma_m,\rho_1\ldots\rho_n} = P_{\sigma_1\ldots\sigma_m,\rho_1\ldots\rho_n} . \tag{4.94}$$

The vanishing of the BV operator then follows from (4.81). □

Finally, let us note that the involution $\omega_{\mathcal{P}}$, $\omega_{\mathcal{P}}^2 = 1$, that exchanges the holomorphic and antiholomorphic generators, i.e., $\omega_{\mathcal{P}}(x_\sigma) = x_{\dot\sigma}$, $\omega_{\mathcal{P}}(x^*_{\dot\sigma}) = x^*_\sigma$ extends to all polyderivations $\mathcal{P}(\mathcal{R}_N)$, such that $\omega_{\mathcal{P}}(\mathcal{P}_+(\mathcal{R}_N)) \cong \mathcal{P}_-(\mathcal{R}_N)$

4.3 $N=3$: The BV Algebra Structure of $\mathcal{P}(\mathcal{R}_3)$

The major motivation for explicitly constructing the BV algebra $\mathcal{P}(\mathcal{R}_N)$ was to better understand the special case, $N=3$, which plays a central role in Chap. 5. We will now specialize the results of Sect. 4.2 to this case.

4.3.1 The Algebra $\mathcal{P}(\mathcal{R}_3)$

Consider the ground ring $\mathcal{R}_3$ as an $\mathfrak{sl}_3$ module, where $\mathfrak{sl}_3 \subset \mathfrak{so}_6$ is the subalgebra defined in (4.93). If (s_1, s_2) denotes an $\mathfrak{sl}_3$ irreducible module with the Dynkin labels s_1 and s_2, respectively, then the branching rule for an $\mathfrak{so}_6$ module $[m;0]$ is given by

$$[m;0] = \bigoplus_{s_1+s_2=m} (s_1, s_2)\,, \tag{4.95}$$

and the following result is an immediate consequence of Theorem 4.13.

Theorem 4.28 *The ground ring $\mathcal{R}_3$ is a model space for the Lie algebra $\mathfrak{sl}_3$, i.e., $\mathcal{R}_3$ is a direct sum of all finite-dimensional irreducible modules of $\mathfrak{sl}_3$, each occurring with multiplicity one.*

In the following, we will often write $\mathfrak{P}$ instead of $\mathcal{P}(\mathcal{R}_3)$ for the space of polyderivations of $\mathcal{R}_3$.

It is worth bringing out the simplicity of this result. The ground ring is generated by[4] x_σ and x^σ with the single relation, $x_\sigma x^\sigma = 0$. Thus the elements of the ring are simply tensors which are independently totally symmetric in their upper and lower indices, and which vanish when an upper index is contracted with a lower index – this is precisely a tensorial presentation of the irreducible representations of $\mathfrak{sl}_3$. The subspace of $\mathcal{R}_3$ spanned by monomials with s_1 factors of x_σ and s_2 factors of $x_{\dot\sigma}$ makes up exactly one irreducible $\mathfrak{sl}_3$ representation (s_1, s_2), i.e., that with highest weight $\Lambda = s_1\Lambda_1 + s_2\Lambda_2$. We will denote this subspace by $\mathcal{R}_3(\Lambda)$ in the following. This further decomposition of $\mathcal{R}_3$ may clearly be considered as the decomposition under $\mathfrak{sl}_3 \oplus (\mathfrak{u}_1)^2$, where the additional $(\mathfrak{u}_1)^2$ generators just count the number of x_σ and $x_{\dot\sigma}$ in a given monomial.

To determine the decomposition of $\mathfrak{P}$ with respect to $\mathfrak{sl}_3$ we need the branching rules,

[4] We recall that $x_\sigma = h_{\sigma\dot\sigma} x^{\dot\sigma}$, $x_{\dot\sigma} = h_{\dot\sigma\sigma} x^\sigma$.

$$
\begin{aligned}
[m;1] &= \bigoplus_{s_1+s_2=m-1} [(s_1,s_2)\oplus(s_1+1,s_2)\oplus(s_1,s_2+1)\oplus(s_1+1,s_2+1)] \\
[m;2] &= \bigoplus_{s_1+s_2=m} [(s_1,s_2)\oplus(s_1-1,s_2-1)] \\
&\oplus \bigoplus_{s_1+s_2=m-1} [2\,(s_1,s_2)\oplus(s_1+2,s_2)\oplus(s_1,s_2+2)]\,.
\end{aligned}
\tag{4.96}
$$

These formulae are valid for $m \geq 1$. The summation runs over $s_1, s_2 \geq 0$, and terms with negative labels are to be omitted. The branching rules for $[m;3]$, $[m;4]$ and $[m;5]$ are obtained using isomorphisms $[m;k] \cong [m;5-k]$, $k = 0,1,2$.

By comparing Table 4.1 with the above branching rules, we find that $\mathfrak{P}$ decomposes into a sum of disjoint "cones" of $\mathfrak{sl}_3$ modules, each cone being a direct sum of modules $(s_1^0+s_1, s_2^0+s_2)$, $s_1, s_2 \geq 0$. In particular, for $n=1$ we find five cones with the tips (s_1^0, s_2^0) equal to $(0,0)$, $(0,0)$, $(0,1)$, $(1,0)$ and $(1,1)$, which correspond to the derivations

$$
C_+ \;=\; x^{\dot\sigma}x^*_{\dot\sigma}\,, \quad C_- \;=\; x^\sigma x^*_\sigma\,, \tag{4.97}
$$

$$
P_{\sigma,\rho} \;=\; \tfrac{1}{2}(x_\sigma x^*_\rho - x_\rho x^*_\sigma)\,, \quad P_{\dot\sigma,\dot\rho} \;=\; \tfrac{1}{2}(x_{\dot\sigma} x^*_{\dot\rho} - x_{\dot\rho} x^*_{\dot\sigma})\,, \tag{4.98}
$$

and

$$
D_{\sigma\dot\sigma} \;=\; x_\sigma x^*_{\dot\sigma} - x_{\dot\sigma} x^*_\sigma - \tfrac{1}{3} h_{\sigma\dot\sigma}(x^{\dot\sigma}x^*_{\dot\sigma} - x^\sigma x^{*\sigma})\,, \tag{4.99}
$$

respectively. In the following we will also find it convenient to define

$$
D_\sigma \;=\; \epsilon_{\sigma\rho\pi}P^{\rho,\pi}\,, \quad D_{\dot\sigma} \;=\; \epsilon_{\dot\sigma\dot\rho\dot\pi}P^{\dot\rho,\dot\pi}\,. \tag{4.100}
$$

Note that while the derivations $D_{\sigma\dot\sigma}$ generate the $\mathfrak{sl}_3$ algebra, C_+ and C_- yield the additional $(\mathrm{u}_1)^2$ discussed earlier. With this representation as derivations it is clear that we have, in fact, a decomposition of all of $\mathfrak{P}$ into $\mathfrak{sl}_3 \oplus (\mathrm{u}_1)^2$ modules. The complete result is summarized in Theorem G.3.

Remark. The model space of $\mathfrak{sl}_3$ can also be realized as the space of polynomial functions on the algebraic variety $A = N_+\backslash SL(3,\mathbb{C})$, where N_+ is the complex subgroup of $SL(3,\mathbb{C})$ generated by the positive root generators [BGG2]. The space A is called the base affine space. In this realization of the ground ring $\mathcal{R}_3$, the algebra of polyderivations $\mathcal{P}(\mathcal{R}_3)$ is nothing but the algebra of polynomial polyvector fields on A. This provides a beautiful geometric interpretation for $\mathcal{P}(\mathcal{R}_3)$, and, in particular, gives a natural explanation of its cone decomposition. We give a detailed discussion of this geometric construction in Sect. 4.6 and Appendix G.

4.4 G Modules and BV Modules

The notion of a G module (BV module) of a G algebra (BV algebra) can be introduced by generalizing the dot and the bracket action (BV operator) on the algebra itself.

Definition 4.29 *Let $(\mathfrak{A}, \cdot, [-,-])$ be a G algebra and $\mathfrak{M} = \bigoplus_{n\in\mathbb{Z}} \mathfrak{M}^n$ a graded module of $\mathfrak{A}$. We call the action of the algebra $\mathfrak{A}$ on $\mathfrak{M}$ the dot action, and thus call $\mathfrak{M}$ a dot algebra module of $\mathfrak{A}$. Then $\mathfrak{M}$ is a G module of $\mathfrak{A}$ if there further exists a bracket map,*

$$[-,-]_M : \mathfrak{A}^m \times \mathfrak{M}^n \longrightarrow \mathfrak{M}^{m+n-1},$$

such that

$$[a\cdot b, m]_M = a\cdot[b,m]_M + (-1)^{|a||b|} b\cdot[a,m]_M, \tag{4.101}$$

$$[a, b\cdot m]_M = [a,b]\cdot m + (-1)^{(|a|-1)|b|} b\cdot[a,m]_M, \tag{4.102}$$

$$[[a,b],m]_M = [a,[b,m]_M]_M - (-1)^{(|a|-1)(|b|-1)}[b,[a,m]_M]_M. \tag{4.103}$$

for all $a, b \in \mathfrak{A}$ and $m \in \mathfrak{M}$.

Remark. Relations (4.102) and (4.103) may be interpreted as the statement that the operators $[a,-]_M$, $a \in \mathfrak{A}$, define a representation of the graded Lie algebra $(\mathfrak{A}, [-,-])$, which acts as a graded derivation of the dot action of $\mathfrak{A}$ on $\mathfrak{M}$.

Definition 4.30 *Let $(\mathfrak{A}, \cdot, \Delta)$ be a BV algebra and $\mathfrak{M} = \bigoplus_{n\in\mathbb{Z}} \mathfrak{M}^n$ a graded module of $\mathfrak{A}$ as a dot algebra. Then $\mathfrak{M}$ is a BV module of $\mathfrak{A}$ if there exists a map*

$$\Delta_M : \mathfrak{M}^n \longrightarrow \mathfrak{M}^{n-1},$$

such that $\Delta_M^2 = 0$ and for any $a, b \in \mathfrak{A}$ and $m \in \mathfrak{M}$,

$$\begin{aligned}\Delta_M(a\cdot b\cdot m) &= \Delta(a\cdot b)\cdot m + (-1)^{|a|} a\cdot\Delta_M(b\cdot m) + (-1)^{(|a|-1)|b|} b\cdot\Delta_M(a\cdot m)\\ &\quad - (\Delta a)\cdot b\cdot m - (-1)^{|a|} a\cdot(\Delta b)\cdot m - (-1)^{|a|+|b|} a\cdot b\cdot\Delta_M(m).\end{aligned} \tag{4.104}$$

Clearly, a BV module $\mathfrak{M}$ of a BV algebra $\mathfrak{A}$ is also a G module of $\mathfrak{A}$ with the bracket defined by

$$[a,m]_M = (-1)^{|a|}\big(\Delta_M(a\cdot m) - (\Delta a)\cdot m - (-1)^{|a|} a\cdot(\Delta_M m)\big), \quad a\in\mathfrak{A}, m\in\mathfrak{M}, \tag{4.105}$$

which measures to what extent Δ_M fails to be a derivation of the dot action of $\mathfrak{A}$ on $\mathfrak{M}$.

Free modules on one generator, ω, provide simplest examples of G modules and/or BV modules. They are spanned by expressions of the form

$$a_1\cdot[a_2,[\ldots[a_{n-1}, a_n\cdot\omega]_M \ldots]_M]_M, \quad a_1,\ldots,a_n \in \mathfrak{A}, \tag{4.106}$$

and

$$a_1 \cdot \Delta_M(a_2 \cdot \Delta_M(\ldots \Delta_M(a_n \cdot \omega) \ldots)), \quad a_1, \ldots, a_n \in \mathfrak{A}, \tag{4.107}$$

subject to the defining relations (4.101)-(4.103) and (4.104), respectively.

4.4.1 Natural G Modules for the G Algebra $(\mathcal{P}(\mathcal{R}), \cdot, [-,-]_S)$

For the G algebra of polyderivations $(\mathcal{P}(\mathcal{R}), \cdot, [-,-]_S)$ of an Abelian algebra $\mathcal{R}$, a natural class of G modules consists of polyderivations $\mathcal{P}(\mathcal{R}, M)$, where M is a suitable module of $\mathcal{R}$.

Theorem 4.31 *Suppose that M is a module of $\mathcal{R}$ on which the Lie algebra $\mathcal{D}(\mathcal{R})$ acts by derivations of the dot product action of $\mathcal{R}$, i.e., the representation $a \mapsto K_a$, $a \in \mathcal{D}(\mathcal{R})$, satisfies (see Sect. 4.1.1)*

$$K_a(K_b(m)) - K_b(K_a(m)) = K_{[a,b]_S}(m), \quad a, b \in \mathcal{D}(\mathcal{R}), m \in M, \tag{4.108}$$

$$K_a(x \cdot m) = a(x) \cdot m + x \cdot K_a(m), \quad a \in \mathcal{D}(\mathcal{R}), x \in \mathcal{R}, m \in M. \tag{4.109}$$

Then the space of polyderivations $\mathcal{P}(\mathcal{R}, M)$ naturally has the structure of a G module of $(\mathcal{P}(\mathcal{R}), \cdot, [-,-]_S)$.

Proof. The proof parallels that of Theorem 4.9. The module structure with respect to the dot product is defined by (4.12), with $a \in \mathcal{P}(\mathcal{R})$ and $b \in \mathcal{P}(\mathcal{R}, M)$. The bracket is constructed by induction setting

$$\begin{aligned} [a,m]_M &= 0, \quad a \in \mathcal{R}, m \in M, \\ [a,m]_M &= K_a(m), \quad a \in \mathcal{P}^1(\mathcal{R}), m \in M, \\ [a,m]_M &= -m(a), \quad a \in \mathcal{R}, m \in \mathcal{P}^1(\mathcal{R}, M), \end{aligned} \tag{4.110}$$

and then using

$$[a,m]_M(x) = [a, m(x)]_M + (-1)^{|m|-1}[a(x), m]_M, \quad x \in \mathcal{R}, \tag{4.111}$$

for $|a| + |m| \geq 2$. □

Remark. We have implicitly assumed that the grading on $\mathcal{P}(\mathcal{R}, M)$ as a G module is the same as the degree of polyderivations, i.e., $\mathfrak{M}^n = \mathcal{P}^n(\mathcal{R}, M)$. If we shift the grading of the module by taking $\mathfrak{M}^n \to \mathfrak{M}^{n+k}$, an obvious modification of the construction above equips $\mathcal{P}(\mathcal{R}, M)$ with another $\mathcal{P}(\mathcal{R})$ G module structure. Clearly, all structures with k respectively even or odd are equivalent, so it makes sense to talk about $\mathcal{P}(\mathcal{R}, M)$ and as an "even" or "odd" G module of $\mathcal{P}(\mathcal{R})$.

4.5 $N=3$: Twisted Modules of $\mathcal{P}(\mathcal{R}_3)$

As an application of this construction we go immediately to the case of interest, $N=3$. Here there is a construction of the twisted $\mathcal{R}_3$ modules which follows naturally from consideration of the hidden symmetry alluded to earlier.

4.5.1 The Hidden Symmetry Structure

As discussed in Sect. 4.2.2 , the ground ring is a module for itself under left multiplication. There is a hidden symmetry algebra, which includes multiplication by ring generators, for which the ground ring is an irreducible module. In the case of $\mathcal{R}_3$ this hidden symmetry algebra is $\mathfrak{so}_8$. Under the chain of embeddings, $\mathfrak{sl}_3 \subset \mathfrak{so}_6 \subset \mathfrak{so}_8$, the adjoint representations of $\mathfrak{so}_6$ and $\mathfrak{so}_8$ decompose with respect to $\mathfrak{sl}_3$ as

$$\mathrm{ad}_{\mathfrak{so}_6} = \mathbf{8} \oplus (\mathbf{3} \oplus \overline{\mathbf{3}}) \oplus \mathbf{1}\,, \tag{4.112}$$

$$\mathrm{ad}_{\mathfrak{so}_8} = \mathbf{8} \oplus (\mathbf{3} \oplus \overline{\mathbf{3}}) \oplus (\mathbf{3} \oplus \overline{\mathbf{3}}) \oplus (\mathbf{3} \oplus \overline{\mathbf{3}}) \oplus \mathbf{1} \oplus \mathbf{1}\,. \tag{4.113}$$

The operators corresponding to the decomposition (4.113) are

$$D_{\sigma\dot{\sigma}}\,, \quad (D_\sigma, D_{\dot{\sigma}})\,, \quad (P_\sigma, U_{\dot{\sigma}})\,, \quad (U_\sigma, P_{\dot{\sigma}})\,, \quad C_+\,, \quad C_-\,, \tag{4.114}$$

where, in addition to the first order derivations given in (4.97)-(4.99), we also have zero and second order differential operators on $\mathcal{R}_3$,

$$P_\sigma = x_\sigma\,, \quad P_{\dot{\sigma}} = x_{\dot{\sigma}}\,, \tag{4.115}$$

and

$$\begin{aligned} U_\sigma &= 2\frac{\partial}{\partial x^\sigma} + x^\rho \frac{\partial}{\partial x^\rho}\frac{\partial}{\partial x^\sigma} + x_\rho \frac{\partial}{\partial x_\rho}\frac{\partial}{\partial x^\sigma} - x_\sigma \frac{\partial}{\partial x_\rho}\frac{\partial}{\partial x^\rho}\,, \\ U_{\dot{\sigma}} &= 2\frac{\partial}{\partial x^{\dot{\sigma}}} + x^{\dot{\rho}} \frac{\partial}{\partial x^{\dot{\rho}}}\frac{\partial}{\partial x^{\dot{\sigma}}} + x_{\dot{\rho}} \frac{\partial}{\partial x_{\dot{\rho}}}\frac{\partial}{\partial x^{\dot{\sigma}}} - x_{\dot{\sigma}} \frac{\partial}{\partial x_{\dot{\rho}}}\frac{\partial}{\partial x^{\dot{\rho}}}\,. \end{aligned} \tag{4.116}$$

The derivations $D_{\sigma\dot{\sigma}}$, D_σ, $D_{\dot{\sigma}}$ and $C_+ - C_-$ generate the $\mathfrak{so}_6$ algebra of Sect. 4.2. As we have seen in Sect. 4.2.3, this symmetry lifts from the ground ring to the polyderivations. But there is no such lift for the operators U, U_σ and $U_{\dot{\sigma}}$, the reason being that they do *not* act on $\mathcal{R}_3$ as derivations.

This structure arose out of the consideration of a particular extension of $\mathfrak{sl}_3$ to $\mathfrak{so}_6 \subset \mathfrak{so}_8$; namely, that for which the pair $(D_\sigma, D_{\dot{\sigma}})$ corresponds to $(\mathbf{3} \oplus \overline{\mathbf{3}})$ in (4.112). From (4.113) it is clear that this extension may be done in three ways, utilising any of the pairs in (4.114). In fact, the existence of three extensions is explained by the triality of $\mathfrak{so}_8$, i.e., the three inequivalent representations of $\mathfrak{so}_8$ of dimension eight. It is interesting to understand the extensions which involve the two remaining pairs in (4.114) since, as explained in the next section, this leads to new modules of the ground ring. Indeed, since for a given choice of the extension to $\mathfrak{so}_6$ there are still two ways to assign a remaining $(\mathbf{3} \oplus \overline{\mathbf{3}})$ pair in (4.113) to the ring generators, this will produce a total of six ground ring modules.

4.5.2 "Twisted" Modules of $\mathfrak{P}$

The $\mathcal{R}_3$ module discussed above – namely, $\mathcal{R}_3$ itself – will be denoted by M_1. We will now explicitly construct the remaining ring modules alluded to there. It will be convenient in the following to denote by M the vector space spanned by monomials in x_σ and $x_{\dot\sigma}$, modulo the constraint $h^{\sigma\dot\sigma}x_\sigma x_{\dot\sigma}=0$. Clearly M carries a representation of $\mathfrak{sl}_3$ as differential operators – in fact, precisely the $D_{\sigma\dot\sigma}$ in (4.99) – and, for $\Lambda = s_1\Lambda_1 + s_2\Lambda_2$, we may introduce (in analogy with $\mathcal{R}_3(\Lambda)$) the subspaces $M(\Lambda)$ spanned by monomials with s_1 factors of x_σ and s_2 factors of $x_{\dot\sigma}$. The space M may also carry a realization of the ring $\mathcal{R}_3$ by differential operators. Indeed, M_1 is M on which the generators act by the (zeroth order) differential operators $(P_\sigma, P_{\dot\sigma})$ in (4.115).

Theorem 4.32 *The six pairs of operators $(P^w_\sigma, P^w_{\dot\sigma})$, $w \in W$, given by*

$$(P_\sigma,P_{\dot\sigma}),\quad (D_\sigma,P_{\dot\sigma}),\quad (P_\sigma,D_{\dot\sigma}),\quad (U_\sigma,D_{\dot\sigma}),\quad (D_\sigma,U_{\dot\sigma}),\quad (U_\sigma,U_{\dot\sigma}), \tag{4.117}$$

for w equal to 1, r_1, r_2, r_{12}, r_{21} and r_3, respectively, define six inequivalent $\mathcal{R}_3$ module structures as differential operators acting on M.

Remark. We will denote by M_w the ground ring module defined by the realization $(P^w_\sigma, P^w_{\dot\sigma})$ on M.

Proof. It is straightforward to verify that, for each w, the differential operators P^w_σ and $P^w_{\dot\sigma}$ commute and satisfy the constraint (4.30), i.e., $h^{\sigma\dot\sigma}P^w_\sigma P^w_{\dot\sigma} = 0$. Thus the module structure is established. Examining the explicit action of the differential operators on monomials in $M(\Lambda)$, one finds immediately that the operators P^w_σ and $P^w_{\dot\sigma}$ act as epimorphisms between vector spaces, and

$$\begin{aligned}(P^w_\sigma) &: M(\Lambda) \longrightarrow M(\Lambda + w^{-1}\Lambda_1)\,,\\ (P^w_{\dot\sigma}) &: M(\Lambda) \longrightarrow M(\Lambda + w^{-1}\Lambda_2)\,.\end{aligned} \tag{4.118}$$

Hence the module structures are inequivalent since they map differently between irreducible $\mathfrak{sl}_3$ modules. □

Remark. For all $w \in W$ the modules M_w are isomorphic to $\mathcal{R}_3$ as $\mathfrak{sl}_3$ modules. The action (4.118) is the motivation for the labelling by Weyl group elements.

Consistent with the comments at the end of the last subsection, the existence of the six modules M_w is equivalent to the existence of six realizations of $\mathfrak{so}_8$ as differential operators on M. We display those realizations in Table 4.2, one on each line labelled by the corresponding $\mathfrak{sl}_3$ Weyl group element, $w \in W$. The fact that the operators in each line of Table 4.2 independently generate $\mathfrak{so}_8$ is shown by explicit computation.

Table 4.2. Six realizations of the generators of $\mathfrak{so}_8$ on M.

w	P^w_σ	$P^w_{\dot\sigma}$	D^w_σ	$D^w_{\dot\sigma}$	$D^w_{\sigma\dot\sigma}$	U^w_σ	$U^w_{\dot\sigma}$	C^w_+	C^w_-
1	P_σ	$P_{\dot\sigma}$	D_σ	$D_{\dot\sigma}$	$D_{\sigma\dot\sigma}$	U_σ	$U_{\dot\sigma}$	C_+	C_-
r_1	D_σ	$P_{\dot\sigma}$	P_σ	$U_{\dot\sigma}$	$D_{\sigma\dot\sigma}$	U_σ	$D_{\dot\sigma}$	$-C_+ - 2$	$C_+ + C_- + 1$
r_2	P_σ	$D_{\dot\sigma}$	U_σ	$P_{\dot\sigma}$	$D_{\sigma\dot\sigma}$	D_σ	$U_{\dot\sigma}$	$C_+ + C_- + 1$	$-C_- - 2$
r_{12}	U_σ	$D_{\dot\sigma}$	P_σ	$U_{\dot\sigma}$	$D_{\sigma\dot\sigma}$	D_σ	$P_{\dot\sigma}$	$-C_+ - C_- - 3$	C_+
r_{21}	D_σ	$U_{\dot\sigma}$	U_σ	$P_{\dot\sigma}$	$D_{\sigma\dot\sigma}$	P_σ	$D_{\dot\sigma}$	C_-	$-C_+ - C_- - 3$
r_3	U_σ	$U_{\dot\sigma}$	D_σ	$D_{\dot\sigma}$	$D_{\sigma\dot\sigma}$	P_σ	$P_{\dot\sigma}$	$-C_- - 2$	$-C_+ - 2$

The main result of this section then follows.

Theorem 4.33 *Let $\mathfrak{P}_w \equiv \mathcal{P}(\mathcal{R}_3, M_w)$ be the algebra of polyderivations of $\mathcal{R}_3$ with values in M_w. Then $\mathfrak{P}_w$ is a G module of $\mathfrak{P}$.*

Proof. In view of Theorem 4.31 it is sufficient to define an action of $\mathfrak{P}^1$ on M_w that satisfies (4.108) and (4.109). Given that the ring generators are realized as differential operators on M_w as in Table 4.2, $P^w_i = (P^w_\sigma, P^w_{\dot\sigma})$, a natural candidate for the generators of $\mathfrak{P}^1$ is the set $P^w_{i,j} = (P^w_{\sigma,\rho}, P^w_{\dot\sigma,\dot\rho}, P^w_{\sigma,\dot\sigma}, C^w)$, constructed from the other generators in the table,

$$\begin{aligned} P^w_{\sigma,\rho} &= \tfrac{1}{2}\epsilon_{\sigma\rho\pi}D^\pi\,, \quad P^w_{\dot\sigma,\dot\rho} = \tfrac{1}{2}\epsilon_{\dot\sigma\dot\rho\dot\pi}D^{\dot\pi}\,, \\ P^w_{\sigma,\dot\sigma} &= 2\,D^w_{\sigma\dot\sigma} + \tfrac{2}{3}h_{\sigma\dot\sigma}(C^w_+ - C^w_-)\,, \quad C^w = C^w_+ + C^w_-\,. \end{aligned} \tag{4.119}$$

It is straightforward to check that these operators satisfy the relations (4.63) and (4.64), where the dot product is realized as the ordinary product of differential operators. Thus, by Theorem 4.17, the space of differential operators spanned by monomials of P^w_i multiplying $P^w_{j,k}$ on the left gives a realization of $\mathfrak{P}^0 \oplus \mathfrak{P}^1$ as a dot algebra. Moreover, we know that a given line of Table 4.2 generates $\mathfrak{so}_8$ under commutation, independent of w. So, (4.108) and (4.109) hold for the generators realized as above. But then they hold for any element of the corresponding realization of $\mathfrak{P}^0 \oplus \mathfrak{P}^1$. Thus we have found, for each w, a realization of $\mathfrak{P}^0 \oplus \mathfrak{P}^1$ as a G algebra, where the bracket operation is simply the commutator of differential operators. □

We will often refer to M_w as the w-twisted module of $\mathcal{R}_3$. While M_w is isomorphic, as an $\mathfrak{sl}_3$ module, with $\mathcal{R}_3$, it will turn out convenient to twist the $(\mathfrak{u}_1)^2$ weights such that $\Lambda' \to w^{-1}(\Lambda' + \rho) - \rho$. This is precisely consistent with Table 4.2 if we use $C^w_\pm$ as the $(\mathfrak{u}_1)^2$ generators in the twisted module. In particular, the identity in M has weights $(0, w^{-1}\rho - \rho)$ when considered as an element of M_w, and will be denoted by Ω_w. From now on we will denote the action of the ring generators on M_w simply by x_σ or $x_{\dot\sigma}$. Also we will use the terminology "twisted polyderivations" for $\mathfrak{P}_w$, $w \neq 1$.

4.5.3 A Classification of Twisted Polyderivations

We now describe in more detail the spaces of twisted polyderivations, $\mathfrak{P}_w$, especially for $w = r_1$ and r_2. As in Sect. 4.2.3, the computation of those spaces may be posed as an algebraic problem of finding all $\Phi \in \mathrm{Hom}(\bigwedge^n \mathcal{R}_3^1, M_w)$, whose coefficients of expansion, $\Phi = \Phi^{i_1\ldots i_n} x^*_{i_1} \ldots x^*_{i_n}$, $\Phi^{i_1\ldots i_n} \in M_w$, satisfy the analogue of (4.36), i.e.,

$$x_i \cdot \Phi^{ii_1\ldots i_{n-1}} = 0\,, \quad i_1,\ldots,i_{n-1} = 1,\ldots,6\,. \tag{4.120}$$

The decomposition of $\mathfrak{P}_w$ with respect to $\mathfrak{sl}_3 \oplus (\mathfrak{u}_1)^2$, whose action is induced from that on $\mathcal{R}_3$ and M_w, is crucial for solving (4.120) by reducing it to separate irreducible components. However, since this symmetry is now smaller than $\mathfrak{so}_6$ in Sect. 4.2.3, the analysis is rather lengthy.[5] For that reason, rather than discussing the general case, let us illustrate the method with the simple example of generalized vector fields with values in M_{r_1}, and then present the complete solution for $\mathfrak{P}_{r_1}$ and $\mathfrak{P}_{r_2}$.

Example. An arbitrary generalized vector field, $\Phi \in \mathfrak{P}^1_{r_1}$, that transforms in an $\mathfrak{sl}_3$ representation with weight Λ, is of the form

$$\Phi = \sum_{\lambda \in P_1(\Lambda)} \Phi^\sigma_\lambda x^*_\sigma + \sum_{\lambda \in P_2(\Lambda)} \Phi^{\dot\sigma}_\lambda x^*_{\dot\sigma}\,, \tag{4.121}$$

where $\Phi^\sigma_\lambda, \Phi^{\dot\sigma}_\lambda \in M_{r_1}(\lambda)$, and $P_i(\Lambda)$ consist of those weights $\lambda \in P_+$ for which $\mathcal{L}(\Lambda)$ arises in the tensor product $\mathcal{L}(\lambda) \otimes \mathcal{L}(\Lambda_i)$, i.e.,

$$\begin{aligned} P_1(\Lambda) &= \{\Lambda + \Lambda_2, \Lambda + \Lambda_1 - \Lambda_2, \Lambda - \Lambda_1\} \cap P_+\,, \\ P_2(\Lambda) &= \{\Lambda + \Lambda_1, \Lambda - \Lambda_1 + \Lambda_2, \Lambda - \Lambda_2\} \cap P_+\,. \end{aligned} \tag{4.122}$$

Since the operator $\mathfrak{d}$ in (4.36) is $\mathfrak{sl}_3$ invariant, we conclude that the component of (4.120) along the representation with weight Λ must vanish,

$$\Big(\sum_{\lambda \in P_1(\Lambda)} x_\sigma \cdot \Phi^\sigma_{\lambda_1} + \sum_{\lambda \in P_2(\Lambda)} x_{\dot\sigma} \cdot \Phi^{\dot\sigma}_{\lambda_2} \Big)_\Lambda = 0\,. \tag{4.123}$$

Now, recall that that the action of the ground ring on M_{r_1} merely amounts to shifting between the following representations, see (4.118),

$$(x_\sigma) : M_{r_1}(\Lambda) \longrightarrow M_{r_1}(\Lambda - \Lambda_1 + \Lambda_2)\,, \quad (x_{\dot\sigma}) : M_{r_1}(\Lambda) \longrightarrow M_{r_1}(\Lambda + \Lambda_2)\,. \tag{4.124}$$

Therefore (4.123) reduces to

$$x_\sigma \cdot \Phi^\sigma_{\Lambda + \Lambda_1 - \Lambda_2} + x_{\dot\sigma} \cdot \Phi^{\dot\sigma}_{\Lambda - \Lambda_2} = 0\,. \tag{4.125}$$

[5] Note that the "standard" $\mathfrak{so}_6$ symmetry of the ground ring yields a decomposition of M_w, $w \neq 1, r_3$, into infinite-dimensional modules. Thus it seems simpler to work with a smaller symmetry algebra, that yields a decomposition into finite-dimensional modules only.

Table 4.3. The r_1-twisted cone decomposition of $\mathfrak{P}^1_{r_1}$

Φ	Λ'	(Λ, Λ')
$\Phi^{\sigma}_{\Lambda+\Lambda_2} x^*_{\sigma}$	$r_1\Lambda - 2\Lambda_1 + \Lambda_2$	$(0, -2\Lambda_1 + \Lambda_2)$
$\Phi^{\dot\sigma}_{\Lambda+\Lambda_1} x^*_{\dot\sigma}$	$r_1\Lambda - 4\Lambda_1 + 2\Lambda_2$	$(0, -4\Lambda_1 + 2\Lambda_2)$
$\Phi^{\sigma}_{\Lambda-\Lambda_1} x^*_{\sigma}$	$r_1\Lambda - \Lambda_1 - \Lambda_2$	$(\Lambda_1, -2\Lambda_1)$
$\Phi^{\dot\sigma}_{\Lambda-\Lambda_1+\Lambda_2} x^*_{\dot\sigma}$	$r_1\Lambda - 2\Lambda_1 + \Lambda_2$	$(\Lambda_1, -3\Lambda_1 + 2\Lambda_2)$
$\Phi^{\sigma}_{\Lambda+\Lambda_1-\Lambda_2} x^*_{\sigma} + \Phi^{\dot\sigma}_{\Lambda-\Lambda_2} x^*_{\dot\sigma}$	$r_1\Lambda - 3\Lambda_1$	$(\Lambda_2, -3\Lambda_1 + \Lambda_2)$

Since the action of $(x_{\dot\sigma})$ on M_{r_1} has no zeros, we may always solve this equation and express the components $\Phi^{\dot\sigma}_{\Lambda-\Lambda_2}$ in terms of $\Phi^{\sigma}_{\Lambda+\Lambda_1-\Lambda_2}$.

The weights Λ, for which a solution arises in the sum (4.121), form a set of cones in P_+. This, in turn, translates into an r_1-twisted cone decomposition of $\mathfrak{P}^1_{r_1}$. We have summarized the classification of $\mathfrak{P}^1_{r_1}$ in Table 4.3, where we give a schematic form of each vector field, its $(\mathfrak{u}_1)^2$ weight Λ' that follows immediately from the coordinate expansion, and the tip of the corresponding cone, (Λ, Λ'), determined by the lowest Λ for which a given component exists.

An extension of this example to higher degree polyderivations is complicated by the fact that (4.120) may have components in several $\mathfrak{sl}_3 \oplus (\mathfrak{u}_1)^2$ irreducible modules. Upon expanding the coordinates of the polyderivations in a basis of $\mathfrak{sl}_3 \oplus (\mathfrak{u}_1)^2$ invariant tensors we obtain a system of linear equations. A simple counting of independent solutions yields the following enumeration of $\mathfrak{P}_1$ and $\mathfrak{P}_2$.

Theorem 4.34 *The space of generalized polyderivations $\mathfrak{P}_{r_i}$, $i = 1, 2$, decomposes into a direct sum of r_i-twisted cones of $\mathfrak{sl}_3 \oplus (\mathfrak{u}_1)^2$ modules,*

$$\mathfrak{P}^n_{r_i} = \bigoplus_{(\Lambda,\Lambda')} \bigoplus_{\lambda \in P_+} \mathcal{L}(\Lambda + \lambda) \otimes \mathbb{C}_{\Lambda' + r_i \lambda}\,, \tag{4.126}$$

where, in the case of $\mathfrak{P}_{r_1}$, the tips of the cones, (Λ, Λ'), satisfy $\Lambda + 2\rho = r_1(\Lambda' + \rho - \sigma\rho)$, with $\Lambda \in P_+$ and $\sigma \in \widetilde{W}$ given in Table 4.4.[6] *The result for $\mathfrak{P}_{r_2}$ is obtained by interchanging the fundamental weights Λ_1 and Λ_2, and letting $r_1 \to r_2$, $\sigma \to w_0 \sigma w_0$, $\sigma \in W$, $\sigma_i \to \sigma_i$, $i = 1, 2$.*

A more explicit description of $\mathfrak{P}_{r_1}$ and $\mathfrak{P}_{r_2}$ is obtained by studying the G module action of $\mathfrak{P}$, and in particular of its chiral subalgebras, $\mathfrak{P}_-$ and $\mathfrak{P}_+$, respectively. Consider $\mathfrak{P}_{r_1}$. Since $x_\sigma \cdot \Omega_{r_1} = 0$, we find one polyderivation of degree three,

$$\Gamma_1 = -\tfrac{1}{432} \epsilon^{\sigma\rho\pi}\, \Omega_{r_1} x^*_\sigma x^*_\rho x^*_\pi\,, \tag{4.127}$$

[6] See also Table H.2 in Appendix H.

Table 4.4. The weights Λ in the r_1-twisted cone decomposition of $\mathfrak{P}_{r_1}$

$\sigma \backslash n$	0	1	2	3	4	5
r_{21}	0	$0,\ \Lambda_1$	Λ_1			
r_2		Λ_1	$\Lambda_1,\ 2\Lambda_1$	$2\Lambda_1$		
r_3		0	0, 0	0		
σ_2		Λ_2	$0,\ \Lambda_2$	0		
σ_1			Λ_1	$\Lambda_1,\ \Lambda_1$	Λ_1	
1			Λ_2	$0,\ \Lambda_1+\Lambda_2$	Λ_1	
r_{12}			Λ_2	$0,\ \Lambda_2$	0	
r_1				Λ_2	$0,\ \Lambda_2$	0

at weight $(0,-2\rho)$. Similarly, we have $\Gamma_2 \in \mathfrak{P}_{r_2}$.

Theorem 4.35 *Twisted polyderivations $\mathfrak{P}_{r_1}$ and $\mathfrak{P}_{r_2}$ are generated as free G modules by $\mathfrak{P}_-$ and $\mathfrak{P}_+$ acting on Γ_1 and Γ_2, respectively.*

Proof. See Appendix H. □

By comparing the decomposition of $\mathfrak{P}_{r_i}$ with that of polyvectors $\mathfrak{P}$, as given, e.g., in Theorem G.3, we conclude that for modules sufficiently deep inside the cones the two decompositions are related by the Weyl reflection of $\Lambda' + \rho - \sigma\rho$. This could have been anticipated by looking at the action of the ground ring on the twisted modules at the level of $\mathfrak{sl}_3 \oplus (\mathfrak{u}_1)^2$ modules. The essential difference between $M_1 \simeq \mathcal{R}_3$ and the twisted modules M_w, $w \neq 1$, is the presence of zeros in the action of some (for $w = r_1$ and r_2) or all (for $w = r_{12}$, r_{21} and r_3) ground ring generators, which explains why the the relation between different $\mathfrak{P}_w$ holds only in the bulk, i.e., sufficiently deep inside the cones.

In the remaining cases of polyderivations with values in $M_{r_{12}}$, $M_{r_{21}}$ and M_{r_3}, the cone decomposition breaks down close to the boundaries of the corresponding Weyl chambers. Once more this is explained by the lines of states in the twisted modules that are annihilated by *all* ground ring generators. The presence of such states results in additional solutions to (4.120), beyond those predicted by a naive counting of equations and components. However, inside the chambers, those special cases cannot arise, and once more we find a similar result to the one in the fundamental Weyl chamber.

Theorem 4.36 *In the bulk the space of generalized polyvectors $\mathfrak{P}_w$ is a direct sum of quartets of $\mathfrak{sl}_3 \oplus (\mathfrak{u}_1)^2$ modules, $\mathcal{L}(\Lambda) \otimes \mathbb{C}_{\Lambda'}$ such that:*

i. Each quartet consists of modules of polyvectors of degree n, $n+1$, $n+1$ and $n+2$, respectively,

ii. The weights $\Lambda \in P_+$ and $\Lambda' \in P$ satisfy $\Lambda + 2\rho = w^{-1}(\Lambda' + \rho - \sigma\rho)$, where $\sigma \in \widetilde{W}$ depends on n and w as given in Table 4.5.

Table 4.5. The dependence of σ on n and w in the quartet decomposition of $\mathfrak{P}^n_w$ in the bulk

$n \backslash w$	1	r_1	r_2	r_{12}	r_{21}	r_3
0	r_3	r_{21}	r_{12}	r_2	r_1	1
1	r_{12}, σ_2, r_{21}	r_3, σ_2, r_2	r_1, σ_2, r_3	$1, \sigma_2, r_{21}$	$r_{12}, \sigma_2, 1$	r_2, σ_2, r_1
2	r_1, σ_1, r_2	$r_{12}, \sigma_1, 1$	$1, \sigma_1, r_{21}$	r_1, σ_1, r_3	r_3, σ_1, r_2	r_{21}, σ_1, r_{12}
3	1	r_1	r_2	r_{12}	r_{21}	r_3

Remark. The assignment of $\sigma = \sigma_1$ and $\sigma = \sigma_2$ in Table 4.5 to particular cones appears to be arbitrary at this point, and our choice was motivated by the results in Sect. 5.4.2.

While the G module structure of $\mathfrak{P}_w$ over $\mathfrak{P}$ is rather obvious, it is less clear whether it arises from some BV module structure. Although we cannot answer this question in general, we would like to point out that in the case of $\mathfrak{P}_{r_i}$, $i = 1, 2$, there is a natural candidate for a BV operator. This operator, Δ, is uniquely determined by the condition

$$\Delta_i \Gamma_i \;=\; 0\,, \quad i = 1, 2\,, \tag{4.128}$$

together with the properties of the bracket. Using the explicit parametrization of $\mathfrak{P}_{r_i}$ as free modules of $\mathfrak{P}_\mp$, given in Appendix H, we then find

$$\Delta_i(\Phi_0 \cdot [\Phi_1, [\ldots, [\Phi_n, \Gamma_i] \ldots]]) \;=\; (-1)^{|\Phi_0|}[\Phi_0, [\Phi_1, [\ldots, [\Phi_n, \Gamma_i] \ldots]]\,, \tag{4.129}$$

where $\Phi_0\,, \ldots, \Phi_n \in \mathfrak{P}_\mp$. Essentially by construction, Δ_1 and Δ_2 turn $\mathfrak{P}_{r_1}$ and $\mathfrak{P}_{r_2}$ into a BV module of $\mathfrak{P}_-$ and $\mathfrak{P}_+$, respectively. We would like to conjecture that in fact Δ_i defines on $\mathfrak{P}_{r_i}$ a BV module structure with respect to the full BV algebra of polyderivations $\mathfrak{P}$. It appears that a direct algebraic proof of this conjecture along the lines of Sect. 4.2.5 is rather cumbersome. In particular, it would require a more explicit enumeration of the bases in $\mathfrak{P}_{r_i}$ beyond the one given in Appendix H.

4.6 BV Algebras on the Base Affine Space $A(G)$

In this section we introduce two BV algebras, $\mathcal{P}(A)$ and BV[$\mathfrak{g}$], associated with the base affine space $A = A(G)$ of a complex Lie group G. In the lowest rank cases, i.e., for $G = SL(n, \mathbb{C})$ with $n = 2$ or 3, the algebras $\mathcal{P}(A)$ are isomorphic with the algebras of polyderivations $\mathcal{P}(\mathcal{C}_2)$ and $\mathcal{P}(\mathcal{R}_3)$, respectively. For other groups, in particular for $G = SL(n, \mathbb{C})$ with $n \geq 4$, they provide new examples of BV algebras beyond those introduced in Sects. 4.2 and 4.3.

The algebra BV[$\mathfrak{g}$] is a natural extension of $\mathcal{P}(A)$. It is introduced in terms of the cohomology with respect to the nilpotent subalgebra $\mathfrak{n}_+$ of $\mathfrak{g}$, where $\mathfrak{g}$ is the Lie algebra of G. It is BV[$\mathfrak{g}$] that will play a central role in the description of the BV structure of $\mathcal{W}$ cohomology in the next chapter.

4.6.1 The Base Affine Space $A(G)$

Let G be a complex finite-dimensional Lie group and $\mathfrak{g}$ its Lie algebra. We assume that $\mathfrak{g}$ is simple and simply-laced with the generators e_A satisfying commutation relations

$$[e_A, e_B]_- = f_{AB}{}^C e_C \,. \tag{4.130}$$

In the following we will fix a Cartan decomposition $\mathfrak{g} \cong \mathfrak{n}_- \oplus \mathfrak{h} \oplus \mathfrak{n}_+ \cong \mathfrak{b}_- \oplus \mathfrak{n}_+$ and choose e_A as the Chevalley generators of $\mathfrak{g}$, denoted by $\{e_{-\alpha}, h_i, e_\alpha\}$, $i = 1, \ldots, \ell \equiv \text{rank}\, \mathfrak{g}$, $\alpha \in \Delta_+$.

The space $\mathcal{E}(G)$ of regular functions on G (i.e., polynomial functions in the matrix elements of $g \in G$) carries the left and right regular representation of G. Explicitly, we have for $f \in \mathcal{E}(G)$ and $g, g' \in G$

$$\Pi^L(g) \cdot f(g') = f(g^{-1}g'), \qquad \Pi^R(g) \cdot f(g') = f(g'g) \,. \tag{4.131}$$

We will denote by Π^L_A and Π^R_A the operators representing the corresponding actions of the generator e_A on $\mathcal{E}(G)$. They span two commuting algebras $\mathfrak{g}_L$ and $\mathfrak{g}_R$, respectively, both isomorphic with $\mathfrak{g}$. A classical result in representation theory is the decomposition of $\mathcal{E}(G)$ into finite-dimensional irreducible modules of $\mathfrak{g}_L \oplus \mathfrak{g}_R$.

Theorem 4.37 [Peter-Weyl] *For any finite-dimensional simple Lie group G, the decomposition of $\mathcal{E}(G)$ under the action of $\mathfrak{g}_L \oplus \mathfrak{g}_R$ is given by*

$$\mathcal{E}(G) \cong \bigoplus_{\Lambda \in P_+} \mathcal{L}(\Lambda^*) \otimes \mathcal{L}(\Lambda) \,. \tag{4.132}$$

Here $\mathcal{L}(\Lambda)$ and $\mathcal{L}(\Lambda^*)$ are finite-dimensional irreducible modules of $\mathfrak{g}$ with highest weights Λ and Λ^*, respectively, and $\Lambda^* = -w_0\Lambda$.

Let N_-, H, and N_+ be the complex subgroups of G generated by the subalgebras in the Cartan decomposition of $\mathfrak{g}$. In particular, for $G = SL(n, \mathbb{C})$ they are given by the lower triangular, diagonal, and upper triangular matrices, respectively. Following [BGG2], we define the base affine space of G as the quotient $A = N_+ \backslash G$. The space of regular functions on A, $\mathcal{E}(A)$, consisting of those functions in $\mathcal{E}(G)$ that are invariant under N_+^L, carries a representation of $\mathfrak{h}_L \oplus \mathfrak{g}_R$. So, from Theorem 4.37, we immediately conclude that

Theorem 4.38 [BGG2] *Under the action of $\mathfrak{h}_L \oplus \mathfrak{g}_R$ we have*

$$\mathcal{E}(A) \cong \bigoplus_{\Lambda \in P_+} \mathbb{C}_{\Lambda^*} \otimes \mathcal{L}(\Lambda) \,, \tag{4.133}$$

where $\mathbb{C}_{\Lambda^*}$ denotes the one-dimensional representation of $\mathfrak{h}_L$ with weight Λ^*. In other words, $\mathcal{E}(A)$ is a model space for $\mathfrak{g}$.

4.6.2 The Algebra $\mathcal{P}(A)$

Geometric objects on A can be studied effectively using standard techniques of induced representations. A good example of this is the description of $\mathcal{E}(A)$ given by Theorem 4.38. In a similar spirit, we will therefore define polyvector fields on A as regular sections of homogenous vector bundles over A, rather than, as would be more natural if we worked in the smooth category, through differential operators acting on $\mathcal{E}(A)$. In the case of vector fields, the equivalence of the two approaches follows immediately from the explicit construction of all differential operators on A in [BGG2,GeKi1,GeKi2]. We will illustrate this on examples in Appendix G.

Since $A = N_+\backslash G$, the tangent space to A at the origin is isomorphic with $\mathfrak{b}_- \cong \mathfrak{n}_+\backslash\mathfrak{g}$. Let[7] Π^{bc} denote the representation of $\mathfrak{n}_+$ on $\mathfrak{b}_-$ arising from the left action of N_+ on G, as well as its extension to

$$\bigwedge\mathfrak{b}_- \cong \bigoplus_{n=0}^{D} \bigwedge^n\mathfrak{b}_- , \qquad D = |\Delta_+| + \ell , \tag{4.134}$$

Definition 4.39 *The space $\mathcal{P}^n(A)$ of polyvectors of order n on A is the space of regular sections of the vector bundle $G \times_{N_+} \bigwedge^n\mathfrak{b}_-$.*

Let $\mathcal{E}(G, \bigwedge^n\mathfrak{b}_-)$ denote the space of regular functions on G with values in $\bigwedge^n\mathfrak{b}_-$. We recall that the total space of the bundle $G \times_{N_+} \bigwedge^n\mathfrak{b}_-$ consists of pairs (g,t), $g \in G$, $t \in \bigwedge^n\mathfrak{b}_-$, subject to an equivalence relation $(g,t) \sim (mg, \Pi^{bc}(m)t)$, $m \in N_+$. Then an n-vector field on A, defined as a section of this bundle, is given by a function $\Phi \in \mathcal{E}(G, \bigwedge^n\mathfrak{b}_-)$ such that

$$\Phi(mg) = \Pi^{bc}(m)\Phi(g) , \qquad m \in N_+ , \quad g \in G . \tag{4.135}$$

Or, in an infinitesimal form,

$$\Pi^L(x)\Phi(g) = -\Pi^{bc}(x)\Phi(g) , \qquad x \in \mathfrak{n}_+ , \quad g \in G . \tag{4.136}$$

Note that the polyvectors of order zero are simply identified with the regular functions on A.

Now, let us turn to the $\mathfrak{g}$ module structure of the polyvectors and a generalization of Theorem 4.38. The (right) action of G on A lifts to $\mathcal{P}(A)$ as $\Pi^R(g')\cdot\Phi(g) = \Phi(gg')$, $g, g' \in G$. When necessary we will write $\mathfrak{g}_R$ as above when talking about the corresponding right action of $\mathfrak{g}$. Since $[\mathfrak{h}, \mathfrak{n}_+]_- \subset \mathfrak{n}_+$, the constraint (4.136) is invariant under the action of $\mathfrak{h}$ (also called $\mathfrak{h}_L$) arising from the left regular action of $\mathfrak{h}$ on $\mathcal{E}(G)$ and the adjoint action on $\bigwedge\mathfrak{b}_-$. Moreover, $\mathfrak{g}_R$ and $\mathfrak{h}_L$ commute.

Theorem 4.40 *The space of n-vectors, $\mathcal{P}^n(A)$, is a completely reducible module of $\mathfrak{h}_L \oplus \mathfrak{g}_R$, with the decomposition given by*

[7] The notation here is motivated by the explicit realization of $\bigwedge\mathfrak{b}_-$ below.

$$\begin{aligned} \mathcal{P}^n(A) &\cong \bigoplus_{\Lambda\in P_+} \mathrm{Hom}_{\mathfrak{g}}(\mathcal{L}(\Lambda), \mathcal{P}^n(A)) \otimes \mathcal{L}(\Lambda) \\ &\cong \bigoplus_{\Lambda\in P_+} \mathrm{Hom}_{\mathfrak{n}_+}(\mathcal{L}(\Lambda), \textstyle\bigwedge^n \mathfrak{b}_-) \otimes \mathcal{L}(\Lambda), \end{aligned} \tag{4.137}$$

where $\mathfrak{h}$ and $\mathfrak{g}$ act on the first and the second factor in the tensor product, respectively.

The decomposition in (4.137) is a well known result called the Frobenius reciprocity. Its proof is outlined in Appendix G, where we also explicitly compute the space of polyvectors on the base affine space of $SL(3,\mathbb{C})$.

Using the isomorphism $\mathcal{E}(G, \bigwedge^n \mathfrak{b}_-) \cong \mathcal{E}(G) \otimes \bigwedge^n \mathfrak{b}_-$, the space of polyvectors may be characterized as the $\mathfrak{n}_+$ invariants in the tensor product. To make it more explicit, it is convenient to first introduce a realization of $\bigwedge \mathfrak{b}_-$ in terms of ghosts.

Consider the Clifford algebra of the vector space $\mathfrak{g} \oplus \mathfrak{g}'$, where $\mathfrak{g}'$ is the dual of $\mathfrak{g}$. In physicists' language this algebra is realized by the ghost operators $c(x)$, $x \in \mathfrak{g}$, and the antighost operators $b(x')$, $x' \in \mathfrak{g}'$. Let us set $c_A = c(e_A)$ and $b^A = b(e^A)$, where e_A and e^A are the dual bases of $\mathfrak{g}$ and $\mathfrak{g}'$, respectively. Then the anticommutation relations between the ghost and the antighost operators are

$$[c_A, c_B]_+ = 0, \quad [b^A, b^B]_+ = 0, \quad [c_A, b^B]_+ = \delta_A{}^B, \quad A, B = 1, \ldots, \dim \mathfrak{g}. \tag{4.138}$$

In the following we will distinguish between the ghosts/antighosts associated with $\mathfrak{b}_-$ and $\mathfrak{n}_+$. The former will always be denoted by c_a and b^a, where $a = (-\alpha, i)$ is the collective $\mathfrak{b}_-$ index, while the latter by $\omega_\alpha \equiv c_\alpha$ and $\sigma^\alpha \equiv b^\alpha$. This is partly to avoid possible confusion, but also to emphasize the different roles played by both sets of operators.

Let F^{bc} be the Fock space of the (c_a, b^a) ghosts, which is the module freely generated from the vacuum $|bc\rangle$ satisfying $b^a|bc\rangle = 0$. Clearly, we may identify $\bigwedge \mathfrak{b}_-$ with F^{bc}, where the $\mathfrak{n}_+$ action is given by[8]

$$\Pi_\alpha^{bc} = f_{\alpha b}{}^c \, c_c \, b^b . \tag{4.139}$$

In particular, this identification induces a graded commutative product on F^{bc}. Moreover, F^{bc} is also an $\mathfrak{h}$ module, where the $\mathfrak{h}$ generators are

$$\Pi_i^{bc} = f_{ib}{}^c \, c_c \, b^b . \tag{4.140}$$

Thus $\mathcal{E}(G) \otimes \bigwedge \mathfrak{b}_-$ has a natural structure of a graded, graded commutative algebra and carries commuting actions of $\mathfrak{g} \oplus \mathfrak{h}$ and $\mathfrak{n}_+$, defined by the operators Π_A^R and $\Pi_i^L + \Pi_i^{bc}$, and $\Pi_\alpha^L + \Pi_\alpha^{bc}$, respectively. Both $\mathfrak{g} \oplus \mathfrak{h}$ and $\mathfrak{n}_+$ act by derivations of the algebra product. In terms of $\mathcal{E}(G) \otimes \bigwedge \mathfrak{b}_-$ the polyvectors $\mathcal{P}(A)$ are simply given by the $\mathfrak{n}_+$-invariant elements, i.e.,

$$\mathcal{P}^n(A) \cong (\mathcal{E}(G) \otimes \textstyle\bigwedge^n \mathfrak{b}_-)^{\mathfrak{n}_+} . \tag{4.141}$$

---

[8] One should not confuse the label bc with the dummy indices b and c.

4.6.3 The Algebra BV[$\mathfrak{g}$]

This last formulation of $\mathcal{P}(A)$ given in (4.141) immediately suggests a rather beautiful generalization. The natural framework for determining invariants of group actions is Lie algebra cohomology, and in this more general context the polyvectors are obtained as the zeroth order cohomology of $\mathfrak{n}_+$ with coefficients in $\mathcal{E}(G) \otimes \bigwedge \mathfrak{b}_-$. One is thus naturally led to consider the algebra BV[$\mathfrak{g}$] defined by the *full* cohomology,

$$\mathrm{BV}[\mathfrak{g}] \equiv H(\mathfrak{n}_+, \mathcal{E}(G) \otimes \textstyle\bigwedge \mathfrak{b}_-) . \tag{4.142}$$

We will now examine BV[$\mathfrak{g}$] more closely, using this as an opportunity to introduce further notation and to derive some elementary results. Consider the ghost Fock space $F^{\sigma\omega}$ with vacuum $|\sigma\omega\rangle$ satisfying $\omega_\alpha|\sigma\omega\rangle = 0$. The $\mathfrak{n}_+$ action on $F^{\sigma\omega}$ is given by

$$\Pi_\alpha^{\sigma\omega} = -f_{\alpha\beta}{}^\gamma \sigma^\beta \omega_\gamma , \tag{4.143}$$

and the $\mathfrak{h}$ action by

$$\Pi_i^{\sigma\omega} = -f_{i\alpha}{}^\alpha \sigma^\alpha \omega_\alpha . \tag{4.144}$$

Again, there is a natural graded commutative product on $F^{\sigma\omega}$.

Definition 4.41 *The cohomology $H(\mathfrak{n}_+, \mathcal{E}(G) \otimes \bigwedge \mathfrak{b}_-)$ is defined as the cohomology of the differential*

$$d = \sigma^\alpha \left(\Pi_\alpha^L + \Pi_\alpha^{bc} + \tfrac{1}{2} \Pi_\alpha^{\sigma\omega} \right) \tag{4.145}$$

acting on the complex $\mathcal{C}(G) \equiv \mathcal{E}(G) \otimes F^{bc} \otimes F^{\sigma\omega}$. It will also be denoted by $H_d(\mathcal{C}(G))$.

The complex $\mathcal{C}(G)$ is bi-graded by the bc ghost and the $\sigma\omega$ ghost numbers

$$\mathrm{gh}(c_a) = -\mathrm{gh}(b^a) = (0,1) , \qquad \mathrm{gh}(\sigma^\alpha) = -\mathrm{gh}(\omega_\alpha) = (1,0) , \tag{4.146}$$

with d of degree $(1,0)$. Clearly, this bi-degree passes to the cohomology. We will write $H^n(\mathfrak{n}_+, \mathcal{E}(G) \otimes \bigwedge \mathfrak{b}_-)$ for the cohomology in *total* ghost number n. Similarly, we will write $\mathrm{BV}^n[\mathfrak{g}]$. When it is important to distinguish the bi-grading, we will write, e.g., $\mathrm{BV}^{(m,n)}[\mathfrak{g}]$.

Combining the above discussions, the complex $\mathcal{C}(G)$ is a $\mathfrak{g} \oplus \mathfrak{h}$ module under Π_A^R and

$$\Pi_i^{\mathrm{tot}} \equiv \Pi_i^L + \Pi_i^{bc} + \Pi_i^{\sigma\omega} . \tag{4.147}$$

Note that with respect to this $\mathfrak{h}$ action, the weights of $b^{-\alpha}$ and ω_α are α, whilst those of σ^α and $c_{-\alpha}$ are $-\alpha$. Since d commutes both with the action of $\mathfrak{g}$ and $\mathfrak{h}$, we have a direct sum decomposition

$$\mathrm{BV}[\mathfrak{g}] = \bigoplus_{\Lambda \in P_+} H(\mathfrak{n}_+, \mathcal{L}(\Lambda^*) \otimes \textstyle\bigwedge \mathfrak{b}_-) \otimes \mathcal{L}(\Lambda) . \tag{4.148}$$

As an $\mathfrak{h}$ module,

$$H(\mathfrak{n}_+, \mathcal{L}(\Lambda) \otimes \textstyle\bigwedge \mathfrak{b}_-) \cong \displaystyle\bigoplus_{\lambda \in P(\mathcal{C}(\Lambda))} H(\mathfrak{n}_+, \mathcal{L}(\Lambda) \otimes \textstyle\bigwedge \mathfrak{b}_-)_\lambda \,, \tag{4.149}$$

where, obviously, $\mathcal{C}(\Lambda) = \mathcal{L}(\Lambda) \otimes F^{bc} \otimes F^{\sigma\omega}$, and $P(V)$ denotes the set of weights of an $\mathfrak{h}$ module V.

The decomposition (4.148) reduces the problem of computing BV[$\mathfrak{g}$] to that of computing the cohomology of finite-dimensional modules.

Theorem 4.42 *Let $\Lambda \in P_+$*

i. The cohomology $H^n(\mathfrak{n}_+, \mathcal{L}(\Lambda) \otimes \bigwedge \mathfrak{b}_-)_{\Lambda'}$ is nontrivial only if there exists a $w \in W$ and $\lambda \in P(\bigwedge^k \mathfrak{b}_-)$ such that

$$\Lambda' = w(\Lambda + \rho) - \rho + \lambda \,, \tag{4.150}$$

and $n = \ell(w) + k$.

ii. For $\Lambda \in P_+$ in the bulk, i.e. $(\Lambda, \alpha_i) \geq N(\mathfrak{g})$ for some $N(\mathfrak{g}) \in \mathbb{N}$ sufficiently large (in particular $N(\mathfrak{sl}_n) = n - 1$), we have

$$\begin{aligned} H(\mathfrak{n}_+, \mathcal{L}(\Lambda) \otimes \textstyle\bigwedge \mathfrak{b}_-) &\cong H(\mathfrak{n}_+, \mathcal{L}(\Lambda)) \otimes \textstyle\bigwedge \mathfrak{b}_- \\ &\cong \left(\displaystyle\bigoplus_{w \in W} \mathbb{C}_{w(\Lambda+\rho)-\rho} \right) \otimes \textstyle\bigwedge \mathfrak{b}_- \,. \end{aligned} \tag{4.151}$$

Proof. Consider the gradation of the complex $\mathcal{C} = \bigoplus_{k \in \mathbb{Z}} \mathcal{C}_k$ given by (ρ, λ^{bc}), where λ^{bc} is the weight corresponding to Π_i^{bc}. With respect to this gradation the differential (4.145) decomposes as

$$d = \sum_{k \geq 0} d_k \,, \tag{4.152}$$

with

$$\begin{aligned} d_0 &= \sigma^\alpha \left(\Pi_\alpha^L + \tfrac{1}{2} \Pi_\alpha^{\sigma\omega} \right) \,, \\ d_k &= \sum_{(\rho,\alpha)=k} \sigma^\alpha \Pi_\alpha^{bc} \,, \qquad k \geq 1 \,. \end{aligned} \tag{4.153}$$

In particular, $d_k \equiv 0$ for $k \geq (\rho, \theta) + 1 = \mathrm{h}^\vee$, where $\mathrm{h}^\vee$ is the dual Coxeter number of $\mathfrak{g}$ (for $\mathfrak{g} = \mathfrak{sl}_n$ we have $\mathrm{h}^\vee = n$). The spectral sequence (E_k, δ_k) corresponding to this gradation converges since the complex is finite-dimensional. The first term in the spectral sequence is given by

$$E_1 = H_{d_0}(\mathcal{C}) \cong H(\mathfrak{n}_+, \mathcal{L}(\Lambda)) \otimes \textstyle\bigwedge \mathfrak{b}_- \,, \tag{4.154}$$

where [Bt,Kt]

$$H(\mathfrak{n}_+, \mathcal{L}(\Lambda)) \cong \bigoplus_{w \in W} \mathbb{C}_{w(\Lambda+\rho)-\rho} \,. \tag{4.155}$$

This proves the first part of the theorem. To prove the second part we will now determine the condition for the spectral sequence to collapse at the first term.

The differential δ_1 on E_1 is simply given by d_1, so we find that a sufficient condition for δ_1 to act trivially is that there exist no $w, w' \in W$, $\ell(w') = \ell(w)+1$, $\lambda, \lambda' \in P(\bigwedge \mathfrak{b}_-)$ such that

$$w(\Lambda+\rho) - w'(\Lambda+\rho) \;=\; \lambda' - \lambda \;=\; \alpha_i\,, \tag{4.156}$$

for some i. Similarly, we find that a sufficient condition for δ_k to act trivially on E_k is that there exist no $w, w' \in W$, $\ell(w') = \ell(w)+1$, $\lambda, \lambda' \in P(\bigwedge \mathfrak{b}_-)$ such that

$$w(\Lambda+\rho) - w'(\Lambda+\rho) \;=\; \lambda' - \lambda \;\in\; \{\beta \in Q_+ \,|\, (\rho,\beta) \le k\}\,. \tag{4.157}$$

So, since $\delta_k \equiv 0$ for $k \ge \mathrm{h}^\vee$, we find that a sufficient condition for $E_\infty \cong \ldots \cong E_2 \cong E_1$ is that there exist no $w, w' \in W$, $\ell(w') = \ell(w)+1$, $\lambda, \lambda' \in P(\bigwedge \mathfrak{b}_-)$ such that

$$w(\Lambda+\rho) - w'(\Lambda+\rho) \;=\; \lambda' - \lambda \;\in\; \{\beta \in Q_+ \,|\, (\rho,\beta) \le \mathrm{h}^\vee - 1\}\,. \tag{4.158}$$

This condition is met when Λ is sufficiently deep inside the fundamental Weyl chamber, i.e., $(\Lambda, \alpha_i) \ge N(\mathfrak{g})$ for some $N(\mathfrak{g}) \in \mathbb{N}$ sufficiently large. □

Remark. The first part of the theorem is a necessary condition that holds for all weights $\Lambda \in P_+$. In particular (4.150) restricts the $\mathfrak{g}$ and $\mathfrak{h}$ weights that may arise in $\mathrm{BV}[\mathfrak{g}]$.

In the cases where Λ lies close to the boundary of P_+ the cohomology is a proper subspace of $H(\mathfrak{n}_+, \mathcal{L}(\Lambda)) \otimes \bigwedge \mathfrak{b}_-$. For $\mathfrak{g} = \mathfrak{sl}_2$ and $\mathfrak{sl}_3$ the complete result is calculated in Appendix I. For later convenience we present here the result in the case $\mathfrak{g} = \mathfrak{sl}_3$ in a slightly different convention, i.e., by letting $w \to w^{-1}$ and $\sigma \to w^{-1}\sigma w_0$ in the results of Appendix I. Note that under this substitution

$$\begin{aligned} \Lambda' + 2\rho \;&=\; \left(w^{-1}(\Lambda+\rho) - \rho\right) + w^{-1}\sigma w_0 \rho - \rho + 2\rho \\ &=\; w^{-1}(\Lambda + \rho - \sigma\rho)\,, \end{aligned} \tag{4.159}$$

and

$$n \;=\; \ell(w^{-1}) + \ell(w^{-1}\sigma w_0) \;=\; 3 - \ell_w(\sigma)\,, \tag{4.160}$$

where we use $w\sigma_i = \sigma_i$, $w \in W$, $i = 1, 2$ (cf. Sect. 2.4.1), and $\sigma_1 w_0 = \sigma_2$, $\sigma_2 w_0 = \sigma_1$ to evaluate the products involving elements σ_1 and σ_2.

Theorem 4.43 *For $\mathfrak{g} \cong \mathfrak{sl}_3$, the cohomology $H(\mathfrak{n}_+, \mathcal{L}(\Lambda) \otimes \bigwedge \mathfrak{b}_-)_{\Lambda'}$ is nontrivial only if there exists a $w \in W$ and $\sigma \in \widetilde{W}$ such that $\Lambda' + 2\rho = w^{-1}(\Lambda + \rho - \sigma\rho)$. The set of of allowed pairs (w, σ) depends on Λ and is given in Table 4.6. For each allowed pair (w, σ) there is a quartet of cohomology states at ghost numbers $n, n+1, n+1$ and $n+2$ where $n = 3 - \ell_w(\sigma)$.*

Table 4.6. Condition on Λ for the pair (w, σ) to be allowed ($m_i = (\Lambda, \alpha_i)$ and "–" means there is no condition on $\Lambda \in P_+$).

$w\backslash\sigma$	1	r_1	r_2	σ_1	r_1r_2	r_2r_1	σ_2	$r_1r_2r_1$
1	$m_1 \geq 1$, $m_2 \geq 1$	$m_2 \geq 2$	$m_1 \geq 2$	–	$m_2 \geq 1$	$m_1 \geq 1$	$m_1 \geq 1$, $m_2 \geq 1$	–
r_1	$m_2 \geq 1$	–	$m_1 \geq 1$	$m_1 \geq 1$	$m_2 \geq 1$	–	$m_2 \geq 1$	–
r_2	$m_1 \geq 1$	$m_2 \geq 1$	–	$m_2 \geq 1$	–	$m_1 \geq 1$	$m_1 \geq 1$	–
r_1r_2	$m_2 \geq 1$	$m_2 \geq 1$	–	$m_1 \geq 1$	–	–	$m_2 \geq 1$	–
r_2r_1	$m_1 \geq 1$	–	$m_1 \geq 1$	$m_2 \geq 1$	–	–	$m_1 \geq 1$	–
$r_1r_2r_1$	–	$m_2 \geq 2$	$m_1 \geq 2$	$m_1 \geq 1$, $m_2 \geq 1$	$m_2 \geq 1$	$m_1 \geq 1$	–	–

4.6.4 BV Algebra Structures on $\mathcal{P}(A)$ and BV[$\mathfrak{g}$]

Now, let us turn to the structure of the algebras BV[$\mathfrak{g}$] and $\mathcal{P}(A)$.

Lemma 4.44 BV[$\mathfrak{g}$] *is a graded, graded commutative algebra with the product "$\cdot$" induced from the product on the underlying complex $\mathcal{C}(G)$.*

Proof. Let $|0\rangle = |bc\rangle \otimes |\sigma\omega\rangle$. Then $\Phi \in \mathcal{C}(G)$ is of the form

$$\Phi = \Phi^{a_1 \ldots a_m}{}_{\alpha_1 \ldots \alpha_n} \sigma^{\alpha_1} \ldots \sigma^{\alpha_n} c_{a_1} \ldots c_{a_m} |0\rangle\,, \qquad \Phi^{a_1 \ldots a_m}{}_{\alpha_1 \ldots \alpha_n} \in \mathcal{E}(G)\,. \tag{4.161}$$

The product of two such element is thus given by

$$\Phi \cdot \Psi = (-1)^{np}\, \Phi^{a_1 \ldots a_m}{}_{\alpha_1 \ldots \alpha_n} \Psi^{b_1 \ldots b_p}{}_{\beta_1 \ldots \beta_q} \sigma^{\alpha_1} \ldots \sigma^{\beta_q} c_{a_1} \ldots c_{b_p} |0\rangle\,. \tag{4.162}$$

We verify that

$$d(\Phi \cdot \Psi) = d\Phi \cdot \Psi + (-1)^{m+n} \Phi \cdot d\Psi\,, \tag{4.163}$$

from which it follows immediately that the product passes to the cohomology. Obviously, it is graded commutative according to the total ghost number. □

The algebra BV[$\mathfrak{g}$] contains the algebra of polyvectors $\mathcal{P}(A)$ as a subalgebra,

$$\mathcal{P}^m(A) \cong BV^{(0,m)}[\mathfrak{g}]\,. \tag{4.164}$$

Let $\imath : \mathcal{P}(A) \longrightarrow$ BV[$\mathfrak{g}$] be the corresponding embedding. From now on we will identify $\mathcal{P}(A)$ with its image in BV[$\mathfrak{g}$]. In particular an element $\Phi \in \mathcal{C}(G)$ is a polyvector provided

$$\omega_\alpha \Phi = 0\,, \quad \Pi^{\rm tot}_\alpha \Phi = 0\,, \qquad \alpha \in \Delta_+\,, \tag{4.165}$$

where

$$\Pi^{\rm tot}_\alpha = [d, \omega_\alpha]_+ = \Pi^L_\alpha + \Pi^{bc}_\alpha + \Pi^{\sigma\omega}_\alpha\,. \tag{4.166}$$

Clearly, the second condition in (4.165) is required by compatibility of the first one with the cohomology.

It will be of principal importance that, as the notation suggests, BV[$\mathfrak{g}$] allows the introduction of a BV algebra structure – and, in fact, one for which the BV operator has trivial homology. This will be demonstrated in two steps. First, we will show that there is a natural BV operator on BV[$\mathfrak{g}$] which preserves the space of polyvectors, but has nontrivial homology. Secondly, we will construct a deformation of that "naive" BV operator such that its homology becomes trivial.

Theorem 4.45 *Consider the operator*

$$\Delta_0 = -b^a(\Pi^L_a + \Pi^{\sigma\omega}_a) + \tfrac{1}{2} f_{ab}{}^c b^a b^b c_c , \tag{4.167}$$

where $\Pi^{\sigma\omega}_a = -f_{a\alpha}{}^{\beta} b^a \sigma^\alpha \omega_\beta$. *Then*

i. $[d, \Delta_0]_+ = 0$,
ii. Δ_0 *is a BV operator on* BV[$\mathfrak{g}$],
iii. $\Delta_0 \imath(\mathcal{P}(A)) \subset \imath(\mathcal{P}(A))$ *and thus* Δ_0 *is a BV operator on* $\imath(\mathcal{P}(A))$.

Remark. We write $\Delta_0\big|_{\imath(\mathcal{P}(A))} = \Delta'_0$.

Proof. Note that $-\Delta_0$ is the differential of $\mathfrak{b}_-$ homology with coefficients in $\mathcal{E}(G) \otimes F^{\sigma\omega}$. Thus $\Delta_0^2 = 0$, a fact easily verified by explicit algebra. Moreover, if we combine the $\mathfrak{b}_-$ and $\mathfrak{n}_+$ ghosts to $\mathfrak{g}$ ghosts, i.e., set $c_A = \{c_a, \omega_\alpha\}$ and $b^A = \{b^a, \sigma^\alpha\}$ then

$$\delta = d - \Delta_0 = b^A \Pi^L_A - \tfrac{1}{2} f_{AB}{}^C b^A b^B c_C , \tag{4.168}$$

is the differential of a twisted cohomology of $\mathfrak{g}$ with coefficients in $\mathcal{E}(G)$. Thus $\delta^2 = 0$ and the assertion (i) follows.

The second order derivation property of Δ_0 is shown as follows: The first term in Δ_0 is a product of first order derivations $\Pi^L_a + \Pi^{\sigma\omega}_a$ on $\mathcal{E}(G) \otimes F^{\sigma\omega}$ and b^a on F^{bc}. The product of such first order derivations acting on the tensor product of spaces is well known to be a second order derivation. Upon normal ordering of the bc ghosts, the second term – with the nontrivial action on F^{bc} only – becomes a sum of terms with one or a product of two b^a's, which are first and second order derivations, respectively. Since we have already shown that $\Delta_0^2 = 0$, this proves (ii).

Part (iii) follows from the observation that on $\imath(\mathcal{P}(A))$ the only term involving the $\sigma\omega$ ghosts, i.e., $b^a \Pi^{\sigma\omega}_a$, vanishes. Then Δ_0 reduces to

$$\Delta'_0 = -b^a \Pi^L_a + \tfrac{1}{2} f_{ab}{}^c b^a b^b c_c , \tag{4.169}$$

and clearly preserves $\imath(\mathcal{P}(A))$. □

Remarks

i. The differential δ defines an equivariant version of the twisted homology introduced in [FeFr1] as an analogue of the semi-infinite homology in the category of finite-dimensional Lie algebras. Here, that semi-infinite character is determined by the choice of the ghost Fock space F.

ii. Since b^{α} acts like a multiplication, rather than a derivation, on F, the full operator δ is not a second order derivation on $\mathcal{E}(G) \otimes F^{\mathrm{gh}}$.

iii. In Appendix G we show that the bracket induced by Δ_0' on $\mathcal{P}(A)$ coincides with the Schouten bracket on polyvectors. Upon the identification $\mathcal{P}(A(SL(3,\mathbb{C}))) \cong \mathcal{P}(\mathcal{R}_3)$ we also find that $\Delta_0' = -\Delta_S$ (cf. Theorem G.6).

We will now seek a deformation of Δ_0 of the form $\Delta = \Delta_0 + \Delta_1$, such that Δ is a BV operator on BV[$\mathfrak{g}$], and commutes with $\mathfrak{g} \oplus \mathfrak{h}$. In particular this requires that $[d, \Delta_1]_+ = 0$, which implies that Δ_1 must be a nontrivial element in the "operator cohomology" of d on $\mathcal{U}(\mathfrak{g}) \otimes \mathcal{U}(bc) \otimes \mathcal{U}(\sigma\omega)$. Here $\mathcal{U}(\cdot)$ denotes the enveloping algebra and d acts by the commutator. Since computing this cohomology is difficult, we make the further simplification that Δ_1 is an element of $\mathcal{U}(bc) \otimes \mathcal{U}(\sigma\omega)$. This assumption is motivated by the naive expectation that the only second order derivation that has degree -1 and acts nontrivially on the $\mathcal{E}(G)$ component in BV[$\mathfrak{g}$] must be of the form $b^a \Pi_a^L$, and thus is already accounted for in Δ_0. Then we have

Theorem 4.46 *Let $\mathfrak{g}$ be a simple, simply laced Lie algebra and $\varepsilon : Q \times Q \longrightarrow \{\pm 1\}$ its asymmetry function with respect to the chosen Chevalley basis. Define*

$$\Delta' = \sum_{\alpha \in \Delta_+} \sum_{i=1}^{\ell} (\alpha, \alpha_i)\, \sigma^{\alpha} b^{-\alpha} b^i - \sum_{\alpha,\beta \in \Delta_+} \varepsilon(\alpha,\beta) \sigma^{\alpha+\beta} b^{-\alpha} b^{-\beta} . \qquad (4.170)$$

Then $\Delta_t = \Delta_0 + t\Delta'$ is a one parameter family of BV operators on BV[$\mathfrak{g}$].

Remark. We recall that the asymetry function for a simply-laced $\mathfrak{g}$ is defined by the following properties

$$\varepsilon(\alpha,0) = \varepsilon(0,\alpha) = \varepsilon(\alpha,\alpha) = 1, \quad \varepsilon(\alpha,\beta)\varepsilon(\beta,\alpha) = (-1)^{(\alpha,\beta)}, \qquad (4.171)$$

and

$$\varepsilon(\alpha+\beta,\gamma) = \varepsilon(\alpha,\gamma)\varepsilon(\beta,\gamma), \quad \varepsilon(\alpha,\beta+\gamma) = \varepsilon(\alpha,\beta)\varepsilon(\alpha,\gamma), \qquad (4.172)$$

where $\alpha, \beta, \gamma \in \Delta$. Moreover, the structure constants of $\mathfrak{g}$ are given by

$$f_{\alpha\beta}{}^{\gamma} = \epsilon(\alpha,\beta), \quad f_{\alpha,-\alpha}{}^{i} = -(\alpha, \Lambda_i), \quad f_{i,\alpha}{}^{\alpha} = -f_{i,-\alpha}{}^{-\alpha} = (\alpha,\alpha_i) . \qquad (4.173)$$

Proof. Obviously Δ' is a second order derivation and satisfies $(\Delta')^2 = 0$. Hence we must only show $[\Delta_0, \Delta']_+ = 0$. Upon evaluating the commutator and using (4.172)–(4.173), we find terms with the following ghost structures:

(i) $\sigma^{\alpha+\beta+\gamma} b^{-\alpha} b^{-\beta} b^{-\gamma}$,
(ii) $\sigma^{\alpha+\beta} b^{-\alpha} b^{-\beta} b^{i}$,
(iii) $\sigma^{\alpha} b^{-\alpha} b^{i} b^{j}$.

In case (i) the coefficient, upon antisymetrization with respect to α, β and γ, is proportional to

$$\begin{aligned}\varepsilon(\alpha,\beta)\epsilon(\beta,\gamma)\epsilon(\gamma,\alpha) &- \varepsilon(\beta,\alpha)\epsilon(\alpha,\gamma)\epsilon(\gamma,\beta)\\ &= \left(1-(-1)^{(\alpha,\beta)+(\beta,\gamma)+(\gamma,\alpha)}\right)\varepsilon(\alpha,\beta)\epsilon(\beta,\gamma)\epsilon(\gamma,\alpha)\,.\end{aligned} \tag{4.174}$$

Note that those terms may arise only if $\alpha+\beta+\gamma$ is a root, which in particular implies $(\alpha,\beta)+(\beta,\gamma)+(\gamma,\alpha)=2$, and thus vanishing of (4.174).

In case (ii) we must have $(\alpha,\beta)=-1$ because $\alpha+\beta$ is a root. In that case the coefficient

$$(\alpha+\beta,\alpha_i)\epsilon(\alpha,\beta)+[\,(\beta,\alpha_i)\epsilon(\alpha,\alpha+\beta)-(\alpha\leftrightarrow\beta)]\,,$$

vanishes.

Finally, in case (iii) the coefficient is proportional to $(\alpha,\alpha_i)(\alpha,\alpha_j)$ and yields zero upon antisymmetrization on ij. □

Remark. We have verified that for $\mathfrak{g}=\mathfrak{sl}_2$ and $\mathfrak{sl}_3$ the generator Δ' of the one parameter family of deformations of Δ_0 is uniquely determined by requiring that it be a nontrivial element of the $\mathfrak{n}_+$ cohomology at ghost number -1. It is an interesting open problem to determine whether this also true for an arbitrary $\mathfrak{g}$.

A simple scaling argument shows that in fact all BV algebra structures on BV[$\mathfrak{g}$] for $t\neq 0$ are equivalent. Indeed, if we let $\sigma^\alpha\to\lambda\sigma^\alpha$, $\omega_\alpha\to\lambda^{-1}\omega_\alpha$ then $d\to\lambda d$, $\Delta_0\to\Delta_0$ while $\Delta'\to\lambda\Delta'$. Since the cohomology classes must scale homogenously with respect to this transformation, we may use it to set $t=1$, and denote $\Delta=\Delta_{t=1}$.

Finally, in Appendix G we show that the homology of Δ_0 on $\mathcal{P}(A)$ is nontrivial and spanned by the volume element $\prod_a c_a|bc\rangle$. To study the homology of Δ_0 and Δ on BV[$\mathfrak{g}$] we proceed similarly.

Lemma 4.47 *The operators Δ_0 and Δ may have nontrivial homologies on* BV[$\mathfrak{g}$] *at most in the subspace of $\mathfrak{g}$ singlets with the $\mathfrak{h}$ weight equal -2ρ.*

Proof. First, we note that c_i, $i=1,\ldots,\ell$ are well defined operators on BV[$\mathfrak{g}$]. This follows from $[d,c_i]_+=0$. Secondly,

$$[\Delta_t,c_i]_+ \;=\; M_i+tN_i\,, \tag{4.175}$$

where

$$M_i \;=\; -\Pi_i^{\rm tot}-(\alpha_i,\rho)\mathbf{1}\,, \tag{4.176}$$

is a diagonal operator while

$$N_i \;=\; \sum_{\alpha\in\Delta_+}(\alpha_i,\alpha)\sigma^\alpha b^{-\alpha}\,, \tag{4.177}$$

a nilpotent operator on BV[$\mathfrak{g}$]. Moreover $[M_i, N_i]_- = 0$. Thus, on the subspace where M_i does not vanish, the operator on the right hand side in (4.175) is invertible and its inverse, $(M_i+tN_i)^{-1}$, commutes with Δ_t. Then $(M_i+tN_i)^{-1}c_i$ is a contracting homotopy for Δ_t. We note (cf. Appendix I) that the $\mathfrak{h}$ weights of states in $\bigwedge\mathfrak{n}_-$ are of the form $\sigma\rho - \rho$, where $\sigma \in \widetilde{W} \equiv W \cup \{\sigma_1, \sigma_2\}$. Together with (4.150) this gives the following condition for the vanishing of the M_i's, $i = 1, \ldots, \ell$,

$$w(\Lambda + \rho) = \sigma\rho\,, \tag{4.178}$$

where $\Lambda \in P_+$, $w \in W$ and $\sigma \in \widetilde{W}$. Clearly, the only solution to this equation is $\Lambda = 0$ and $w = \sigma$. □

In the two cases $\mathfrak{g} = \mathfrak{sl}_2$ and $\mathfrak{sl}_3$ where BV[$\mathfrak{g}$] has been computed, we may determine the homology of Δ_0 and Δ by explicitly evaluating the action of Δ_t on BV$[\mathfrak{g}]_{-2\rho}$. In both cases, as we show explicitly below, the homology of Δ is trivial.

Example. The subspace BV$[\mathfrak{sl}_2]_{-2\rho}$ is spanned by

$$v_1 = \sigma^\alpha|0\rangle\,, \qquad v_2 = c_1 c_{-\alpha}|0\rangle\,, \tag{4.179}$$

and we have

$$\Delta_t v_1 = 0\,, \qquad \Delta_t v_2 = 2tv_1\,. \tag{4.180}$$

This shows that the homology of Δ_0 is spanned by v_1 and v_2, while that of Δ is trivial.

Example. BV$[\mathfrak{sl}_3]_{-2\rho}$ is spanned by

$$\begin{aligned}
v_3 &= 2\sigma^{\alpha_1}\sigma^{\alpha_2}\sigma^{\alpha_3}|0\rangle\,,\\
v_4 &= (\sigma^{\alpha_1}\sigma^{\alpha_2}c_{-\alpha_1}c_{-\alpha_2} - \sigma^{\alpha_1}\sigma^{\alpha_3}c_{-\alpha_2}c_1 + \sigma^{\alpha_2}\sigma^{\alpha_3}c_{-\alpha_1}c_2)|0\rangle\,,\\
\widetilde{v}_4 &= (\sigma^{\alpha_3}c_{-\alpha_3}c_1c_2 - \sigma^{\alpha_1}c_{-\alpha_2}c_{-\alpha_3}c_2 - \sigma^{\alpha_2}c_{-\alpha_1}c_{-\alpha_3}c_1\\
&\quad - \tfrac{1}{2}\sigma^{\alpha_1}c_{-\alpha_2}c_{-\alpha_3}c_1 - \tfrac{1}{2}\sigma^{\alpha_2}c_{-\alpha_1}c_{-\alpha_3}c_2\\
&\quad - \tfrac{1}{2}\sigma^{\alpha_3}c_{-\alpha_1}c_{-\alpha_2}c_2 + \tfrac{1}{2}\sigma^{\alpha_3}c_{-\alpha_1}c_{-\alpha_2}c_1)|0\rangle\,,\\
v_5 &= c_{-\alpha_1}c_{-\alpha_2}c_{-\alpha_3}c_1c_2|0\rangle\,.
\end{aligned} \tag{4.181}$$

A somewhat lengthy algebra yields

$$\Delta_t v_5 = 2t\widetilde{v}_4\,, \qquad \Delta_t v_4 = 2tv_3\,, \tag{4.182}$$

which shows that Δ_0 vanishes on BV$[\mathfrak{sl}_3]_{-2\rho}$, and thus the entire space comprises its homology, while the homology of Δ is trivial.

These explicit examples are actually those which are of principal interest in this book. Based on them, we feel confident in conjecturing that (under the previous restrictions) for all $\mathfrak{g}$, BV[$\mathfrak{g}$] is acyclic with respect to Δ.

4.6.5 The Description of BV[$\mathfrak{g}$] as a $\mathcal{P}(A)$ Module

The total ghost number zero subspace $\mathcal{R}[\mathfrak{g}] \equiv \mathrm{BV}^{(0,0)}[\mathfrak{g}]$ is an Abelian algebra which we call the ground ring. We will now describe the structure of BV[$\mathfrak{g}$] in terms of algebras of polyderivations that are associated with an Abelian algebra as discussed in Sects. 4.1.4, 4.1.4 and 4.4.1. In particular, we will generalize the notion of twisted modules of a ground ring from Sect. 4.5.2.

By the earlier discussion we have

$$\mathcal{R}[\mathfrak{g}] \cong \mathcal{P}^0(A) \cong \mathcal{E}(A)\,. \tag{4.183}$$

At this point we have two natural BV algebras associated with $\mathcal{R}[\mathfrak{g}]$:

(i) $(\mathcal{P}(\mathcal{A}),\,\cdot\,,\Delta_0)$ (see Appendix G)
(ii) $(\mathcal{P}(\mathcal{R}[\mathfrak{g}]),\,\cdot\,,\Delta_S)$ (see Sect. 4.1.4).

In Sect. 1.3.3 and Appendix G (Theorem G.6) we have shown that for $\mathfrak{g} = \mathfrak{sl}_2$ and $\mathfrak{sl}_3$, where $\mathcal{R}[\mathfrak{sl}_2] \cong \mathcal{C}_2$ and $\mathcal{R}[\mathfrak{sl}_3] \cong \mathcal{R}_3$, respectively, the algebra of polyvectors in (i) is isomorphic with the algebra of polyderivations in (ii). It is reasonable to expect that this result should also hold for a general $\mathfrak{g}$, i.e., $\mathcal{P}(\mathcal{R}[\mathfrak{g}]) \cong \mathcal{P}(A)$.

We may now consider the general construction in Sect. 4.1.4 of a G algebra homomorphism, π, from a given G algebra to the polyderivations of its ground ring. In the present context we have the BV algebra (BV[$\mathfrak{g}$], $\cdot$, Δ), with G algebra bracket induced by (4.5) from Δ given in Sect. 4.6.4. As is clear from the form of Δ, BV[$\mathfrak{g}$] is a G algebra only under grading by *total* ghost number.

Theorem 4.48 *Let* $\pi : \mathrm{BV}[\mathfrak{g}] \to \mathcal{P}(\mathcal{R}[\mathfrak{g}])$ *be the homomorphism defined in Sect. 4.1.4. Upon identification* $\mathcal{P}(\mathcal{R}[\mathfrak{g}]) \cong \mathcal{P}(A)$, *the homomorphism* π *becomes the projection onto the* $\sigma\omega$ *ghost number zero component, i.e.,*

$$\pi(\Phi) = \delta^{n0}\Phi^{a_1\ldots a_m} c_{a_1}\ldots c_{a_m}|0\rangle\,, \tag{4.184}$$

where $\Phi \in \mathrm{BV}[\mathfrak{g}]$ *is given in (4.161).*

Proof. Let $x \in \mathcal{P}(A)$. Since $\Delta x = 0$, (4.5) gives for $[x,\Phi]$ just two terms with a relative minus sign. The pure ghost terms of Δ then cancel, and we are left with

$$\begin{aligned}[x,\Phi] &= (-1)^n\sigma^{\alpha_1}\ldots\sigma^{\alpha_n}\big[\,\Pi^L_a(\Phi^{a_1\ldots a_m}{}_{\alpha_1\ldots\alpha_n}\,x) \\ &\qquad - (\Pi^L_a\Phi^{a_1\ldots a_m}{}_{\alpha_1\ldots\alpha_n})\,x\,\big]\,b^a c_{a_1}\ldots c_{a_m}|0\rangle \\ &= m(-1)^{m+n-1}\sigma^{\alpha_1}\ldots\sigma^{\alpha_n}\Phi^{a_1\ldots a_m}{}_{\alpha_1\ldots\alpha_n}(\Pi^L_{a_m}x)\,c_{a_1}\ldots c_{a_{m-1}}|0\rangle\,,\end{aligned} \tag{4.185}$$

since Π^L_a acts as a derivation. By iterating this formula we find the general form of the bracket action of BV[$\mathfrak{g}$] on its ground ring. In particular, the multiple bracket $[x_{i_1},\ldots,[x_{i_k},\Phi]\ldots]$, $x_{i_1},\ldots,x_{i_k} \in \mathcal{E}(A)$, vanishes for $k > m$, while for $k = m$ we have

$$\begin{aligned}[x_{i_1},\ldots,[x_{i_k},\Phi]\ldots] &= \Phi^{a_1\ldots a_m}{}_{\alpha_1\ldots\alpha_n}\sigma^{\alpha_1}\ldots\sigma^{\alpha_n} \\ &\qquad\times\,\Pi^L_{a_1}\wedge\ldots\wedge\Pi^L_{a_m}(x_{i_1}\otimes\ldots\otimes x_{i_m})\,.\end{aligned} \tag{4.186}$$

The theorem now follows by setting $k = m + n$, which yields a nonvanishing result only for $n = 0$, and recalling the identification between polyvectors $\mathcal{P}^n(A)$ and differential operators on $\mathcal{E}(A)^{\otimes n}$

$$\Phi^{a_1 \ldots a_n} c_{a_1} \ldots c_{a_n} |0\rangle \quad \longleftrightarrow \quad \Phi^{a_1 \ldots a_n} \Pi^L_{a_1} \wedge \ldots \wedge \Pi^L_{a_n}\,, \quad \Phi^{a_1 \ldots a_n} \in \mathcal{E}(G)\,, \tag{4.187}$$

which is discussed in Appendix G. □

We have also seen, in Sect. 4.6, that Δ reduces to $\Delta_0 = -\Delta_S$ under (4.184). Thus π is actually a BV homomorphism, and we have that

$$0 \longrightarrow \mathcal{I} \longrightarrow \mathrm{BV}[\mathfrak{g}] \stackrel{\pi}{\longrightarrow} \mathcal{P}(A) \longrightarrow 0 \tag{4.188}$$

is an exact sequence of BV algebras. Moreover, $\mathcal{P}(A)$ has an obvious natural embedding into $\mathrm{BV}[\mathfrak{g}]$ as a dot algebra.

The remaining problem is to describe $\mathcal{I} \equiv \operatorname{Ker} \pi$, the cohomology at nonzero $\sigma\omega$ ghost number. We will first identify the twisted ground ring modules in terms of the zero bc ghost number cohomology.

Theorem 4.49

i. The bc ghost number zero cohomology is given by

$$\mathrm{BV}^{(n,0)}[\mathfrak{g}] = \bigoplus_{\substack{w \in W \\ \ell(w)=n}} \mathrm{BV}_w\,, \tag{4.189}$$

where BV_w is an $\mathfrak{h} \oplus \mathfrak{g}$ module,

$$\mathrm{BV}_w \cong \bigoplus_{\Lambda \in P_+} \mathbb{C}_{w(\Lambda^* + \rho) - \rho} \otimes \mathcal{L}(\Lambda)\,. \tag{4.190}$$

ii. For each $w \in W$, BV_w is a "twisted" module of the ground ring $\mathcal{R}[\mathfrak{g}]$, with the action of $\mathcal{R}[\mathfrak{g}]$ on BV_w, at the level of $\mathfrak{h} \oplus \mathfrak{g}$ modules, given by

$$\mathcal{R}[\mathfrak{g}](\Lambda) \cdot \mathrm{BV}_w(\Lambda') \sim \mathrm{BV}_w(\Lambda' + w_0 w^{-1} w_0 \Lambda)\,. \tag{4.191}$$

Here $\mathcal{R}[\mathfrak{g}](\Lambda)$ and $\mathrm{BV}_w(\Lambda)$ denote irreducible components transforming as $\mathbb{C}_{\Lambda^*} \otimes \mathcal{L}(\Lambda)$ and $\mathbb{C}_{w(\Lambda^*+\rho)-\rho} \otimes \mathcal{L}(\Lambda)$, respectively, in the decomposition with respect to $\mathfrak{h} \oplus \mathfrak{g}$.

Remark. Note that the $\mathfrak{h}$ weights for different BV_w lie in disjoint "cones" in P_+.

Proof. Since $\mathrm{BV}^{(n,0)}[\mathfrak{g}] \equiv H^n(\mathfrak{n}_+, \mathcal{E}(G))$, the first part follows from (4.132), (4.148) and (4.155). The $\mathfrak{h} \oplus \mathfrak{g}$ modules that can arise in the decomposition into irreducible modules of the left hand side in (4.191) must be of the form $BV_{w'}(\Lambda'')$, where $\ell(w') = \ell(w)$ and $\Lambda'' \in P_+$. By examining the $\mathfrak{h}$ weight of the product in (4.191) we find that

$$\Lambda + w_0 w^{-1} w_0 (\Lambda' + \rho) = w_0 w'^{-1} w_0 (\Lambda'' + \rho)\,, \tag{4.192}$$

which, as we prove in Lemma 4.50 below, implies $w = w'$. By solving (4.192) we get $\Lambda'' = \Lambda' + w_0 w^{-1} w_0 \Lambda$. If $\Lambda'' \in P_+$, we find that $\mathcal{L}(\Lambda'')$ arises precisely once in the decomposition of $\mathcal{L}(\Lambda) \otimes \mathcal{L}(\Lambda')$, which proves (ii). Otherwise the product must vanish. □

Lemma 4.50 *If $\ell(w') = \ell(w'')$, $w', w'' \in W$, and there exist weights $\Lambda, \Lambda', \Lambda'' \in P_+$ such that*

$$\Lambda + w'(\Lambda' + \rho) \;=\; w''(\Lambda'' + \rho) \tag{4.193}$$

then $w' = w''$.

Proof. Define $\Phi_w = \Delta_- \cap w\Delta_+$, $w \in W$. The sets Φ_w are in one to one correspondence with the elements of the Weyl group. Indeed if $\Phi_{w'} = \Phi_{w''}$ then the well known relation [BGG1] $\sum_{\alpha \in \Phi_w} \alpha = w\rho - \rho$ implies $w'\rho = w''\rho$ and thus $w' = w''$. Now, suppose we have a solution to (4.193) such that $w' \neq w''$, $\ell(w') = \ell(w'')$. Then, according to the above, we may choose a root $\alpha \in \Delta_+$ such that $w'^{-1}\alpha \in \Delta_+$ and $w''^{-1}\alpha \in \Delta_-$. Taking the product with α on the both sides of (4.193) we obtain

$$(\alpha, \Lambda) + (w'^{-1}\alpha, \Lambda' + \rho) \;=\; (w''^{-1}\alpha, \Lambda'' + \rho)\,. \tag{4.194}$$

Then the left hand side is found to be strictly positive while the right hand side strictly negative, which shows that we cannot have $w' \neq w''$. □

Example. For $\mathfrak{g} = \mathfrak{sl}_3$ we can identify BV_w with the twisted module $M_{w_0 w w_0}$ introduced in Sect. 4.5.2. In particular, the states Ω_w corresponding to the tip of the weight cone for all the BV_w are easily seen to be

$$1\,, \quad \sigma^{\alpha_1}\,, \quad \sigma^{\alpha_2}\,, \quad \sigma^{\alpha_1}\sigma^{\alpha_3}\,, \quad \sigma^{\alpha_2}\sigma^{\alpha_3}\,, \quad \sigma^{\alpha_1}\sigma^{\alpha_2}\sigma^{\alpha_3}\,, \tag{4.195}$$

acting on the ghost vacuum, for w equal to 1, r_1, r_2, r_{12}, r_{21}, and r_3, respectively.

Theorems 4.48 and 4.49 provide us with descriptions of $\mathrm{BV}^{(m,n)}[\mathfrak{g}]$ for two different boundary values of ghost numbers. Loosely speaking, they tell us that in order to increase the bc ghost number we need to consider higher order derivations, while to increase the $\sigma\omega$ ghost number we must take twisted modules corresponding to Weyl group elements of longer length. This leads us to consider (twisted) polyderivations, $\mathcal{P}(\mathcal{R}[\mathfrak{g}], \mathrm{BV}_w)$, of the ground ring with values in the twisted modules, cf. Sect. 4.4.1. Since, in the bulk, $\mathrm{BV}[\mathfrak{g}]$ as described in Theorem 4.42 (ii) displays the Weyl group symmetry, our immediate goal is to relate the (Weyl) sectors in (4.151) to twisted polyderivations.

Take $\Phi \in \mathrm{BV}^{(n,m)}[\mathfrak{g}]$ and consider the multiple bracket

$$\pi_n(\Phi) = [\ldots[[\Phi, x_{i_1}], \ldots], x_{i_n}]\,. \tag{4.196}$$

The calculation in (4.186) shows that $\pi_n(\Phi)$ must an element of BV_{w}, with $\ell(w) = n$. To determine which particular w arises, one can examine the $\mathfrak{h}$ weight of $\pi_n(\Phi)$. One should note that it is possible for $\pi_n(\Phi)$ to vanish, even though its

$\mathfrak{h}$ weight lies in the allowed region, and we will discuss examples to that effect in the next chapter in a somewhat different setting. To summarize, we have

Theorem 4.51 *In the bulk there is an isomorphism of $\mathfrak{h} \oplus \mathfrak{g}$ modules*

$$\mathrm{BV}^n[\mathfrak{g}] \approx \bigoplus_{w \in W} \mathcal{P}^{n-\ell(w)}(\mathcal{R}[\mathfrak{g}], \mathrm{BV}_w)), \tag{4.197}$$

with natural maps

$$\pi_w : \mathrm{BV}^{n+\ell(w)}[\mathfrak{g}] \longrightarrow \mathcal{P}^n(\mathcal{R}[\mathfrak{g}], \mathrm{BV}_w), \tag{4.198}$$

defined by

$$\pi_w(\Phi)(x_{i_1}, \ldots, x_{i_n}) \equiv \begin{cases} [\ldots[[\Phi, x_{i_1}], \ldots], x_{i_n}], & \Phi \in \mathrm{BV}^{(\ell(w),n)}[\mathfrak{g}] \\ 0, & \textit{otherwise.} \end{cases} \tag{4.199}$$

Remark. In the case $\mathfrak{g} = \mathfrak{sl}_3$ one may verify (4.197) directly by comparing the $\mathfrak{sl}_3 \oplus (\mathfrak{u}_1)^2$ decomposition of twisted polyderivations $\mathfrak{P}_w$ in the bulk (Theorem 4.36) with the similar decomposition of $\mathrm{BV}[\mathfrak{sl}_3]$ (Theorem 4.43).

5 The BV Algebra of the $\mathcal{W}_3$ String

5.1 Introduction

In this section we will study, with various degrees of rigor, the algebraic structure of $H(\mathcal{W}_3, \mathfrak{C})$ that is induced by the VOA structure of the underlying complex $\mathfrak{C}$.

5.1.1 General Results

Let $\mathfrak{H}$ denote the cohomology $H(\mathcal{W}_3, \mathfrak{C})$ considered as an operator algebra. A straightforward application of the results in [Gt,LiZu2,PeSc,WiZw,WuZh] yields the following general theorem:

Theorem 5.1 *The cohomology $\mathfrak{H}$ carries a structure of a BV algebra $(\mathfrak{H}, \cdot, b_0)$, where the product "$\cdot$" is induced from the normal ordered product in $\mathfrak{C}$, while the BV operator, $b_0 \equiv b_0^{[2]}$, is the zero mode of the Virasoro antighost field $b^{[2]}(z)$.*

Proof. Using (3.9) and (3.10)we find that $\mathfrak{C}$ carries an action of the Virasoro algebra via $T^{\text{tot}}(z) = [d, b^{[2]}(z)]_+$, with diagonalizable energy operator L_0^{tot}. Thus the complex $(\mathfrak{C}, d)$ is an example of a topological chiral algebra, and the theorem follows from the discussion in [LiZu2], Sect. 3.9.4. □

For completeness let us recall some explicit formulae (see [LiZu2,Wi2]). The dot product of two operators in $\mathfrak{H}$ is given by

$$(\mathcal{O} \cdot \mathcal{O}')(z) = \frac{1}{2\pi i} \oint_{C_z} \frac{dw}{w-z} \mathcal{O}(w)\, \mathcal{O}'(z)\,. \tag{5.1}$$

It is graded commutative according to the ghost number of the operators. Since all nontrivial cohomology states are annihilated by L_0^{tot} (see Lemma 3.7), all singular terms in the OPE on the right hand side are trivial in cohomology. Thus (5.1) is also equivalent, in cohomology, to

$$(\mathcal{O} \cdot \mathcal{O}')(z) = \lim_{w \to z} \mathcal{O}(w)\mathcal{O}'(z)\,. \tag{5.2}$$

The action of the BV operator is

$$(b_0\mathcal{O})(z) = \oint_{C_z} \frac{dw}{2\pi i} (w-z)\, b^{[2]}(w)\mathcal{O}(z)\,. \tag{5.3}$$

The associativity and graded commutativity of the product at the level of cohomology, as well as the required properties of the BV operator (see Definition 4.2), follow immediately [PeSc]. Moreover, one finds that the corresponding bracket[1] is simply obtained as

$$[\mathcal{O},\mathcal{O}'](z) = (-1)^{gh(\mathcal{O})} \oint_{C_z} \frac{dw}{2\pi i} (b_{-1}^{[2]}\mathcal{O})(w)\,\mathcal{O}'(z). \tag{5.4}$$

The action of $\mathfrak{sl}_3 \oplus (\mathfrak{u}_1)^2$ commutes with the BV operator b_0, and acts as a derivation of the dot product. The latter follows from the distributivity of the normal ordered product with respect to the horizontal algebra defined by zero modes of spin one currents. Thus we also have a refinement of Theorem 3.11.

Theorem 5.2

i. The symmetry algebra $\mathfrak{sl}_3 \oplus (\mathfrak{u}_1)^2$ acts on $\mathfrak{H}$ by (infinitesimal) BV algebra automorphisms.

ii. $\mathfrak{H}$ is a direct sum of irreducible finite dimensional modules of $\mathfrak{sl}_3 \oplus (\mathfrak{u}_1)^2$.

In determining the explicit structure of the BV algebra $(\mathfrak{H}, \cdot, b_0)$ we will distinguish between two types of arguments. The first type, referred to as "kinematics," involves arguments based on general properties of the cohomology such as, in particular, the dimensions of the cohomology spaces at various matter and Liouville momenta, and their $\mathfrak{sl}_3 \oplus (\mathfrak{u}_1)^2$ content. The second type of argument is based on explicit computations. Those have been mostly carried out using the algebraic manipulation program Mathematica™ together with the CFT package OPEdefs [Th].

5.1.2 More Notation

In the following we will need a convenient parametrization of the operators in $\mathfrak{H}$. By examining the decomposition (3.67), we find that whenever $(\Lambda^M, -i\Lambda^L)$ satisfies (3.68) with $\sigma \in W$ (i.e., $\sigma \notin \{\sigma_1, \sigma_2\}$) there is precisely one quartet of $\mathfrak{sl}_3 \oplus (\mathfrak{u}_1)^2$ modules with those weights. Clearly the modules at the "bottom" and the "top" of the quartet are unique, and we will denote them by

$$\Psi^{(n)}_{\Lambda^M,-i\Lambda^L} \quad \text{and} \quad \Psi^{(n+2)}_{\Lambda^M,-i\Lambda^L}, \tag{5.5}$$

respectively. In cases where $\sigma \in \{\sigma_1, \sigma_2\}$ there is still only one module with the lowest and the highest ghost number, but the difference between those two ghost numbers is 3 rather than 2. To resolve ambiguity in the remaining cases, where an $\mathfrak{sl}_3 \oplus (\mathfrak{u}_1)^2$ module in cohomology is not characterized uniquely by $(\Lambda^M, -i\Lambda^L)$ and n, we will use either additional letters or labels.

We will often use the same notation as in (5.5) for the operator corresponding to the highest weight in a given module. Of course, in this case Λ^M and Λ^L are the momenta of the operator. Since $\mathfrak{sl}_3 \oplus (\mathfrak{u}_1)^2$ acts by automorphisms of the BV algebra, it is convenient to express most of the results at the level of modules rather than operators. To emphasize this we will then write "$\sim$" instead of the usual equality sign.

[1] Note that in this notation the bracket does *not* denote the commutator.

5.2 A Preliminary Survey of $\mathfrak{H}$

5.2.1 The Ground Ring $\mathfrak{H}^0$

We have seen in Sect. 3.5.3 that the cohomology at ghost number 0 is concentrated in the fundamental Weyl chamber, and consists of a single cone with tip $(0,0)$. Let us examine the lowest lying modules in $\mathfrak{H}^0$. First, there is the unit operator

$$\mathbf{1}(z) = \Psi^{(0)}_{0,0}(z)\,. \tag{5.6}$$

At Liouville weights $i\Lambda_1$ and $i\Lambda_2$ there is a triplet $\Psi^{(0)}_{\Lambda_1,\Lambda_1}$ and an anti-triplet $\Psi^{(0)}_{\Lambda_2,\Lambda_2}$. Let us denote their elements by $\widehat{x}_\sigma$ and $\widehat{x}^\sigma$, $\sigma \in \{1,2,3\}$, respectively, with

$$\widehat{x}_1(z) = \Psi^{(0)}_{\Lambda_1,\Lambda_1}(z)\,, \quad \widehat{x}^3(z) = \Psi^{(0)}_{\Lambda_2,\Lambda_2}(z)\,. \tag{5.7}$$

Explicit expressions for those two operators were first computed in [BLNW2]. They are given in Appendix J.

Lemma 5.3 [BMP5] *In cohomology, the operators $\widehat{x}_\sigma$ and $\widehat{x}^\sigma$, $\sigma \in \{1,2,3\}$, satisfy the constraint*

$$\widehat{x}_\sigma \cdot \widehat{x}^\sigma = 0\,. \tag{5.8}$$

Proof. The left hand side in (5.8) is a cohomology class of ghost number zero, with Liouville momentum $-i\Lambda^L = \Lambda_1 + \Lambda_2$, which transforms as a singlet under $\mathfrak{sl}_3$. However, by Theorem 3.19, the only nontrivial cohomology at this Liouville momentum is in the adjoint representation of $\mathfrak{sl}_3$, which implies (5.8). □

Remark. It is rather straightforward to check by direct computation of dimensions of the cohomology that there can be no cohomology at weights $\Lambda^M = 0$, $-i\Lambda^L = \Lambda_1+\Lambda_2$, and ghost number 0 (see Appendix D). However, verifying (5.8) directly, using the explicit representatives of the ground ring generators given in Appendix J, is clearly a formidable computation (which we have not attempted to perform). Recently, another verification of (5.8) has been given in [Zh].

Theorem 5.4 *The associative abelian algebra generated by $\mathbf{1}$, $\widehat{x}_\sigma$ and $\widehat{x}^\sigma$, $\sigma \in \{1,2,3\}$, is isomorphic with $\mathcal{R}_3$, i.e.,*

$$(\Psi^{(0)}_{\Lambda_1,\Lambda_1})^m \cdot (\Psi^{(0)}_{\Lambda_2,\Lambda_2})^n \sim \Psi^{(0)}_{m\Lambda_1+n\Lambda_2,m\Lambda_1+n\Lambda_2}\,. \tag{5.9}$$

Proof. In view of Lemma 5.3, and the discussion in Sect. 4.2.1, we must only show that the product of the highest weight operators in (5.9) does not vanish in cohomology. This can be done by examining explicitly the representatives (J.1) and (J.2) together with the BRST current (3.12), as we now demonstrate.

Define $\phi^{\pm,i} = \phi^{M,i} \pm i\phi^{L,i}$, $i = 1,2$. In terms of those fields the highest weight operators have the form

$$\widehat{x}_1(z) = P_{x_1}\, e^{i\Lambda_1\cdot\phi^+}, \quad \widehat{x}^3(z) = P_{x^3}\, e^{i\Lambda_2\cdot\phi^+}, \tag{5.10}$$

where the operator-level 2 prefactors P_{x_1} and P_{x^3} are

$$\begin{aligned} P_{x_1} &= -\tfrac{\sqrt{3}}{4}\,(2\sqrt{3}\partial\phi^{-,1}\partial\phi^{-,1} + 5\partial\phi^{-,1}\partial\phi^{-,2} + \partial\phi^{-,2}\partial\phi^{-,1}) + \dots, \\ P_{x^3} &= -\tfrac{3}{4}(\partial\phi^{-,1}\partial\phi^{-,1} - 3\partial\phi^{-,2}\partial\phi^{-,2}) + \dots. \end{aligned} \tag{5.11}$$

The dots in (5.11) stand for terms with $c^{[j]}$, $b^{[j]}$, $\partial\phi^{+,i}$, and their derivatives, as well as the derivatives $\partial^n\phi^{-,i}$, $n > 1$; the terms that have been written explicitly are the only ones which depend solely on $\partial\phi^{-,i}$

We will refer to a polynomial in $\partial\phi^{-,i}$ as a "leading term" of an operator. Remarkably, the leading terms in (5.11) do *not* depend on the choice of a representative in the cohomology class of $\widehat{x}_1$ and $\widehat{x}^3$. Indeed, the only ghost number -1 operators at the same weights and operator-level 2 are $b^{[2]}e^{i\Lambda_i\cdot\phi^+}$, and the assertion follows by examining the residue of the first order pole in the OPE with the BRST current.

Consider the normal ordered product

$$\begin{aligned} (\widehat{x}_1)^m(\widehat{x}^3)^n &\equiv (\widehat{x}_1 \dots (\widehat{x}_1(\widehat{x}^3 \dots (\widehat{x}^3\widehat{x}^3)\dots))\dots) \\ &= P_{mn}e^{i(m\Lambda_1+n\Lambda_2)\cdot\phi^+}. \end{aligned} \tag{5.12}$$

The prefactor P_{mn} has operator-level $2m+2n$ and ghost number zero. The proof of the theorem now reduces to showing that

i. The leading term in P_{mn} is the product of the leading terms of all factors in (5.12).
ii. This term does not depend on the choice of operator in the cohomology class of the product.

Recall that the normal ordered product of two operators is just the first nonsingular term in their OPE. Since

$$\phi^{\pm,i}(z)\phi^{\pm,j}(w) \sim \text{regular}, \quad \phi^{+,i}(z)\phi^{-,j}(w) \sim -2\delta^{ij}\log(z-w) + \text{regular}, \tag{5.13}$$

there are no contractions between the exponentials in computing (5.12), as all of them depend only on $\phi^{+,i}$. Contributions to the normal ordered product that would yield additional leading terms beyond the product of those in (5.11) can arise only after the Taylor expansion cancels pole terms arising from contractions between the prefactors and the exponentials, and between the prefactors. However, a moment's thought reveals that all such terms in which $\partial\phi^{-,i}$ is present must also have as factors either other fields, or derivatives of $\partial\phi^{-,i}$. This proves claim (i).

To show (ii), we must examine contributions to the residue of the first order pole in the OPE of the BRST current with an arbitrary operator that has ghost number -1, and the same momenta and level as $(\widehat{x}_1)^m(\widehat{x}^3)^n$. A similar argument to the one above shows that none of the terms arising via Taylor expansion can yield a polynomial in $\partial\phi^{-,i}$. Thus the only terms we need to be concerned with are obtained through a single contraction of $b^{[i]}$ with $c^{[i]}$, as otherwise we would have either higher order poles or uncancelled ghost operators. In fact, the only possibility is that the BRST current has a term of the form $c^{[i]} \times P[\partial\phi^{-,i}]$ which upon contraction with a term of the form $b^{[i]} \times \widetilde{P}[\partial\phi^{-,i}]$ would contribute to the leading term. The result (ii) then follows as a simple consequence of the fact that, as read from (3.9), the BRST current has no such term when expanded in $\phi^{\pm,i}$. □

As a direct consequence of Theorems 3.17 and 5.4 we obtain the following isomorphism.

Theorem 5.5 *The ground ring $\mathfrak{H}^0$, of the BV algebra $\mathfrak{H}$, is isomorphic to $\mathcal{R}_3$. The isomorphism $\pi : \mathfrak{H}^0 \to \mathcal{R}_3$ is explicitly given by*

$$\pi(\widehat{x}_\sigma) = x_\sigma\,, \quad \pi(\widehat{x}_{\dot\sigma}) = x_{\dot\sigma}\,, \quad \sigma, \dot\sigma \in \{1,2,3\}\,. \tag{5.14}$$

Since $\mathfrak{H}^n \cong 0$ for $n < 0$ (see Theorem 3.20), we can extend π to a G algebra homomorphism $\pi : \mathfrak{H} \to \mathfrak{P} \equiv \mathcal{P}(\mathcal{R}_3)$, as discussed in Sect. 4.1.4. Our immediate goal is to use π to establish a precise relation between $\mathfrak{H}$ and $\mathfrak{P}$ as BV algebras. First, however, we need to further study the explicit operator cohomology states at higher ghost numbers.

5.2.2 $\mathfrak{H}^1$: The $\mathfrak{sl}_3 \oplus (\mathfrak{u}_1)^2$ Symmetry of $\mathfrak{H}$ Revisited

Since $\mathfrak{H}$ is a BV algebra with BV operator b_0, there is a Lie algebra action of $\mathfrak{H}^1$ on $\mathfrak{H}$ defined by $\Phi \mapsto [\Psi, \Phi]$, $\Psi \in \mathfrak{H}^1$, $\Phi \in \mathfrak{H}$ (see Sect. 4.1.4). Moreover, by Lemma 4.5, the derivation $[\Psi, -]$ commutes with the BV operator if $b_0\Psi$ has vanishing bracket with $\mathfrak{H}$. We will now show that the algebra $\mathfrak{sl}_3 \oplus (\mathfrak{u}_1)^2$, as introduced in Sect. 3.3.3 , does in fact arise as a subalgebra of $\mathfrak{H}^1$ in this way.

Consider $\mathfrak{H}^1$ at the Liouville weight $\Lambda^L = 0$. From Theorem 3.19, or simply Table D.2, we find that it consists of three $\mathfrak{sl}_3$ modules: the adjoint and two singlets.

The highest weight operator in the adjoint representation is

$$\Psi^{(1)}_{\Lambda_1+\Lambda_2,0}(z) = \left(-c^{[2]} - \sqrt{\tfrac{3}{2}}\,i\partial\phi^{M,1}c^{[3]} + \tfrac{1}{\sqrt{2}}\,i\partial\phi^{M,2}c^{[3]} + b^{[2]}\partial c^{[3]}c^{[3]}\right)\mathcal{V}_{\Lambda_1+\Lambda_2,0}\,. \tag{5.15}$$

It satisfies

$$(b^{[2]}_{-1}\Psi^{(1)}_{\Lambda_1+\Lambda_2,0})(z) = -\mathcal{V}_{\Lambda_1+\Lambda_2,0}(z)\,; \tag{5.16}$$

i.e., its action on $\mathfrak{H}$, defined by the bracket (5.4), is the same as that of the $\mathfrak{sl}_3$ automorphism. Clearly, the same holds for the remaining operators in the octet. Let us denote them by $\widehat{D}_{\sigma\dot\sigma}(z)$, where

$$\widehat{D}_{1\dot3}(z) = \Psi^{(1)}_{\Lambda_1+\Lambda_2,0}(z)\,. \tag{5.17}$$

Theorem 5.6 *The operators $\widehat{D}_{\sigma\dot\sigma}$ close under the bracket onto the $\mathfrak{sl}_3$ algebra.*

Proof. Recall that

$$(\mathbf{8}\otimes\mathbf{8})_a = \mathbf{8}\oplus\mathbf{10}\oplus\overline{\mathbf{10}}. \tag{5.18}$$

Since at $\Lambda^L = 0$ and ghost number $n = 1$ there is no cohomology in either the **10** or the $\overline{\mathbf{10}}$ of $\mathfrak{sl}_3$, the $\widehat{D}_{\sigma\dot{\sigma}}$'s span a subspace in $\mathfrak{H}^1$ which is closed under the bracket. The theorem now follows by noting that the action of this algebra on the ground ring coincides with $\mathfrak{sl}_3$. □

In fact, we also have

$$\pi(\widehat{D}_{\sigma\dot{\sigma}}) = D_{\sigma\dot{\sigma}}\,,\quad \sigma,\dot{\sigma}\in\{1,2,3\}\,. \tag{5.19}$$

The two singlets at $-i\Lambda^L = 0$ can be understood as a part of the quartet associated with the identity operator. This quartet consists of $\mathbf{1}(z)$, $C^{[2]}(z)$, $C^{[3]}(z)$ and $C^{[23]}(z)$, where $C^{[23]}(z) = (C^{[2]}\cdot C^{[3]})(z)$. The ghost number one operators $C^{[2]}(z)$ and $C^{[3]}(z)$ are given in (J.4) and (J.5), respectively. Note that neither of them depends on the matter fields, $\phi^{M,i}$, so they indeed transform as singlets under $\mathfrak{sl}_3$.

From the explicit formulae we find

$$\begin{aligned} C^{[2]} &= -\ 4\partial c^{[2]} - (\alpha_1+\alpha_2)\cdot\partial\phi^L c^{[2]} + \ldots\,,\\ C^{[3]} &= -\,(\alpha_1-\alpha_2)\cdot\partial\phi^L c^{[2]} + \ldots\,, \end{aligned} \tag{5.20}$$

where the dots stand for terms without $c^{[2]}$ or its derivatives. Thus

$$(b_{-1}^{[2]}C^{[2]})(z) = -(\alpha_1+\alpha_2)\cdot\partial\phi^L(z)\,,\quad (b_{-1}^{[2]}C^{[3]})(z) = -(\alpha_1-\alpha_2)\cdot\partial\phi^L(z)\,, \tag{5.21}$$

which shows that $C^{[2]}$ and $C^{[3]}$ are the $(\mathrm{u}_1)^2$ generators we are looking for. Moreover, if we set

$$\widehat{C}(z) = C^{[2]}(z)\,,\quad \widehat{C}_\pm(z) = \tfrac{1}{2}(C^{[2]}(z)\pm C^{[3]}(z))\,, \tag{5.22}$$

then (compare to (4.97)),

$$\pi(\widehat{C}) = C\,,\quad \pi(\widehat{C}_\pm) = C_\pm\,. \tag{5.23}$$

As another straightforward consequence of (5.20), we will obtain explicit formulae for the action of the BV operator b_0.

Theorem 5.7 *Let $\Phi\in\mathfrak{H}$ be an arbitrary operator with Liouville momentum $-i\Lambda^L = t_1\Lambda_1 + t_2\Lambda_2$ satisfying $b_0\Phi = 0$. Then*

$$b_0(C^{[2]}\cdot\Phi) = -(4+t_1+t_2)\Phi\,,\quad b_0(C^{[3]}\cdot\Phi) = -(t_1-t_2)\Phi\,, \tag{5.24}$$

$$b_0(C^{[23]}\cdot\Phi) = -(t_2-t_1)C^{[2]}\cdot\Phi - (4+t_1+t_2)C^{[3]}\Phi\,. \tag{5.25}$$

Proof. For Φ of the form (3.18), the action of b_0 in (5.3) simply amounts to setting to zero all the terms in the polynomial prefactor P that do not contain $\partial c^{[2]}$ as a factor, and removing $\partial c^{[2]}$ from all the terms in which it is present.

Thus, $b_0\Phi = 0$ implies that $\partial c^{[2]}$ is absent from all the terms in P. But then $(\partial c^{[2]}\Phi)(z) \neq 0$, and from (5.20) we have

$$(C^{[2]} \cdot \Phi) \;=\; -(4+t_1+t_2)(\partial c^{[2]}\Phi)+\Phi', \quad (C^{[3]} \cdot \Phi) \;=\; -(t_1-t_2)(\partial c^{[2]}\Phi)+\Phi'', \tag{5.26}$$

where $b_0\Phi' = b_0\Phi'' = 0$. These equations imply (5.24), and (5.25) is then obtained using the second order derivation property of b_0, see (4.4). □

Corollary 5.8

$$\tfrac{1}{4}(b_0 C^{[2]})(z) \;=\; -\mathbf{1}(z)\,, \quad (b_0 C^{[3]})(z) \;=\; 0\,. \tag{5.27}$$

Proof. Take $\Phi = \mathbf{1}$ in (5.24). □

Note that by kinematics we must have $b_0\widehat{D}_{\sigma\dot{\sigma}} = 0$, which, together with Corollary 5.8, proves directly that the algebra generated by $\widehat{D}_{\sigma\dot{\sigma}}$ and $\widehat{C}_{\pm}$ commutes with the BV operator on $\mathfrak{H}$.

Theorem 5.9 *The G algebra homomorphism π is equivariant with respect to the action of $\mathfrak{sl}_3 \oplus (\mathfrak{u}_1)^2$ on $\mathfrak{H}$ and $\mathfrak{P}$.*

Proof. Since π is a G algebra homomorphism, the equivariance of π with respect to $\mathfrak{sl}_3 \oplus (\mathfrak{u}_1)^2$ follows from (5.19) and (5.23). Indeed, for any $\Psi \in \mathfrak{H}$ we have

$$\pi([\,\widehat{D},\Psi]) \;=\; [\,\pi(\widehat{D}),\pi(\Psi)\,]\,, \tag{5.28}$$

where $\widehat{D} = \widehat{D}_{\sigma\dot{\sigma}}$, $\widehat{C}_+$ or $\widehat{C}_-$. □

5.2.3 More $\mathfrak{H}^1$

The operators at the tips of the two remaining $w = 1$ cones in Table 3.2 have Liouville momenta $-i\Lambda^L = -\Lambda_1 + \Lambda_2$ and $\Lambda_1 - \Lambda_2$, and transform under $\mathfrak{sl}_3$ as the triplet and the anti-triplet, respectively. We denote them by antisymmetric tensors $\widehat{P}_{\dot{\rho},\dot{\sigma}}$ and $\widehat{P}_{\rho,\sigma}$. The highest weight operators are

$$\widehat{P}_{\dot{2},\dot{3}}(z) \;=\; \Psi^{(1)}_{\Lambda_1,-\Lambda_1+\Lambda_2}(z)\,, \quad \widehat{P}_{1,2}(z) \;=\; \Psi^{(1)}_{\Lambda_2,\Lambda_1-\Lambda_2}(z)\,, \tag{5.29}$$

and their explicit expressions can be found in Appendix J.

Now consider the action of $\widehat{P}_{\rho,\sigma}$ on the ground ring. First we must have $[\widehat{P}_{\mu,\rho}, \widehat{x}^{\dot{\sigma}}] = 0$, because at the total Liouville momentum of this operator, given by $-i\Lambda^L = 2\Lambda_1 - \Lambda_2$, there is no cohomology with ghost number zero. Similarly, the other bracket, $[\widehat{P}_{\mu,\rho}, \widehat{x}^{\sigma}]$, must be a linear combination of the generators in the triplet of $\mathfrak{sl}_3$. By an explicit computation we verify that in fact

$$[\widehat{P}_{\mu,\rho}, \widehat{x}^{\sigma}] \;=\; -\tfrac{1}{2}(\delta^{\sigma}_{\mu}\widehat{x}_{\rho} - \delta^{\sigma}_{\rho}\widehat{x}_{\mu})\,. \tag{5.30}$$

A similar result also holds for $\widehat{P}_{\dot{\sigma},\dot{\rho}}$, and we conclude that

$$\pi(\widehat{P}_{\rho,\sigma}) = P_{\rho,\sigma}, \quad \pi(\widehat{P}_{\dot{\rho},\dot{\sigma}}) = P_{\dot{\rho},\dot{\sigma}}. \tag{5.31}$$

At this point we have considered all cones in $\mathfrak{H}^1$, except for the two twisted cones with $w = r_1$ and $w = r_2$ in Table 3.2. By comparing the weights of operators in the twisted cones with those of the polyvectors, in Table G.1, we conclude that all these operators must act trivially on the ground ring. In other words the bracket between those operators and the elements of the ground ring must vanish.

The operators at the tips of the two twisted cones will be denoted by

$$\widehat{\Omega}_{r_1}(z) = \Psi^{(1)}_{0,-2\Lambda_1+\Lambda_2}(z), \quad \widehat{\Omega}_{r_2}(z) = \Psi^{(1)}_{0,\Lambda_1-2\Lambda_2}(z). \tag{5.32}$$

They are given in (J.32) and (J.33), respectively.

5.2.4 An Extension of $\mathfrak{so}_6$

Now, we wish to examine whether the $\mathfrak{so}_6$ symmetry of the ground ring is realized by operators in $\mathfrak{H}^1$. First let us set

$$\widehat{P}_{\sigma,\dot{\sigma}} = -\widehat{P}_{\dot{\sigma},\sigma} = \tfrac{1}{2}\widehat{D}_{\sigma\dot{\sigma}} + \tfrac{1}{6}h_{\sigma\dot{\sigma}}\,(\widehat{C}_+ - \widehat{C}_-), \tag{5.33}$$

and combine them with $\widehat{P}_{\rho,\sigma}$ and $\widehat{P}_{\dot{\rho},\dot{\sigma}}$ to $\widehat{P}_{i,j}$, $i,j = 1,\ldots,6$. Then, from (5.19), (5.23) and (5.31), we find that $\pi(\widehat{P}_{i,j}) = P_{i,j}$, or, simply, that the $\widehat{\Lambda}_{ij} = 2\widehat{P}_{i,j}$ act on the ground ring as the $\mathfrak{so}_6$ algebra.

However, when we consider the bracket between the operators $\widehat{\Lambda}_{ij}$ in $\mathfrak{H}^1$, we find that they do not form a Lie subalgebra isomorphic with $\mathfrak{so}_6$, but rather generate an (infinite dimensional) extension of $\mathfrak{so}_6$. In particular we find

Lemma 5.10

$$\begin{aligned} [\widehat{P}_{\dot{\mu},\dot{\nu}}, \widehat{P}_{\dot{\rho},\dot{\sigma}}] &= -\tfrac{1}{48}\left(\epsilon_{\dot{\mu}\dot{\nu}\dot{\rho}}\widehat{x}_{\dot{\sigma}} - \epsilon_{\dot{\mu}\dot{\nu}\dot{\sigma}}\widehat{x}_{\dot{\rho}}\right)\cdot\widehat{\Omega}_{r_1}, \\ [\widehat{P}_{\mu,\nu}, \widehat{P}_{\rho,\sigma}] &= -\tfrac{1}{48}\left(\epsilon_{\mu\nu\rho}\widehat{x}_{\sigma} - \epsilon_{\mu\nu\sigma}\widehat{x}_{\rho}\right)\cdot\widehat{\Omega}_{r_2}. \end{aligned} \tag{5.34}$$

Proof. By kinematics, the general form of the first bracket at the level of modules is

$$[\Psi^{(1)}_{\Lambda_1,-\Lambda_1+\Lambda_2}, \Psi^{(1)}_{\Lambda_1,-\Lambda_1+\Lambda_2}] \sim n\Psi^{(1)}_{\Lambda_2,-2\Lambda_1+2\Lambda_2}, \tag{5.35}$$

where $n = 0$ or 1. The operator on the right hand side turns out to be a product

$$\Psi^{(1)}_{\Lambda_1,-2\Lambda_1+2\Lambda_2} \sim \Psi^{(0)}_{\Lambda_2,\Lambda_2}\cdot\Psi^{(1)}_{0,-2\Lambda_1+\Lambda_2}. \tag{5.36}$$

The general form in (5.34) then follows by the $\mathfrak{sl}_3 \oplus (\mathfrak{u}_1)^2$ covariance, and the overall normalization factor is fixed by explicitly evaluating the bracket between a single pair of operators. □

Let us state, without further detail, that all other brackets between the operators $\widehat{P}_{i,j}$ close as expected, thus the $\mathfrak{so}_6$ commutation rules are violated only by (5.34).

5.2.5 A Summary for $\mathfrak{H}^1$

We may summarize the structure of $\mathfrak{H}^1$ as follows.

Theorem 5.11 *Let us set, according to the cone decomposition in Table 3.2,*

$$\mathfrak{H}^1 = \mathfrak{H}^1_1 \oplus \mathfrak{H}^1_{r_1} \oplus \mathfrak{H}^1_{r_2} . \tag{5.37}$$

Then $\mathfrak{H}^1_1 \cong \pi(\mathfrak{H}^1_1) \cong \mathfrak{P}^1$ *and* $\operatorname{Ker}\pi \cong \mathfrak{H}^1_{r_1} \oplus \mathfrak{H}^1_{r_2}$.

Proof. We have shown that all the generators of $\mathfrak{P}^1$ as a module over the ground ring, are obtained as the image under π of the tips of cones in $\mathfrak{H}^1$. Thus, since π is a dot algebra homomorphism, $\pi(\mathfrak{H}^1) = \mathfrak{P}^1$. By comparing the $\mathfrak{sl}_3 \oplus (\mathfrak{u}_1)^2$ content in Tables 3.2 G.1 we see that π must be an isomorphism between $\mathfrak{H}^1_1$ and $\mathfrak{P}^1$, and vanish on $\mathfrak{H}^1_{r_1}$ and $\mathfrak{H}^1_{r_2}$. □

Corollary 5.12 *The subspace* $\mathfrak{H}^1_1$ *is generated as an* $\mathfrak{H}^0$ *dot module by* $\widehat{C} = \widehat{C}_+ + \widehat{C}_-$ *and* $\widehat{P}_{i,j}$.

5.3 The Relation Between $\mathfrak{H}$ and $\mathfrak{P}$

In the following subsections we will give a complete proof of the following theorem which summarizes the structure of $\mathfrak{H}$ using the homomorphism π.

Theorem 5.13

i. The map $\pi : \mathfrak{H} \to \mathfrak{P}$ *is a BV algebra homomorphism between* $(\mathfrak{H}, \cdot, b_0)$ *and* $(\mathfrak{P}, \cdot, \Delta_S)$.

ii. Let $\mathfrak{I} \equiv \operatorname{Ker}\pi$ *be a BV ideal of* $\mathfrak{H}$. *We have an exact sequence of BV algebras*

$$0 \longrightarrow \mathfrak{I} \longrightarrow \mathfrak{H} \xrightarrow{\pi} \mathfrak{P} \longrightarrow 0 . \tag{5.38}$$

There exists a dot algebra homomorphism $\imath : \mathfrak{P} \to \mathfrak{H}$, *such that* $\pi \circ \imath = \mathrm{id}$, *i.e., the sequence splits as a sequence of* $\imath(\mathfrak{P})$ *dot modules.*

5.3.1 $\pi(\mathfrak{H}) = \mathfrak{P}$

Consider the unique ghost number 5 singlet with the Liouville momentum $-i\Lambda^L = -2\Lambda_1 - 2\Lambda_2$,

$$\widehat{X}(z) = \Psi^{(5)}_{0,-2\Lambda_1-2\Lambda_2} , \tag{5.39}$$

given in (J.30). Using the $\mathfrak{so}_6$ notation, define

$$\widehat{P}_{i_1,i_2\ldots i_m} = \frac{1}{(6-m)!}\frac{6}{m}\,\epsilon_{j_1\ldots j_{6-m}i_1\ldots i_m}[\widehat{x}^{j_{6-m}},\ldots,[\widehat{x}^{j_1},\widehat{X}]\ldots] , \tag{5.40}$$

where $2 \leq m \leq 5$. It is easy to verify by an explicit computation that, for $m = 2$, (5.40) agrees with the previous definition of the operators $\widehat{P}_{i,j}$.

A further explicit computation shows that, for $m = 1$, (5.40) extends to

$$\widehat{P}_i = \widehat{x}_i , \tag{5.41}$$

which implies

$$\pi(\widehat{X}) = X . \tag{5.42}$$

In fact, (5.41) implies a stronger result:

Lemma 5.14 *For all* $1 \leq m \leq 5$,

$$\pi(\widehat{P}_{i_1,i_2 \ldots i_m}) = P_{i_1,i_2 \ldots i_m} . \tag{5.43}$$

Therefore, we have

Theorem 5.15

$$\pi(\mathfrak{H}) = \mathfrak{P} . \tag{5.44}$$

Proof. Lemma 5.14, together with (5.42), shows that all generators of the dot algebra $\mathfrak{P}$, given in Theorem 4.17, are in the image of π. □

In Appendix J we have listed the complete set of operators $\widehat{P}_{i_1,i_2 \ldots i_m}$ corresponding to the highest weights of all $\mathfrak{sl}_3$ modules.

5.3.2 π is a BV Algebra Homomorphism

Using Theorem 4.12, the first part of Theorem 5.13 is proved by the following lemma.

Lemma 5.16 *For* $\Psi \in \mathfrak{H}^1$,

$$\pi(b_0 \Psi) = \Delta_S \pi(\Psi) . \tag{5.45}$$

Proof. By kinematics we have $b_0 \Psi = 0$, for all $\Psi \in \mathfrak{H}^1_{r_1} \oplus \mathfrak{H}^1_{r_2} \cong \operatorname{Ker} \pi$, so (5.45) holds there. In $\mathfrak{H}^1_1$ we can use Corollary 5.12 to conclude that, since b_0 and Δ_S are second order derivations and π is a dot algebra homomorphism, it is sufficient to verify (5.45) on $\widehat{C}$ and $\widehat{P}_{i,j}$, and on their products with a single ground ring generator. Note that by kinematics, together with (5.33) and (5.27), we must have $b_0 \widehat{P}_{i,j} = 0 = \Delta_S P_{i,j}$. Similarly, (4.88) and (5.27) show that (5.45) holds for $\widehat{C}$ and C. Then

$$\pi(b_0(\widehat{x}_i \cdot \widehat{P}_{i,j})) = \pi([\widehat{x}_i, \widehat{P}_{j,k}]) = [x_i, P_{j,k}] = \Delta_S(x_i \cdot P_{j,k}) . \tag{5.46}$$

The last case, $\widehat{C} \cdot \widehat{x}_i$, is proved using (4.89) and Theorem 5.7. □

5.3.3 An Embedding $\imath : \mathfrak{P} \to \mathfrak{H}$

We have seen in Sect. 5.2, particularly Sect. 5.2.5, that a simple kinematical analysis yields a unique embedding of $\mathfrak{P}^0 \oplus \mathfrak{P}^1$ into $\mathfrak{H}$. However, this is not the case at higher ghost numbers, where at some momenta there are more states in the cohomology than in the corresponding polyderivations.

The simplest example is at ghost number two along the boundaries of the fundamental Weyl chamber. Indeed, by comparing Table G.1 with Table 3.1, or Table 3.2, we find that the $\mathfrak{sl}_3 \oplus (\mathfrak{u}_1)^2$ modules with highest weights $(\Lambda_1 + n\Lambda_2, -\Lambda_1 + (n+1)\Lambda_2)$ and $(n\Lambda_1 + \Lambda_2, (n+1)\Lambda_1 - \Lambda_2)$, $n \geq 0$, are doubly degenerate in $\mathfrak{H}^2$ but nondegenerate in $\mathfrak{P}^2$. The same phenomenon is present at higher ghost numbers.

The problem then is to find an embedding $\imath : \mathfrak{P} \to \mathfrak{H}$ which preserves as much of the BV algebra structure of $\mathfrak{P}$ as possible. We have already seen (Sect. 5.2.4) that at ghost number one the image of $\mathfrak{P}^1$ in $\mathfrak{H}^1$ is not closed under the bracket. Thus, one can at most expect to embed $\mathfrak{P}$ as dot algebra. In that case, although $\mathfrak{P}$ is generated as a BV algebra by the ground ring generators x_i, C, and the "volume element" X – all of which embed uniquely into $\mathfrak{H}$ – it is necessary to define the embedding of the remaining dot algebra generators $P_{i_1,i_2\ldots i_m}$, $m = 2,\ldots,5$.

Theorem 5.17 *Let us define*

$$\imath(C) \;=\; \widehat{C}\,, \quad \imath(X) \;=\; \widehat{X}\,, \tag{5.47}$$

$$\imath(P_{i_1,i_2\ldots i_m}) \;=\; \widehat{P}_{i_1,i_2\ldots i_m}\,, \quad 1 \leq m \leq 5\,. \tag{5.48}$$

Then $\imath$ extends uniquely to a dot algebra embedding of $\mathfrak{P}$ into $\mathfrak{H}$.

Proof. Clearly, it is sufficent to prove that the elements $\widehat{C}$, $\widehat{X}$, and $\widehat{P}_{i_1,i_2\ldots i_m}$, $m = 1,\ldots,5$, satisfy (4.62)-(4.66) in Theorem 4.17. The last relation,

$$\widehat{C}\cdot\widehat{X} \;=\; 0\,, \tag{5.49}$$

is easily verified by using (J.8), (J.4), (J.5) and (J.33). It also follows by kinematics, as there is no cohomology with $-i\Lambda^L = -2\Lambda_1 - 2\Lambda_2$ and $n = 6$ (see Table D.2). Thus we must show, in addition to our previous result (5.8), that

$$\widehat{x}_{[i}\cdot\widehat{P}_{i_1,i_2\ldots i_m]} \;=\; 0\,, \tag{5.50}$$

$$\widehat{x}^i\cdot\widehat{P}_{i,j_1\ldots j_m} \;=\; -\tfrac{m}{m+1}\,\widehat{C}\cdot\widehat{P}_{j_1,j_2\ldots j_m}\,, \tag{5.51}$$

$$\widehat{P}_{i_1,i_2\ldots i_m}\cdot\widehat{P}_{j_1,j_2\ldots j_n} \;=\; (-1)^{m-1}\tfrac{m+n-1}{n}\,\widehat{x}_{[i_1}\cdot\widehat{P}_{i_2,i_3\ldots i_m]j_1\ldots j_n}\,, \tag{5.52}$$

where $m,n = 1,\ldots,5$ and $\widehat{P}_{i_1,i_2\ldots i_6} = \epsilon_{i_1 i_2\ldots i_6}\widehat{X}$.

Using the complete antisymmetry of the multiple bracket, which follows immediately from (iv) in Definition 4.1, we may invert (5.40) as

$$[\widehat{x}^{j_{6-m}},[\widehat{x}^{j_{5-m}}\ldots,[\widehat{x}^{j_1},\widehat{X}]\ldots]] = \frac{1}{6(m-1)!}\epsilon^{j_1\ldots j_{6-m}i_1\ldots i_m}\widehat{P}_{i_1,i_2\ldots i_m}, \quad m=1,\ldots,5. \tag{5.53}$$

This implies (see (F.12)) that

$$[\widehat{x}^i,\widehat{P}_{i_1,i_2\ldots i_m}] = (m-1)\,\delta^i{}_{[i_1}\widehat{P}_{i_2,i_3\ldots i_m]}\,. \tag{5.54}$$

Now, for arbitrary $\Psi \in \mathfrak{H}$ we have

$$\widehat{x}_i\cdot[\widehat{x}^i,\Psi] = \tfrac{1}{2}[\widehat{x}_i\cdot\widehat{x}^i,\Psi] = 0, \tag{5.55}$$

so (5.50) follows from (5.54) after multiplication by $\widehat{x}_i$.

The second relation (5.51) is proved by induction on m. First we have, using (5.49), (5.23) and (4.72),

$$0 = [\widehat{x},\widehat{C}\cdot\widehat{X}] = -\widehat{C}\cdot[x^i,\widehat{X}]+x^i\cdot\widehat{X}, \tag{5.56}$$

which by (5.40) is equivalent to (5.51) for $m=5$. Now suppose that (5.51) is true for $m=5$, 4, ..., $n+1$, with $1<n<5$. Then[2]

$$\begin{aligned}
\widehat{C}\cdot\widehat{P}_{\{n-1\}} &= \tfrac{n+1}{n(6-n)}\widehat{C}\cdot[x^i,P_{i,\{n-1\}}]\\
&= \tfrac{n+1}{m(6-n)}\left(-[\widehat{x}^i,\widehat{C}\cdot\widehat{P}_{i,\{n-1\}}]+[\widehat{x}^i,\widehat{C}]\cdot\widehat{P}_{i,\{m-1\}}\right)\\
&= \tfrac{n+2}{n(6-n)}[\widehat{x}^i,\widehat{x}^j\cdot\widehat{P}_{j,i\{n-1\}}]-\tfrac{n+1}{n(6-n)}\widehat{x}^i\cdot\widehat{P}_{i,\{n-1\}}\\
&= \tfrac{n+1}{n}\left(-\tfrac{6-n-1}{6-n}-\tfrac{1}{6-n}\right)\widehat{x}^i\cdot\widehat{P}_{i,\{n-1\}}\\
&= -\tfrac{n+1}{n}\widehat{x}^i\cdot\widehat{P}_{i,\{n-1\}}\,.
\end{aligned} \tag{5.57}$$

Finally, let us consider the last relation (5.52) . We have

$$\widehat{P}_{\{m_1\}}\cdot\widehat{P}_{\{m_2\}} = 0 \quad\text{if}\quad m_1+m_2\geq 6\,. \tag{5.58}$$

This can be proved by noting that for all but two pairs (m_1,m_2) there are simply no operators in the complex $\mathfrak{C}$ with the Liouville momentum and the ghost number of the product on the left hand side in (5.58). The two exceptions, (2,4) and (3,3), can be reduced to the other cases using

$$[\widehat{x}^i,\widehat{P}_{i,\{4\}}\cdot\widehat{P}_{\{2\}}] = \tfrac{5}{6}\widehat{P}_{\{4\}}\cdot\widehat{P}_{\{2\}}+\widehat{P}_{i,\{4\}}\cdot[\widehat{x}^i,\widehat{P}_{\{2\}}], \tag{5.59}$$

$$[\widehat{x}^i,\widehat{P}_{i,\{3\}}\cdot\widehat{P}_{\{3\}}] = \tfrac{8}{5}\widehat{P}_{\{3\}}\cdot\widehat{P}_{\{3\}}+\widehat{P}_{i,\{3\}}\cdot[\widehat{x}^i,\widehat{P}_{\{3\}}], \tag{5.60}$$

which follow from the distributivity of the bracket and (5.54).

On the one hand, (5.58) proves (5.52) for $m+n\geq 8$. On the other hand, (5.52) clearly is true if $m=1$ and n arbitrary. Then the complete proof of (5.52) is obtained by induction on $m+n$ and m, using (5.54) and (5.40). This completes the proof of Theorem 5.17, and thus also of Theorem 5.13. □

[2] We use a shorthand notation for index structure that is either obvious or irrelevant, and write $\{m-1\}$ for m indices; e.g., $\widehat{P}_{\{m-1\}}$ for the ghost number $m-1$ operator $\widehat{P}_{i_1,i_2\ldots i_m}$.

5.4 The Bulk Structure of $\mathfrak{H}$

We have seen in the previous section that the action of the BV algebra $\mathfrak{H}$ on its ground ring $\mathfrak{H}^0$ leads to a projection π from $\mathfrak{H}$ onto polyderivations, $\mathfrak{P}$. For a given ghost number n cohomology class, the components of the projection are simply the ring elements isomorphic to its n-times iterated bracket with the ground ring generators. For elements in the kernel of π, there is clearly some point at which this iteration of brackets vanishes, though in general there will be a nontrivial result after some number of iterations less than n. Identifying this last nontrivial stage will allow us to refine our study of the kernel of π. In fact, this construction yields a homomorphism from $\mathfrak{H}^n$ into polyderivations $\mathcal{P}(\mathcal{R}_3, \mathfrak{H}^{n-k})$, $k \leq n$, the homomorphism π corresponding to the maximal case $k = n$. One observation that we make is that in the bulk, i.e., for Liouville momenta sufficiently deep inside Weyl chambers, the cohomology $H(\mathcal{W}_3, \mathfrak{C})$ admits a description in terms of "twisted polyderivations" associated with twisted modules of the ground ring of the type introduced in Sect. 4.5.2. In particular, this result gives then a partial proof of Conjecture 3.23, in the sense that it establishes the lower bound on the cohomology.

Most of the results below are obtained by a combination of kinematical arguments and explicit computations. While a more rigorous treatment along the lines of the discussion in Sect. 4.2.3 or the proof of Theorem 5.4 could be given, the details of such proofs are rather cumbersome, at least in comparison with their counterparts in the fundamental Weyl chamber. We thus mainly limit our discussion to a general summary of the results.

5.4.1 Twisted Modules of $\mathfrak{H}^0$

An examination of the pattern of cohomology states (see Table 3.2, or, more conveniently, the figures in Appendix E) reveals that in each Weyl chamber the cohomology with the lowest ghost number forms precisely one (twisted) cone, $\widehat{M}_w$, of $\mathfrak{sl}_3 \oplus (\mathfrak{u}_1)^2$ modules with highest weights $(\Lambda, w^{-1}(\Lambda+\rho)-\rho)$, $\Lambda \in P_+$, $w \in W$. The operators at the tips of those cones,

$$\widehat{\Omega}_w(z) = \Psi^{(\ell(w))}_{0,w^{-1}\rho-\rho}, \tag{5.61}$$

can be found in Appendix J.4. Note that for $w = 1$ we have the identity operator, while for $w = r_1$ and r_2 these are exactly the two ghost number one operators which already appeared in Sect. 5.2.4.

Now we would like to understand the dot action of the ground ring on each of the cones $\widehat{M}_w$. The simple fact that there is only one $\mathfrak{sl}_3$ module at each Liouville momentum in $\widehat{M}_w$ allows us to determine most of ground ring action by a purely kinematical analysis.

Theorem 5.18 *The twisted cones, $\widehat{M}_w$, $w \in W$, are closed under the dot product action of the ground ring, i.e., $\mathfrak{H}^0 \cdot \widehat{M}_w \subset \widehat{M}_w$, and as $\mathfrak{H}^0$ modules they are isomorphic to the corresponding twisted $\mathcal{R}_3$ modules, M_w, introduced in Sect. 4.5.2.*

Proof. Clearly, the decomposition of each cone into $\mathfrak{sl}_3$ modules is that of a model space, and thus identical with that of the ground ring, $\mathfrak{H}^0$. In fact, for $w=1$, $\widehat{M}_1 \cong \mathfrak{H}^0$. More interesting are $w=r_1$ and r_2, where we observe that, as $\mathfrak{sl}_3$ modules, $\widehat{M}_{r_1} \cong \mathfrak{H}^1_{r_1}$ and $\widehat{M}_{r_2} \cong \mathfrak{H}^1_{r_2}$, respectively. We now outline the main steps of the proof for those two cases.

First consider $\widehat{M}_{r_1}$. By acting with the ground ring generators on the tip of this cone we obtain

$$\Psi^{(0)}_{\Lambda_1,\Lambda_1}\cdot\Psi^{(1)}_{0,-2\Lambda_1+\Lambda_2} \sim 0\,,\quad \Psi^{(0)}_{\Lambda_2,\Lambda_2}\cdot\Psi^{(1)}_{0,-2\Lambda_1+\Lambda_2} \sim \Psi^{(1)}_{\Lambda_2,-2\Lambda_1+2\Lambda_2}\,. \tag{5.62}$$

In fact, it is not too difficult to verify, by examining the leading terms in the first product, that subsequent action of the anti-triplet of ground ring generators always yields a nonvanishing result. This proves that the operators along the boundary, $(n\Lambda_2, -2\Lambda_1+(n+1)\Lambda_2)$, of $\widehat{M}_{r_1}$ are

$$\widehat{x}_{\dot\sigma_1}\cdot\ldots\cdot\widehat{x}_{\dot\sigma_n}\cdot\widehat{\Omega}_{r_1}\,,\qquad \dot\sigma_1,\ldots,\dot\sigma_n\in\{1,2,3\}\,,\quad n\geq 0\,. \tag{5.63}$$

To obtain the remaining operators in the cone we must study the bracket action of $\mathfrak{H}^1$ on $\widehat{M}_{r_1}$, in particular those of

$$\widehat{\mathcal{D}}_\sigma = \epsilon_{\sigma\mu\rho}[\,\widehat{P}^{\mu,\rho},-]\,,\quad \widehat{\mathcal{D}}_{\dot\sigma} = \epsilon_{\dot\sigma\dot\mu\dot\rho}[\,\widehat{P}^{\dot\mu,\dot\rho},-]\,. \tag{5.64}$$

Note that, when acting on the ground ring, $\widehat{\mathcal{D}}_\sigma$ and $\widehat{\mathcal{D}}_{\dot\sigma}$ are the first order differential operators $D^{(1)}_\sigma$ and $D^{(1)}_{\dot\sigma}$ given in (G.18). Once more we verify explicitly that

$$[\Psi^{(1)}_{\Lambda_1,-\Lambda_1+\Lambda_2},\Psi^{(0)}_{0,-2\Lambda_1+\Lambda_2}] \sim \Psi^{(1)}_{\Lambda_1,-3\Lambda_1+2\Lambda_2}\,,\quad [\Psi^{(1)}_{\Lambda_2,\Lambda_1-\Lambda_2},\Psi^{(0)}_{0,-2\Lambda_1+\Lambda_2}] \sim 0\,, \tag{5.65}$$

which suggests that the other boundary of the cone, $(n\Lambda_1, -(n+2)\Lambda_1+(n+1)\Lambda_2)$, is realized by

$$\widehat{\mathcal{D}}_{\sigma_1}\ldots\widehat{\mathcal{D}}_{\sigma_n}\widehat{\Omega}_{r_1}\,,\quad \sigma_1,\ldots,\sigma_n\in\{1,2,3\}\,,\quad n\geq 0\,. \tag{5.66}$$

Since

$$[\widehat{\Omega}_{r_i},\widehat{\Omega}_{r_j}] = 0\,,\quad i,j=1,2\,, \tag{5.67}$$

we find, by repeatedly using (5.34), (5.65), and the Jacobi identity for the bracket, that the $\mathfrak{sl}_3$ tensor in (5.66) is completely symmetric in $\sigma_1\ldots,\sigma_n$. The "leading term" type argument shows that those operators span the required $\mathfrak{sl}_3$ module.

Combining (5.63) with (5.66), and using the fact that the actions of $\widehat{\mathcal{D}}_\sigma$ and $\widehat{x}_{\dot\sigma}$ commute, we find that an explicit basis in $\widehat{M}_{r_1}$ consists of elements $\widehat{x}_{\dot\sigma_1}\cdot\ldots\cdot\widehat{x}_{\dot\sigma_m}\cdot\widehat{\mathcal{D}}_{\sigma_1}\ldots\widehat{\mathcal{D}}_{\sigma_n}\widehat{\Omega}_{r_1}$, $m,n\geq 0$. Moreover, since $\widehat{x}^\sigma\widehat{\mathcal{D}}_\sigma=0$, this basis also gives an explicit isomorphism $\pi_{r_1}:\widehat{M}_{r_1}\to M_{r_1}$ of $\mathfrak{sl}_3\oplus(\mathfrak{u}_1)^2$ modules,

$$\pi_{r_1}(\widehat{x}_{\dot\sigma_1}\cdot\ldots\cdot\widehat{x}_{\dot\sigma_m}\cdot\widehat{\mathcal{D}}_{\sigma_1}\ldots\widehat{\mathcal{D}}_{\sigma_n}\widehat{\Omega}_{r_1}) = x_{\dot\sigma_1}\cdot\ldots\cdot x_{\dot\sigma_m}\cdot\mathcal{D}_{\sigma_1}\ldots D_{\sigma_n}\Omega_{r_1}\,,\quad m,n\geq 0\,. \tag{5.68}$$

Using (5.62), or equivalently, $\widehat{x}_\sigma\cdot\widehat{\Omega}_{r_1}=0$, it is straightforward to evaluate the action of the triplet of the ground ring generators on the basis elements (5.68), with the result precisely that given in (4.118). Thus π_{r_1} is also an isomorphism of $\widehat{M}_{r_1}$ and M_{r_1} as ground ring modules.

The proof in the case of $\widehat{M}_{r_2}$ is similar. In the remaining three cones, $w = r_1 r_2$, $r_2 r_1$ and r_3, one cannot construct explicit bases of $\widehat{M}_w$ in terms of polyvectors acting on the corresponding operators at the tips of the cones. (However, it is easy to verify that the elements of the form $\widehat{\mathcal{D}}_{\sigma_1} \ldots \widehat{\mathcal{D}}_{\sigma_n} \widehat{\Omega}_{r_2 r_1}$ and, similarly, $\widehat{\mathcal{D}}_{\dot{\sigma}_1} \ldots \widehat{\mathcal{D}}_{\dot{\sigma}_n} \widehat{\Omega}_{r_1 r_2}$ span one of the boundaries in the respective cones.) In those cases our claim is based on first noting that by kinematics the action of the ground ring, if nontrivial, must be of the twisted type as stated in the theorem, and then verifying it by evaluating the products of the ground ring generators with the operators lying close to the tips of the cones. □

Theorem 5.19 *The isomorphisms $\pi_w : \widehat{M}_w \to M_w$ are equivariant with respect to the Lie algebra action of $\mathfrak{H}^1_1 \cong \imath(\mathfrak{P}^1)$ and $\mathfrak{P}^1$ on $\widehat{M}_w$ and M_w, respectively.*

Proof. The proof of this theorem is similar to the one above. □

5.4.2 Interpretation of $\mathfrak{H}$ in Terms of Twisted Polyderivations

We have found that the lowest ghost number subspaces of $\mathfrak{H}$ in each of the Weyl chambers may be identified with the twisted modules of the ground ring. The problem is then to extend the isomorphism π_w to a map between the higher ghost number cohomology and twisted polyderivations $\mathfrak{P}_w \equiv \mathfrak{P}(\mathcal{R}_3, M_w)$ of the ground ring. The result may be summarized as follows.

Theorem 5.20 *There is a natural map, π_w, that identifies $\widehat{M}_w$ and M_w, and maps $\Phi \in \mathfrak{H}^{\ell(w)+n}$, with $-i\Lambda^L + 2\rho$ sufficiently deep inside $w^{-1}P_+$, onto a generalized polyderivation $\pi_w(\Phi) \in \mathfrak{P}^n_w$, given by*

$$\pi_w(\Phi)(x_{i_1}, \ldots, x_{i_n}) \;=\; \pi_w([\ldots[[\Phi, \widehat{x}_{i_1}], \ldots], \widehat{x}_{i_n}]) \,. \tag{5.69}$$

Proof. Clearly π_1 is just the homomorphism π. In the other cases, although the right hand side in (5.69) is well defined for any Φ, the restriction on the Liouville weight is imposed to ensure that the multiple bracket lies in $\widehat{M}_w$. For such Φ, the proof that $\pi_w(\Phi)$ is a twisted polyderivation requires that we check the conditions in Lemma 4.7, which follow immediately using elementary properties of the dot product and the bracket. □

A more interesting question is to what extent π_w is an isomorphism between generalized polyvectors, $\mathfrak{P}_w$, and a subspace of $\mathfrak{H}$. In this respect a comparison of Theorem 4.34, which gives an enumeration of all twisted polyderivations in the bulk, with Theorem 3.25 for the cohomology, leads to the conclusion that in the bulk,

$$\mathfrak{H}^n \;\approx\; \bigoplus_{w \in W} \mathcal{P}^{n-\ell(w)}(\mathcal{R}_3, M_w) \,. \tag{5.70}$$

In fact, we should interpret this equality as a lower bound for the cohomology, and thus a partial proof of Conjecture 3.23.

The description of the cohomology in terms of twisted polyderivations in Theorem 5.20 breaks down close to the origin of the lattice of shifted Liouville momenta, because of the presence of operators that have vanishing brackets with some or even all ground ring generators, and therefore cannot be "detected" by (5.69). A particularly interesting example is the "special operator"

$$\Psi^{(2)}_{0,-\Lambda_1-\Lambda_2}(z) = c^{[2]}c^{[3]}\,\mathcal{V}_{0,-\Lambda_1-\Lambda_2}\,. \tag{5.71}$$

By explicit evaluation of all products and all brackets of this operator with the generators of $\imath(\mathfrak{P})$ we find

Lemma 5.21 *The doublet of operators* $(\Psi^{(2)}_{0,-\Lambda_1-\Lambda_2}, \widehat{C}\cdot\Psi^{(2)}_{0,-\Lambda_1-\Lambda_2})$ *is invariant under dot product and bracket with the elements of* $\imath(\mathfrak{P})$.

In particular, Lemma 5.21 implies that the dot products and the brackets of the special doublet with all ground ring generators vanish.

5.5 Towards the Complete Structure of $\mathfrak{H}$

It remains an open problem to understand how the bulk regions of cohomology, parametrized in terms of twisted polyderivations of the ground ring, are "glued" together. We present here a partial answer to this question that is essentially based on explicit computations of products and brackets between the low lying operators in $\mathfrak{H}$. A more complete understanding, which builds on these results, is discussed in the next section.

As in the analogous problem for the Virasoro cohomology, which has been exhaustively discussed in [LiZu2] and was summarized in Sect. 1.3, the starting point is to understand the action of the BV operator b_0 on $\mathfrak{H}$. The next step will be to unravel the structure of $\mathfrak{H}$ as a module of $\mathfrak{P}$.

5.5.1 The BV Operator b_0

As a simple application of the results in Sect. 5.2.2 we have

Theorem 5.22 *The cohomology of* b_0 *on* $\mathfrak{H}$ *is trivial.*

Proof. Suppose that $b_0\Psi = 0$, where $\Psi \in \mathfrak{H}$ has Liouville momentum $-i\Lambda^L = t_1\Lambda_1 + t_2\Lambda_2$. From (5.24) we find that unless $t_1 = t_2 = -2$, either $\widehat{C}_+$ or $\widehat{C}_-$ yield a contracting homotopy for b_0. In the exceptional case we find that there is simply a quartet of operators in the complex, $\mathfrak{C}$, all of which are nontrivial in cohomology (see Table D.7). Those are

$$\begin{aligned} T_{0,-2\Lambda_1-2\Lambda_2} &= c^{[2]}\partial c^{[3]} c^{[3]} \mathcal{V}_{0,-2\Lambda_1-2\Lambda_2}\,, \\ T^{[2]}_{0,-2\Lambda_1-2\Lambda_2} &= \partial c^{[2]} c^{[2]} \partial c^{[3]} c^{[3]} \mathcal{V}_{0,-2\Lambda_1-2\Lambda_2}\,, \\ T^{[3]}_{0,-2\Lambda_1-2\Lambda_2} &= c^{[2]}\partial^2 c^{[3]} \partial c^{[3]} c^{[3]} \mathcal{V}_{0,-2\Lambda_1-2\Lambda_2}\,, \\ T^{[23]}_{0,-2\Lambda_1-2\Lambda_2} &= \partial c^{[2]} c^{[2]} \partial^2 c^{[3]} \partial c^{[3]} c^{[3]} \mathcal{V}_{0,-2\Lambda_1-2\Lambda_2}\,. \end{aligned} \tag{5.72}$$

They form two doublets under b_0,

$$b_0\, T^{[2]}_{0,-2\Lambda_1-2\Lambda_2} = T_{0,-2\Lambda_1-2\Lambda_2}\,, \quad b_0\, T^{[23]}_{0,-2\Lambda_1-2\Lambda_2} = T^{[3]}_{0,-2\Lambda_1-2\Lambda_2}\,, \tag{5.73}$$

which shows that indeed the cohomology of b_0 is trivial. □

Remark. Note that $T_{0,-2\Lambda_1-2\Lambda_2}$ is the tachyon operator, proportional to $\widehat{\Omega}_{w_0}$, while $T^{[23]}_{0,-2\Lambda_1-2\Lambda_2} = 1728\sqrt{3}\widehat{X}$. More generally, the tachyon operators arise at momenta $(\Lambda, w\Lambda - 2\rho)$, $\Lambda \in P_+$, $w \in W$ [BLNW1]. The quartet of cohomology operators associated with each tachyon is then given by (5.72), but with $\mathcal{V}_{\Lambda,w\Lambda-2\rho}$. It decomposes into two doublets under the action of b_0, as in (5.73).

An immediate consequence of Theorem 5.22 is that all cohomology states are paired into doublets. This does not yet explain the quartet structure, which one might want to associate with the presence of another BV type operator. A naive candidate for such an operator is $b_0^{[3]}$. It turns out, however, that the latter is not a well defined operator on $\mathfrak{H}$, as is easily seen in the following example:

$$b_0^{[3]} C^{[2]}(z) = 8(b^{[2]}c^{[3]})(z)\,. \tag{5.74}$$

The operator on the right hand side is not annihilated by d.

Another consequence of Theorem 5.22 is that the image of polyderivations, $\imath(\mathfrak{P})$, in $\mathfrak{H}$ is not closed under b_0. Indeed if it were, then, given (5.45), this would contradict Theorem 4.26. An obvious example of an operator that is mapped by b_0 outside $\imath(\mathfrak{P})$ is $\widehat{X}$, the image of the Δ-homology class X. Let us denote $\widehat{\Gamma} = b_0\widehat{X}$. It follows from (5.73) that this operator is nonzero.

The nonclosure of $\imath(\mathfrak{P})$ under b_0 also implies nonclosure under the bracket, and we have seen an example to that effect in Sect. 5.2.4.

5.5.2 The Dual Decomposition of $\mathfrak{H}$

The description of $\mathfrak{H}$ in terms of polyderivations $\mathfrak{P}_w$, for $w = r_1$ and r_2, may be generalized to also include the states at the boundaries of those regions. Together with the duality of $\mathfrak{H}$, this will allow an explicit description of the dot module structure of $\mathfrak{H}$ over $\mathfrak{P}$.

Consider $\mathfrak{I} = \operatorname{Ker}\pi$. By Theorem 5.13, $\mathfrak{I}$ is a BV ideal. Thus it is also a BV module of $\mathfrak{H}$ provided we set $\Delta_M = \Delta\big|_{\mathfrak{I}}$. Moreover, we have $\mathfrak{I}^n = 0$ for $n < 1$, and $\mathfrak{I}^1 \cong \widehat{M}_{r_1} \oplus \widehat{M}_{r_2}$. Consider $\mathfrak{I}^1$ as a Lie algebra.

Lemma 5.23 *$\mathfrak{I}^1$ is an Abelian Lie algebra; i.e., the bracket $[-\,,-]$ vanishes when restricted to $\mathfrak{I}^1$.*

Proof. The vanishing of the bracket on $\widehat{M}_{r_1}$ and $\widehat{M}_{r_2}$ follows by kinematics. Indeed, for $\Phi \in \widehat{M}_{r_1}(\Lambda)$ and $\Phi' \in \widehat{M}_{r_1}(\Lambda')$ the bracket, $[\Phi, \Phi']$, has Liouville weight

$$r_1(\Lambda + \Lambda' + 2\rho) - 2\rho \;=\; r_1\left((\Lambda + \Lambda' + \alpha_1) + \rho\right) - \rho\,, \tag{5.75}$$

and thus must vanish because the irreducible representation with highest weight $\Lambda + \Lambda' + \alpha_1$ cannot arise in the tensor product $\mathcal{L}(\Lambda) \otimes \mathcal{L}(\Lambda')$. As for the bracket between $\widehat{M}_{r_1}$ and $\widehat{M}_{r_2}$, we start with (5.67). We then proceed by induction, using the explicit bases in $\widehat{M}_{r_i}$ constructed in Sect. 5.4.1, the vanishing relations

$$\widehat{x}_\sigma \cdot \widehat{\Omega}_{r_1} \;=\; 0\,, \quad \widehat{\mathcal{D}}_{\dot\sigma} \widehat{\Omega}_{r_1} \;=\; 0\,, \quad \widehat{x}_{\dot\sigma} \cdot \widehat{\Omega}_{r_2} \;=\; 0\,, \quad \widehat{\mathcal{D}}_\sigma \widehat{\Omega}_{r_2} \;=\; 0\,, \tag{5.76}$$

and the properties of the bracket. □

Remark. Similar arguments show that, given $\widehat{\Omega}_{r_i} \cdot \widehat{\Omega}_{r_j} = 0$, $i, j = 1, 2$, we must have $\widehat{M}_{r_i} \cdot \widehat{M}_{r_j} = 0$ as well.

Since $\mathfrak{I}^1$ is a ground ring module, as well as the lowest ghost number subspace in $\mathfrak{I}$, it is natural to repeat the construction of Sect. 5.3. Namely, consider the map

$$\pi' \equiv \pi_{r_1} \oplus \pi_{r_2} \;:\; \mathfrak{I}^n \longrightarrow \mathcal{P}^{n-1}(\mathcal{R}_3, \mathfrak{I}^1) \cong \mathfrak{P}_{r_1}^{n-1} \oplus \mathfrak{P}_{r_2}^{n-1}\,, \tag{5.77}$$

which is equal to the identity on $\mathfrak{I}^1$, while for $n \geq 2$ it is given by the multiple brackets (5.69). (Since $\mathfrak{I}$ is a BV ideal, all brackets (5.69) lie in $\mathfrak{I}$ for all $\Phi \in \mathfrak{I}$, the map π' is well-defined on $\mathfrak{I}$, which of course includes the bulk region in Theorem 5.18.) It is straightforward to verify, by induction on the ghost number (as in Sect. 4.1.4), that

$$\pi'(\Phi \cdot \Psi) \;=\; \pi(\Phi) \cdot \pi'(\Psi)\,, \qquad \Phi \in \mathfrak{H}\,, \quad \Psi \in \mathfrak{I}\,, \tag{5.78}$$

where the product on the right hand side corresponds to the dot action of $\mathfrak{P}$ on $\mathfrak{P}_{r_1} \oplus \mathfrak{P}_{r_2}$. In fact, the latter space is a G module of $\mathfrak{P}$ (see Theorem 4.33), and the following stronger result holds:

Theorem 5.24 *The map π' is a G morphism between the G module $\mathfrak{I}$ of $\mathfrak{H}$ and the G module $\mathfrak{P}_{r_1} \oplus \mathfrak{P}_{r_2}$ of $\mathfrak{P}$; i.e., in addition to (5.78) we also have*

$$\pi'([\Phi, \Psi]) \;=\; [\pi(\Phi), \pi'(\Psi)]_M\,, \qquad \Phi \in \mathfrak{H}\,, \quad \Psi \in \mathfrak{I}\,. \tag{5.79}$$

Proof. Let $\Phi \in \mathfrak{H}^m$ and $\Psi \in \mathfrak{I}^n$. Once more the proof follows by induction on $m + n$. In particular, for $m = 0$ and $n = 1$ both sides of (5.79) vanish – the left side because $\mathfrak{I}^0 \cong 0$, and the right side by the definition of the bracket action of $\mathfrak{P}$ on $\mathfrak{P}_{r_1} \oplus \mathfrak{P}_{r_2}$. Next take $m = 1$ and $n = 1$. Using the decomposition (5.37) and Lemma 5.23, the only case in which both sides do not vanish automatically is for $\Phi \in \mathfrak{H}_1^1 \cong \imath(\mathfrak{P})$. Then the equality follows from the isomorphism of $\widehat{M}_{r_i}$ and M_{r_i} as G modules. The general step of the induction is now completed similarly as in the proof of Theorem 5.13, using the definition of the bracket action of $\mathfrak{P}$. □

Conjecture 5.25 *Consider* $\mathfrak{P}_{r_1} \oplus \mathfrak{P}_{r_2}$ *as a BV module of* $\mathfrak{P}$, *with the (conjectured) BV operator* $\Delta = \Delta_1 \oplus \Delta_2$ *defined in (4.128) and (4.129). Then* π' *is a BV morphism between BV modules.*

The ideal $\mathfrak{I}$ at weights $(0, -2\Lambda_1 - 2\Lambda_2)$ is spanned by the operators $T_{0,-2\Lambda_1-2\Lambda_2}$, $T^{[2]}_{0,-2\Lambda_1-2\Lambda_2}$ and $T^{[3]}_{0,-2\Lambda_1-2\Lambda_2} \sim \widehat{\Gamma}$. By an explicit computation we verify that, while

$$\pi'(\widehat{\Gamma}) = \Gamma_1 + \Gamma_2, \tag{5.80}$$

the other two operators are mapped to zero. This shows that π' has a nontrivial cokernel. In fact, by examining the $\mathfrak{sl}_3 \oplus (\mathfrak{u}_1)^2$ decomposition of $\mathfrak{I}$ and $\mathfrak{P}_{r_1} \oplus \mathfrak{P}_{r_2}$, as well as a number of explicit checks, we conclude that π' is onto except at the weight $(0, -2\Lambda_1 - 2\Lambda_2)$. Let us denote the image $\pi'(\mathfrak{I}) = \mathfrak{P}'$.

Let $\mathfrak{I}' = \operatorname{Ker} \pi'$. Using (5.78) and (5.79) we show that $\mathfrak{I}'$ is a G ideal (conjecturally, a BV ideal) of $\mathfrak{H}$. It will turn out convenient to factor out from $\mathfrak{I}'$ the doublet, $\mathfrak{D}_{sp}$, of special states introduced in Lemma 5.21, and write

$$\mathfrak{I}' \cong \mathfrak{H}_- \oplus \mathfrak{D}_{sp}. \tag{5.81}$$

Consider the quotient $\mathfrak{H}_+ \cong \mathfrak{H}/\mathfrak{H}_-$. Note that as a vector space $\mathfrak{H}_+$ is isomorphic with $\mathfrak{P} \oplus \mathfrak{P}' \oplus \mathfrak{D}_{sp}$. By examining the $\mathfrak{sl}_3 \oplus (\mathfrak{u}_1)^2$ decomposition of $\mathfrak{H}_-$ and $\mathfrak{H}_+$ we concluded that each comprises precisely "one half of the cohomology" in the following sense.

Conjecture 5.26 *Let* $\langle -, - \rangle_{\mathfrak{H}}$ *be the nondegenerate bilinear form on* $\mathfrak{H}$, *introduced in Sect. 3.3.4. Then*

i. *The form* $\langle -, - \rangle_{\mathfrak{H}}$ *vanishes identically on* $\mathfrak{H}_-$.
ii. *As a vector space,* $\mathfrak{H}_+$ *is isomorphic with the dual subspace to* $\mathfrak{H}_-$ *in* $\mathfrak{H}$ *with respect to this form.*

Most of this conjecture follows from the $\mathfrak{sl}_3 \oplus (\mathfrak{u}_1)^2$ decomposition. Only the cases where at the same Liouville weight there are states both in $\mathfrak{H}_+$ and $\mathfrak{H}_-$ require a more detailed analysis. We have explicitly checked some of those cases for low lying weights. The extension to the general case is then consistent with the expected module structure of both spaces with respect to the dot action of polyderivations to be discussed shortly.

The question now is whether one can construct $\mathfrak{H}_+$ as a (natural) subspace in $\mathfrak{H}$. In other words we would like to find an extension of the embedding $\imath : \mathfrak{P} \to \mathfrak{H}$ to $\mathfrak{P}'$ and $\mathfrak{D}_{sp}$. The embedding of the special doublet is unambiguous. However, simple kinematics shows that such an embedding on $\mathfrak{P}'$ is ambiguous in the overlap regions with $\imath(\mathfrak{P})$ and $\mathfrak{H}_-$. To deal with this problem we may proceed as in the proof of Theorem 5.13, and use the explicit parametrization of $\mathfrak{P}'$ in terms of free modules of the chiral subalgebras given in Appendix H. Thus we set

$$\imath(\Gamma_i) = \widehat{\Gamma}\,, \tag{5.82}$$

and then require that $\imath(\mathfrak{P}_{r_i})$ is freely generated from $\widehat{\Gamma}$ in $\mathfrak{H}$ as a G submodule of the respective holomorphic subalgebra $\imath(\mathfrak{P}_-)$ or $\imath(\mathfrak{P}_+)$. More explicitly, this construction yields

$$\begin{aligned} \imath(\Phi_0 \cdot [\Phi_1,[\ldots,[\Phi_n,\Gamma_i]\ldots]]) &= \imath(\Phi_0)\cdot[\imath(\Phi_1),[\ldots,[\imath(\Phi_n),\widehat{\Gamma}]\ldots]]\,, \\ &\qquad \Phi_0,\ldots,\Phi_n \in \mathfrak{P}_{\mp}\,. \end{aligned} \tag{5.83}$$

From now on we will identify $\mathfrak{H}_+$ with the image $\imath(\mathfrak{P}\oplus\mathfrak{P}'\oplus\mathfrak{D}_{sp}) \subset \mathfrak{H}$.

To summarize, we have constructed an explicit decomposition

$$\mathfrak{H} \cong \mathfrak{H}_- \oplus \mathfrak{H}_+\,, \tag{5.84}$$

where $\mathfrak{H}_\pm$ is completely isotropic with respect to the bilinear form on $\mathfrak{H}$, and $\mathfrak{H}_+$ is dual to $\mathfrak{H}_-$.

The duality between $\mathfrak{H}_-$ and $\mathfrak{H}_+$, due to the "hermiticity" of the ground ring generators with respect to the bilinear form (which can be proved using explicit expressions in Appendix J.1), holds as the duality of ground ring modules.

It follows from Conjecture 5.25 that $\mathfrak{H}_-$ should be a BV ideal in $\mathfrak{H}$. A combination of kinematics and explicit checks suggest that $\mathfrak{H}_+ \subset \mathfrak{H}$ is a submodule with respect to the dot action of the subalgebra $\imath(\mathfrak{P}) \subset \mathfrak{H}$.

Let us briefly compare the result above with the one for the BV algebra associated with the Virasoro string.[3] The decomposition (5.84) of the algebra as a ground ring module is an analogue of the similar decomposition of $H(\mathcal{W}_2,\mathfrak{C})$ [LiZu2]. However, unlike in the Virasoro case, now $\mathfrak{H}_+$ is much larger than the algebra of polyderivations of the ground ring.

We will not pursue this line of thought any further, but now turn to a more complete description.

5.6 The Complete Structure of $\mathfrak{H}$

In the previous sections we have seen that the cohomology $\mathfrak{H} \equiv H(\mathcal{W}_3,\mathfrak{C})$ possesses the structure of a BV algebra (Theorem 5.1), and that part of $\mathfrak{H}$ can be identified with the BV algebra $\mathcal{P}(\mathcal{R}_3)$ of polyderivations of the Abelian algebra $\mathcal{R}_3$ (Sect. 4.2). More precisely, that we have a BV epimorphism $\pi : \mathfrak{H} \to \mathcal{P}(\mathcal{R}_3)$ (Theorem 5.13). Furthermore, in Sects. 5.4 and 5.5.1 we have seen that (most of the) remaining part of $\mathfrak{H}$ can be modeled on the algebra of twisted polyderivations of $\mathcal{R}_3$.

At the same time, we have shown that $\mathcal{P}(\mathcal{R}_3)$ can be canonically identified with the BV algebra $\mathcal{P}(A)$ of polyvector fields on the base affine space A of $SL(3,\mathbb{C})$ (Sect. 4.6). Extrapolating from the representation as $\mathfrak{n}_+$ invariants, i.e.,

[3] See the summary in Sect. 1.3.

$$\mathcal{P}(A) \cong (\mathcal{E}(SL(3,\mathbb{C})) \otimes \bigwedge \mathfrak{b}_-)^{\mathfrak{n}_+} = H^0(\mathfrak{n}_+, \mathcal{E}(SL(3,\mathbb{C})) \otimes \bigwedge \mathfrak{b}_-), \quad (5.85)$$

we have introduced the natural BV extension

$$\mathrm{BV}[\mathfrak{sl}_3] \equiv H(\mathfrak{n}_+, \mathcal{E}(SL(3,\mathbb{C})) \otimes \bigwedge \mathfrak{b}_-). \quad (5.86)$$

On comparing Tables 3.2 (Sect. 3.5.3) and 4.6 (Sect. 4.6.3) one immediately sees that $\mathfrak{H}$ and BV[$\mathfrak{sl}_3$] are isomorphic as $\mathfrak{sl}_3 \oplus \mathfrak{h}$ modules. Further, in Sect. 4.6.5 we have shown that BV[$\mathfrak{sl}_3$] may be decomposed in terms of twisted polyderivations in direct parallel with $\mathfrak{H}$.

The abovementioned evidence strongly supports the following

Claim 5.27 *We have an isomorphism of BV algebras*

$$H(\mathcal{W}_3, \mathfrak{C}) \cong \mathrm{BV}[\mathfrak{sl}_3] \equiv H(\mathfrak{n}_+, \mathcal{E}(SL(3,\mathbb{C})) \otimes \bigwedge \mathfrak{b}_-). \quad (5.87)$$

Note that the fact that BV[$\mathfrak{sl}_3$] is acyclic with respect to the BV operator Δ (Theorem 4.47 and below) is crucial for the validity of the claim, i.e., for making the identification $\Delta = b_0$.

We have proved that the analogous claim is true for $\mathfrak{g} \cong \mathfrak{sl}_2$, i.e., for the Virasoro algebra, where the cohomology on the left hand side of (5.87) was computed in [BMP3,LiZu1] and its BV structure was unraveled in [LiZu2,WuZh]. This result is summarized in the introduction to this book (Sect. 1.3.3). Based on the above, it is now natural to conjecture that the analogue of Claim 5.27 will hold for all algebras $\mathcal{W}[\mathfrak{g}]$ based on a finite-dimensional simple simply-laced Lie algebra $\mathfrak{g}$.

Note that the result 5.27 gives a precise characterization of the chiral operator algebra of the $\mathcal{W}_3$ string in terms of a very manageable and geometrically-realized BV algebra. A closely related claim has recently appeared in [LiZu3].

5.7 Concluding Remarks and Open Problems

To conclude this book we briefly discuss several interesting avenues for further research which have arisen in our study. It is quite likely that some of them will require qualitatively new insights beyond the present work; where possible, we indicate what we believe to be the most promising approach.

1. The proof of the cohomology for $-i\Lambda^L + 2\rho \notin P_+ \cup w_0 P_+$.

Despite the obvious complexity of the problem, it still seems surprising that, unlike in the case of the Virasoro algebra, the cohomology of the $\mathcal{W}_3$ algebra with values in the tensor product of two Fock modules cannot be computed without resorting to an indirect procedure. While the formal origin of this difficulty is clear – the nature of the quartic terms in the differential precludes any simple spectral sequence argument – one might hope that by a suitable field redefinition the problem could become tractable.

Alternatively, one could try to implement a proof more along the lines of that which works in the fundamental Weyl chamber. For this, one must construct a new class of highest weight modules of the $\mathcal{W}_3$ algebra that are "dual" to the $c^L = 98$ Fock modules $F(\Lambda^L, 2i)$, $-i\Lambda^L + 2\rho \notin P_+ \cup w_0 P_+$, in the sense of the reduction theorem (Theorem 3.8), just as contragredient Verma modules are "dual" to Fock modules when $-i\Lambda^L + 2\rho \in P_+$. (This cohomological "construction" of modules is largely motivated by the analogous problem in the representation theory of affine Lie algebras, where the corresponding cohomology is that with respect to the (twisted) nilpotent subalgebra – see [FeFr1], and also [BMP4], for further details.) By constructing resolutions of $c^M = 2$ irreducible modules in terms of those new modules one could compute the cohomology in a straightforward manner.

2. Is $\mathfrak{H}$ generated by $\imath(\mathfrak{P})$ as a BV algebra?

The operator algebra $\mathfrak{H}$ could be further elucidated. Let us briefly recall the broad structure.

We have seen that $\imath(\mathfrak{P})$ is not closed under b_0. In fact the subspace generated by the bracket and dot action of $\imath(\mathfrak{P})$ on itself contains at least the subspace $\mathfrak{H}_+$. This follows from the discussion in Sect. 5.5.2, and the observation that the special doublet, $\mathfrak{D}_{\rm sp}$, lies in the subspace spanned by elements of the form $[\widehat{x}_\sigma, [\widehat{x}_{\dot\sigma}, b_0\widehat{X}]$ and $\widehat{C} \cdot [\widehat{x}_\sigma, [\widehat{x}_{\dot\sigma}, b_0\widehat{X}]$.

Moreover, while $\imath(\mathfrak{P})$ is closed under the dot product this is not the case for $\mathfrak{H}_+$: dot products of some elements in $\mathfrak{H}_+ \cap \mathfrak{I}$ lie in $\mathfrak{H}_-$.[4] For instance, the "square" of the special state yields

$$\Psi^{[2]}_{0,-\Lambda_1-\Lambda_2} \cdot \Psi^{[2]}_{0,-\Lambda_1-\Lambda_2} \sim T^{[2]}_{0,-2\Lambda_1-2\Lambda_2} . \tag{5.88}$$

Similarly, further products between $\mathfrak{H}_+ \cap \mathfrak{I}$ and $\mathfrak{I}'$ are nonvanishing. A good example is given by the products of the tips of twisted cones $\widehat{\Omega}_w$, which lie in $\mathfrak{I}'$ for $w = r_{12}$, r_{21} and r_3. We find

$$\widehat{\Omega}_{r_1} \cdot \widehat{\Omega}_{r_{12}} \sim \widehat{\Omega}_{r_3} , \quad \widehat{\Omega}_{r_2} \cdot \widehat{\Omega}_{r_{21}} \sim \widehat{\Omega}_{r_3} . \tag{5.89}$$

To gain some insight into the full structure of $\mathfrak{H}$ we have studied the BV subalgebra $\mathfrak{H}_{\rm singl}$, consisting of all elements of $\mathfrak{H}$ transforming as singlets under $\mathfrak{sl}_3$. This algebra is finite dimensional and is spanned by the 19 quartets, as is easily read from Table 3.2. The elements of $\imath(\mathfrak{P})$ form a quartet at the Liouville weight 0, three doublets (with respect to b_0) at $\Lambda_1 - 2\Lambda_2$, $-2\Lambda_1 + \Lambda_2$ and $-\Lambda_1 - \Lambda_2$, and a single element, $\widehat{X}$, at $-2\Lambda_1 - 2\Lambda_2$. It appears that those elements generate the entire $\mathfrak{H}_{\rm singl}$ as a BV algebra.

At this point it is tempting to conjecture that also $\mathfrak{H}$ is generated from $\imath(\mathfrak{P})$. Unfortunately, we were not able to calculate any nontrivial example, beyond the singlet subalgebra, that would support such conjecture. If this conjecture turned

[4] Note that in the Virasoro case the dot product on $\mathfrak{I}$ is always zero.

out false, one would have to understand what is the significance of the (proper) subalgebra generated by $\imath(\mathfrak{P})$ inside $\mathfrak{H}$.

Further, or perhaps alternatively, it is interesting in this context to pursue the decomposition of $\mathrm{BV}[\mathfrak{sl}_3]$ in terms of subsequent kernels of G morphisms as discussed in Sect. 5.6. A direct proof of the corresponding structure in $\mathfrak{H}$ would illuminate the geometrical nature of the result. This leads to the obvious need for

3. The proof of Claim 5.27.

We consider Claim 5.27, and its conjectured extension to arbitrary $\mathcal{W}$ gravities, perhaps the most important result in this book. As such, a complete proof would be highly desirable. It may be that this proof can be completed along the lines of this chapter, in particular the proof of Theorem 5.13, but one can only hope that eventually a proof will be found that does not require such an explicit verification for the generators and their relations. However, the result itself suggests a deep relation to the geometry of Lie groups and cosets, and in this vein we think it likely that such a proof may be found in the context of Drinfel'd-Sokolov reduction. If such a proof does become available, it would at the same time answer question 1; i.e., it would give a complete proof of Theorem 3.25 for the cohomology $\mathfrak{H}$.

4. Further geometrical interpretation of the result.

One could pursue the decomposition of $\mathrm{BV}[\mathfrak{sl}_3]$ in terms of subsequent kernels of the G morphisms π_n introduced in Sect. 4.6.5, extrapolating the parallel discussion in Sect. 5.5.2. A direct proof of the corresponding structure in $\mathfrak{H}$ would illuminate the geometrical nature of the result.

In this context, it may be interesting to find an explicit representation of the image of the π_n as *generalized* polyvector fields on the base affine space A, building on the discussion in Sect. 4.6.5. Of course, to begin one needs a definition of such objects along the lines of the definition (4.136) for regular polyvector fields. A rather obvious conjecture would be that, for bulk weights, a representative of BV can be chosen so that the projection under π_w satisfies

$$\Pi^L(w(x))\Phi = -\Pi^{bc}(x)\Phi\,, \quad x \in \mathfrak{n}_+\,. \tag{5.90}$$

This is consistent with the expected Weyl twisting of the result in the fundamental Weyl chamber. Further, by explicit construction, it seems that the leading term of the projection – in the grading of Theorem 4.42 – satisfies this equation to leading order. This seems to hint at a kind of "Hodge-theoretic" understanding of the cohomology, and is most intriguing.

5. Application to string theory.

At this stage it is possible to contemplate possible physical applications of this work. Already for the Virasoro case it would be very interesting if there is a geometrical description of the *closed* string (the operator algebra of the semirelative cohomology) along the lines of the present work. The relation of various string deformations to deformations of this geometry could be fascinating. The extension of this work to the $\mathcal{W}$ generalizations would then be conceivable. An exciting possibility in this regard would be an understanding of regions in the Calabi-Yau moduli space of string compactifications, in a similar way to how the 2D string describes the conifold points [GoVa]. Further, it should be possible to analyze the tachyon amplitudes along the same line as in the Virasoro case.

These problems could be of interest to both mathematicians and physicists, and we look forward to any progress in these directions.

Appendix A Verma Modules at $c = 2$

A.1 Primitive Vectors

In Tables A.1–A.4 we list the low-lying primitive vectors in the Verma modules $M^{(i)}[s_1, s_2]$ for $i = 1, 2$. We denote by

$$(v_{s_1 s_2}) \equiv (v_{s_1 s_2}, u_{s_1 s_2}), \qquad (w_{s_1 s_2}) \equiv (w_{s_1 s_2}, v_{s_1 s_2}, u_{s_1 s_2}),$$

a doublet and a triplet of states, respectively (cf. Sect. 2.3.2).

Table A.1. Primitive vectors in $M[s_1, s_2]$ (triality 0)

h	$M[0,0]$	$M[1,1]$	$M[3,0]$	$M[0,3]$	$M[2,2]$	$M[4,1]$	$M[1,4]$	$M[3,3]$	$\cdots$
0	u_{00}								
1	(v_{11})	u_{11}							
3	u_{30}	u_{30}	u_{30}						
	u_{03}	u_{03}		u_{03}					
4	(w_{22})	(v_{22})	u_{22}	u_{22}	u_{22}				
7	$\vdots$	(v_{41})	(v_{41})	u_{41}	u_{41}	u_{41}			
	$\vdots$	(v_{14})	u_{14}	(v_{14})	u_{14}		u_{14}		
9	$\vdots$	(w_{33})	(v_{33})	(v_{33})	(v_{33})	u_{33}	u_{33}	u_{33}	
$\vdots$	$\vdots$	$\vdots$	$\vdots$	$\vdots$	$\vdots$	$\vdots$	$\vdots$	$\vdots$	$\vdots$

Table A.2. Primitive vectors in $M^{(2)}[s_1, s_2]$ (triality 0)

h	$M^{(2)}[1,1]$	$M^{(2)}[2,2]$	$M^{(2)}[4,1]$	$M^{(2)}[1,4]$	$M^{(2)}[3,3]$	$\cdots$
1	(v_{11})					
3	u_{30}, u'_{30}					
	u_{03}, u'_{03}					
4	(w_{22}), u'_{22}	(v_{22})				
7	$\vdots$	(v_{41})	(v_{41})			
	$\vdots$	(v_{14})		(v_{14})		
9	$\vdots$	(w_{33}), u'_{33}	(v_{33})	(v_{33})	(v_{33})	
$\vdots$	$\vdots$	$\vdots$	$\vdots$	$\vdots$	$\vdots$	$\vdots$

Table A.3. Primitive vectors in $M[s_1, s_2]$ (triality 1)

h	$M[1,0]$	$M[0,2]$	$M[2,1]$	$M[1,3]$	$M[4,0]$	$M[3,2]$	$M[0,5]$	$M[2,4]$	$\cdots$
1/3	u_{10}								
4/3	u_{02}	u_{02}							
7/3	(v_{21})	u_{21}	u_{21}						
13/3	(v_{13})	(v_{13})	u_{13}	u_{13}					
16/3	u_{40}	u_{40}	u_{40}		u_{40}				
19/3	(w_{32})	(v_{32})	(v_{32})	u_{32}	u_{32}	u_{32}			
25/3	$\vdots$	u_{05}	u_{05}	u_{05}			u_{05}		
28/3	$\vdots$	(w_{24})	(v_{24})	(v_{24})	u_{24}	u_{24}	u_{24}	u_{24}	
$\vdots$	$\vdots$	$\vdots$	$\vdots$	$\vdots$	$\vdots$	$\vdots$	$\vdots$	$\vdots$	$\vdots$

Table A.4. Primitive vectors in $M^{(2)}[s_1, s_2]$ (triality 1)

h	$M^{(2)}[2,1]$	$M^{(2)}[1,3]$	$M^{(2)}[3,2]$	$M^{(2)}[2,4]$	$\cdots$
7/3	(v_{21})				
13/3	(v_{13})	(v_{13})			
16/3	u_{40}, u'_{40}				
19/3	(w_{32}), u'_{32}	(v_{32})	(v_{32})		
25/3	$\vdots$	u_{05}, u'_{05}			
28/3	$\vdots$	(w_{24}), u'_{24}	(v_{24})	(v_{24})	
$\vdots$	$\vdots$	$\vdots$	$\vdots$	$\vdots$	$\vdots$

A.2 Irreducible Modules

In Tables A.5 and A.6 we give the dimension of $L[s_1, s_2]_{(h)}$ for small h.

Table A.5. dim $L[s_1, s_2]_{(h)}$ (triality 0)

$h\backslash[s_1, s_2]$	[0, 0]	[1, 1]	[3, 0]	[2, 2]	[4, 1]	[3, 3]
0	1					
1	0	1				
2	1	2				
3	2	3	1			
4	3	6	1	1		
5	4	10	3	2		
6	8	16	5	5		
7	10	27	9	8	1	
8	17	42	14	16	2	
9	24	64	25	26	4	1
10	36	98	37	45	8	2

Table A.6. dim $L[s_1, s_2]_{(h)}$ (triality 1)

$h\backslash[s_1, s_2]$	[1, 0]	[0, 2]	[2, 1]	[1, 3]	[4, 0]	[3, 2]	[0, 5]	[2, 4]
1/3	1							
4/3	1	1						
7/3	2	1	1					
10/3	3	3	2					
13/3	6	4	4	1				
16/3	9	8	7	2	1			
19/3	15	12	13	4	1	1		
22/3	22	21	21	8	3	2		
25/3	35	31	35	14	5	5	1	
28/3	51	50	55	24	10	9	1	1
31/3	77	73	87	40	15	17	3	2

A.3 Verma Modules

Tables A.7–A.9 give the dimensions of some submodules generated by sets, S, of primitive vectors in the Verma modules $M[s_1, s_2]$.

Table A.7. $\dim M(S)_{(h)}$ for $S \subset M[0,0]$

$h \backslash S$	$\{u_{00}\}$	$\{v_{11}\}$	$\{u_{11}, w_{22}\}$	$\{u_{11}\}$	$\{u_{30}, u_{03}, v_{22}\}$	$\{u_{30}, u_{03}\}$	$\{u_{30}, v_{22}\}$	$\{u_{30}\}$	$\{v_{22}\}$	$\{u_{22}\}$
0	1									
1	2	2	1	1						
2	5	4	2	2						
3	10	8	5	5	2	2	1	1		
4	20	17	11	10	4	3	3	2	2	1
5	36	32	22	20	10	8	7	5	4	2
6	65	57	41	36	20	15	15	10	10	5

Table A.8. $\dim M(S)_{(h)}$ for $S \subset M[1,0]$

$h \backslash S$	$\{u_{10}\}$	$\{v_{21}\}$	$\{v_{21}, u_{20}\}$
1/3	1		
4/3	2		1
7/3	5	2	3
10/3	10	4	7
13/3	20	10	14
16/3	36	19	27
19/3	65	38	50
22/3	110	*	*

Table A.9. $\dim M(S)_{(h)}$ for $S \subset M[1,1]$

$h \backslash S$	$\{u_{11}\}$	$\{u_{30}, u_{03}, v_{22}\}$	$\{u_{30}, u_{03}\}$	$\{u_{30}, u_{03}, w_{33}\}$	$\{u_{03}, w_{33}, v_{41}\}$	$\{u_{30}\}$	$\{v_{22}\}$
1	1						
2	2						
3	5	2	2	2	1	1	
4	10	4	3	3	2	2	2
5	20	10	8	8	5	5	4
6	36	20	15	15	10	10	10
7	65	38	30	30	21	20	20
8	110	68	52	52	*	36	40
9	185	121	94	95	*	65	71
10	300	202	155	157	*	110	128

Appendix B Vertex Operator Algebras Associated to Root Lattices

In this appendix we explicitly construct a VOA, in the chiral algebra $\mathfrak{V}$ of two free scalar fields with momenta lying on the root lattice of $\mathfrak{sl}_3$, which includes the currents of the affine Lie algebra $\widehat{\mathfrak{sl}_3}$. As discussed under (2.56), the principal condition which we must account for is the "statistics" of the VOA – that under interchange of order the OPEs of any two fields in the VOA are related by analytic continuation.

Let Q be the root lattice of a simple simply-laced Lie algebra $\mathfrak{g}$, and let c_α be a set of (momentum dependent) operators on Q. The VOA associated to the lattice Q involves, in particular, the assignment of a vertex operator $\mathcal{V}_\alpha(z) = V_\alpha(z)c_\alpha$ to each $\alpha \in Q$, where $V_\alpha(z) = e^{i\alpha\cdot\phi(z)}$ and c_α is chosen such that $\hat{c}_\alpha \equiv e^{iq\cdot\alpha}c_\alpha$ satisfies

$$\begin{aligned} \hat{c}_\alpha \hat{c}_\beta &= e^{i\pi(\alpha,\beta)}\hat{c}_\beta \hat{c}_\alpha \,, \\ \hat{c}_0 &= 1 \,, \end{aligned} \tag{B.1}$$

for all $\alpha, \beta \in Q$. Note that (B.1) is precisely required to implement the statistics condition, since for any two exponential operators, $e^{i\lambda\cdot\phi(z)}$ and $e^{i\lambda'\cdot\phi(z)}$,

$$e^{i\lambda\cdot\phi(z)}e^{i\lambda'\cdot\phi(w)} = (z-w)^{\lambda\cdot\lambda'}e^{i\lambda\cdot\phi(z)+i\lambda'\cdot\phi(w)} \,. \tag{B.2}$$

The extension of the statistics condition to the exponential operators corresponding to the rest of the root lattice is discussed later. It is enough to just consider the purely exponential operators in $\mathfrak{V}$ since contributions to the OPE from the polynomial field prefactors are clearly meromorphic and satisfy the condition automatically.

We may interpret (B.1) as the statement that $\hat{c}_\alpha$ defines a central extension of Q by the group $\mathbb{Z}/2\mathbb{Z} \cong \{\pm 1\}$ [FrKa]. Such central extensions are uniquely specified by a 2-cocycle $\epsilon : Q \times Q \longrightarrow \{\pm 1\}$, satisfying

$$\epsilon(\alpha,\beta)\epsilon(\alpha+\beta,\gamma) = \epsilon(\alpha,\beta+\gamma)\epsilon(\beta,\gamma) \,, \tag{B.3}$$

$$\epsilon(\alpha,\beta) = e^{i\pi(\alpha,\beta)}\epsilon(\beta,\alpha) \,, \tag{B.4}$$

$$\epsilon(\alpha,0) = 1 \,, \tag{B.5}$$

for all $\alpha, \beta, \gamma \in Q$, through[1]

$$\hat{c}_\alpha \hat{c}_\beta = \epsilon(\alpha,\beta)\hat{c}_{\alpha+\beta} \,. \tag{B.6}$$

---

[1] Given a 2-cocycle $\epsilon(\alpha,\beta)$, the construction of $\hat{c}_\alpha$ is outlined in Sect. 5 of [GNOS]. Despite the slight abuse of language we will call the c_α "phase-cocycles."

Clearly, the consistency of (B.6) implies the 2-cocycle condition (B.3), i.e., the fact that $\epsilon \in H^2(Q, \mathbb{Z}/2\mathbb{Z})$, while (B.4) and (B.5) follow from (B.1).

A 2-cocycle ϵ, satisfying (B.3)-(B.5), is easily constructed as follows [FrKa]. In addition to (B.3)-(B.5) we may impose a bilinearity condition

$$\begin{aligned} \epsilon(\alpha+\beta,\gamma) &= \epsilon(\alpha,\gamma)\epsilon(\beta,\gamma)\,, \\ \epsilon(\alpha,\beta+\gamma) &= \epsilon(\alpha,\beta)\epsilon(\alpha,\gamma)\,. \end{aligned} \tag{B.7}$$

Then ϵ is completely specified by its values $\epsilon(\alpha_i, \alpha_j)$, where $1 \le i \le j \le \ell$ and $\{\alpha_i\}_{i=1}^{\ell}$ is a basis of Q (i.e., a simple root system).

In our case, where $\mathfrak{g} \cong \mathfrak{sl}_3$, we may simply choose

$$\epsilon(\alpha_1,\alpha_1) = \epsilon(\alpha_2,\alpha_2) = \epsilon(\alpha_1,\alpha_2) = 1\,, \tag{B.8}$$

from which it follows, e.g.

$$\epsilon(\alpha_1,-\alpha_1) = \epsilon(\alpha_2,-\alpha_2) = 1\,, \tag{B.9}$$

while

$$\epsilon(\alpha_2,\alpha_1) = \epsilon(\alpha_3,-\alpha_3) = -1\,. \tag{B.10}$$

In fact, for arbitrary $\alpha, \beta \in Q$ we then have

$$\epsilon(\alpha,\beta) = e^{i\pi(\Lambda_2,\alpha)(\Lambda_1,\beta)}\,. \tag{B.11}$$

A c_α, satisfying (B.1) and (B.6) for the 2-cocycle (B.11) is explicitly given by

$$c_\alpha(p) = e^{i\pi p\cdot\xi(\alpha)}\,, \tag{B.12}$$

where

$$\xi(\alpha) = (\Lambda_2,\alpha)\Lambda_1\,. \tag{B.13}$$

Now, if we restrict ourselves to $\alpha \in \Delta$ the modes of the vertex operators $\mathcal{V}_\alpha(z)$ will provide a realization of $\widehat{\mathfrak{sl}_3}$ on $\bigoplus_{\alpha\in Q} F(\alpha,0)$ ismorphic to $L(\Lambda_0)$ (i.e.the so-called basic representation), albeit not in the "conventional" form. In particular we would like to have $\epsilon(\alpha_3,-\alpha_3) = 1$. Clearly, a 2-cocycle ϵ satisfying (B.3)-(B.5) is not unique, but can be modified by a coboundary $\delta(\eta)$, $\eta : Q \longrightarrow \{\pm 1\}$, i.e.

$$\epsilon'(\alpha,\beta) = \epsilon(\alpha,\beta)\eta(\alpha)\eta(\beta)\eta(\alpha+\beta)\,. \tag{B.14}$$

This corresponds to a change

$$\hat{c}'_\alpha = \eta(\alpha)\hat{c}_\alpha\,. \tag{B.15}$$

(We need to take $\eta(0) = 1$ to preserve (B.1) or, equivalently, (B.5).) We can use this 'gauge freedom' to choose $\epsilon(\alpha,\beta)$ such that

$$\epsilon(\alpha,-\alpha) = 1\,, \tag{B.16}$$

for all $\alpha \in Q$ or, equivalently,

$$\hat{c}_\alpha\hat{c}_{-\alpha} = 1\,. \tag{B.17}$$

For example we can take

$$\eta(\alpha) = \begin{cases} 1 & \text{for } (\Lambda_1, \alpha) \geq 0 \\ e^{i\pi(\Lambda_1,\alpha)(\Lambda_2,\alpha)} & \text{for } (\Lambda_1, \alpha) < 0\,. \end{cases} \tag{B.18}$$

Note that with this choice of cocycle we automatically have

$$\begin{aligned} \overline{\omega}_{\mathcal{A}}(\hat{c}'_\alpha) &= \overline{\omega}_{\mathcal{A}}(\eta(\alpha) e^{iq\cdot\alpha} e^{i\pi\alpha\cdot\xi(\alpha)}) \\ &= \eta(\alpha)\eta(-\alpha) e^{i\pi\alpha\cdot\xi(\alpha)} \hat{c}'_{-\alpha} \\ &= \hat{c}'_{-\alpha} \end{aligned} \tag{B.19}$$

such that

$$\overline{\omega}_{\mathcal{A}}(\mathcal{V}_\alpha(z)) = \left(\frac{1}{z}\right)^{\alpha^2} \mathcal{V}_{-\alpha}(\frac{1}{z})\,, \tag{B.20}$$

i.e., this choice makes the realization of $\widehat{\mathfrak{sl}_3}$ unitary with respect to the Hermitean form defined by $\overline{\omega}_{\mathcal{A}}$.

Appendix C Tables for Resolutions of $c = 2$ Irreducible Modules

Table C.1. Dimensions for $L[0,0]$ resolution

h	$L[0,0]$	$\mathcal{M}^{(0)}$	$\mathcal{I}^{(0)}$	$\mathcal{M}^{(-1)}$	$\mathcal{I}^{(-1)}$	$\mathcal{M}^{(-2)}$	$\mathcal{I}^{(-2)}$	$\mathcal{M}^{(-3)}$	$\mathcal{I}^{(-3)}$	$\mathcal{M}^{(-4)}$
0	1	1								
1	0	2	2	2						
2	1	5	4	4						
3	2	10	8	10	2	2				
4	3	20	17	20	3	4	1	2	1	1
5	4	36	32	40	8	10	2	4	2	2
6	8	65	57	72	15	20	5	10	5	5
7	10	110	100	130	30	40	10	20	10	10

Table C.2. Dimensions for $L[1,0]$ resolution

h	$L[1,0]$	$\mathcal{M}^{(0)}$	$\mathcal{I}^{(0)}$	$\mathcal{M}^{(-1)}$	$\mathcal{I}^{(-1)}$	$\mathcal{M}^{(-2)}$	$\mathcal{I}^{(-2)}$	$\mathcal{M}^{(-3)}$	$\mathcal{I}^{(-3)}$	$\mathcal{M}^{(-4)}$
1/3	1	1								
4/3	1	2	1	1						
7/3	2	5	3	4	1	1				
10/3	3	10	7	9	2	2				
13/3	6	20	14	20	6	7	1	1		
16/3	9	36	27	40	13	15	2	2		
19/3	15	65	50	76	26	32	6	7	1	1
22/3	22	110	88	137	49	61	12	14	2	2

Table C.3. Dimensions for $L[1,1]$ resolution

h	$L[1,1]$	$\mathcal{M}^{(0)}$	$\mathcal{I}^{(0)}$	$\mathcal{M}^{(-1)}$	$\mathcal{I}^{(-1)}$	$\mathcal{M}^{(-2)}$	$\mathcal{I}^{(-2)}$	$\mathcal{M}^{(-3)}$	$\mathcal{I}^{(-3)}$	$\mathcal{M}^{(-4)}$
1	1	1								
2	2	2								
3	3	5	2	2						
4	6	10	4	6	2	2				
5	10	20	10	14	4	4				
6	16	36	20	30	10	10				
7	27	65	38	60	22	24	2	2		
8	42	110	68	112	44	48	4	4		
9	64	185	121	202	81	92	11	12	1	1
10	98	300	202	350	148	170	22	24	2	2

Table C.4. Dimensions for $L[2,0]$ resolution

h	$L[2,0]$	$\mathcal{M}^{(0)}$	$\mathcal{I}^{(0)}$	$\mathcal{M}^{(-1)}$	$\mathcal{I}^{(-1)}$	$\mathcal{M}^{(-2)}$	$\mathcal{I}^{(-2)}$	$\mathcal{M}^{(-3)}$	$\mathcal{I}^{(-3)}$	$\mathcal{M}^{(-4)}$
4/3	1	1								
7/3	1	2	1	1						
10/3	3	5	2	2						
13/3	4	10	6	7	1	1				
16/3	8	20	12	14	2	2	1	1		
19/3	12	36	24	30	6	7	1	1		
22/3	21	65	44	56	12	14	2	2		
25/3	31	110	79	105	26	31	5	5		
28/3	50	185	135	182	47	58	11	12	1	1

Appendix D Summary of Explicit Computations

D.1 Introduction

In this appendix we summarize the results of explicit computations of the cohomologies $H(\mathcal{W}_3, F(\Lambda^M, 0) \otimes F(\Lambda^L, 2i))$ that are required to determine tips of all cones in the proof of Theorem 3.25. Given a Liouville weight, Λ^L, the corresponding matter weight, Λ^M, is chosen to be the lowest lying positive weight such that $(\Lambda^M, \Lambda^L) \in L$. This assures that the cohomology will include states from all irreducible $\mathfrak{sl}_3 \oplus (\mathfrak{u}_1)^2$ modules $\mathcal{L}(\Lambda) \otimes \mathbb{C}_{-i\Lambda^L}$ that may arise at this particular Liouville momentum. The number of states in each irreducible module is given by the multiplicity $m^{\Lambda}_{\Lambda^M}$.

For a given (Λ^M, Λ^L) the cohomology may arise only in the finite dimensional subcomplex that is annihilated by $L_0^{\rm tot}$. All operators in this subcomplex are of the form $P\mathcal{V}_{\Lambda^M, -i\Lambda^L}(z)$, where the prefactor P is a polynomial in all the fields (see Sects.3.2 and 3.3.3), whose dimension is equal to

$$h = \tfrac{1}{2}|-i\Lambda^L + 2\rho|^2 - \tfrac{1}{2}|\Lambda^M|^2 - 4\,. \tag{D.1}$$

Thus the number, $d(h,n)$, of linearly independent operators at ghost number n is given by expanding the partition function

$$q^{-4}\prod_{m=0}^{\infty}(1+tq^m)^2(1+t^{-1}q^{m+1})^2(1-q^n)^{-4} = \sum_{h,n\in\mathbb{Z}} d(h,n)q^h t^n\,. \tag{D.2}$$

To compute the action of the differential on the complex it is necessary to determine the OPE of the BRST current with all operators in a basis. Because of the algebraic complexity of this computation, we have used the algebraic manipulations program Mathematica™ together with the CFT package OPEdefs [Th]. As a result we obtain at each ghost number, n, a $d(h,n) \times d(h,n+1)$ complex matrix, $(d_{n,n+1})$, of the differential. The dimension of the kernel of d is then found as the number of zero eigenvalues of the $d(h,n) \times d(h,n)$ hermitian matrix $(d_{n,n+1})^\dagger (d_{n,n+1})$. Here the product of matrices is computed exactly, but the eigenvalues in most cases are found using a numerical routine.

Because of symmetry that exchanges the fundmental weights Λ_1 and Λ_2, it is sufficient to compute only "half" of the cases. As a consistency check, we have included, however, the results that could be deduced using duality (3.51).

The results are summarized in Sect. D.2. The tables are arranged according to the value of the level, h, defined in (D.1). Given h, we first determine which ghost numbers, n, may arise, and what are the dimensions, $\dim C^n = d(h,n)$, of the corresponding subspaces in the complex. Then for various choices of $(\Lambda^M, -i\Lambda^L)$ we list dimensions, $\dim K^n$, of the kernels and dimensions, $\dim H^n$, of cohomologies. The latter are computed using

$$\dim H^n = \dim K^n - (\dim C^{n-1} - \dim K^{n-1}) . \tag{D.3}$$

The cones can be identified by matching dimensions of the cohomologies with the multiplicities of the modules that could be present. Starting with the low (shifted) Liouville weights, this gives a systematic way of determining the boundaries of all the cones.

As an illustration, let us verify Theorem 3.25 at weights $(0,0)$. The representations and the corresponding multiplicities are given in Table D.1.

Table D.1. Multiplicities m_0^Λ

Λ	0	$\Lambda_1+\Lambda_2$	$3\Lambda_1$	$3\Lambda_2$	$2\Lambda_1+2\Lambda_2$	$\cdots$
m_0^Λ	0	2	1	1	3	$\cdots$

The contribution from each cone to the cohomology is read off from Table 3.2. This yields the result in the table below, which agrees with an explicit computation.

Table D.2. $H(\mathcal{W}_3, F(0,0) \otimes F(0,2i))$

$\Lambda\backslash n$	0	1	2	3	4	5	6	7	8
0	1	2	1						
$\Lambda_1+\Lambda_2$		2	4	2					
$3\Lambda_1$			1	2	1				
$3\Lambda_2$			1	2	1				
$\Lambda_1+\Lambda_2$			2	4	2				
$2\Lambda_1+2\Lambda_2$				3	6	3			
$\dim H^n$	1	4	9	13	10	3	0	0	0

All other cases are analyzed similarly.

D.2 The Tables

Table D.3. $h = 0$

$(\Lambda^M, -i\Lambda^L)$	n	0	1	2	3	4	5	6	7	8
	dim C^n	2	39	208	513	684	513	208	39	2
$(0, 0)$	dim K^n	1	6	43	178	345	342	171	37	2
	dim H^n	1	4	9	13	10	3	0	0	0
$(\Lambda_1, -\Lambda_2)$	dim K^n	0	3	41	176	344	342	171	37	2
	dim H^n	0	1	5	9	7	2	0	0	0
$(\Lambda_2, 2\Lambda_1 - 3\Lambda_2)$	dim K^n	0	3	41	176	344	342	171	37	2
	dim H^n	0	1	5	9	7	2	0	0	0
$(0, 2\Lambda_1 - 4\Lambda_2)$	dim K^n	0	2	41	180	348	343	171	37	2
	dim H^n	0	0	4	13	15	7	1	0	0
$(\Lambda_1, 2\Lambda_1 - 5\Lambda_2)$	dim K^n	0	2	38	176	348	344	171	37	2
	dim H^n	0	0	1	6	11	8	2	0	0
$(\Lambda_1, \Lambda_1 - 6\Lambda_2)$	dim K^n	0	2	39	177	347	343	171	37	2
	dim H^n	0	0	2	8	11	6	1	0	0
$(0, -6\Lambda_2)$	dim K^n	0	2	38	177	351	346	171	37	2
	dim H^n	0	0	1	7	15	13	4	0	0
$(\Lambda_2, -\Lambda_1 - 6\Lambda_2)$	dim K^n	0	2	37	173	347	346	172	37	2
	dim H^n	0	0	0	2	7	9	5	1	0
$(\Lambda_2, -3\Lambda_1 - 5\Lambda_2)$	dim K^n	0	2	37	173	347	346	172	37	2
	dim H^n	0	0	0	2	7	9	5	1	0
$(0, -4\Lambda_1 - 4\Lambda_2)$	dim K^n	0	2	37	174	349	348	174	38	2
	dim H^n	0	0	0	3	10	13	9	4	1

Table D.4. $h = -1$

$(\Lambda^M, -i\Lambda^L)$	n	1	2	3	4	5	6	7
	dim C^n	8	56	152	208	152	56	8
$(0, \Lambda_1 - 2\Lambda_2)$	dim K^n	1	11	51	105	104	48	8
	dim H^n	1	4	6	4	1	0	0
$(0, \Lambda_1 - 5\Lambda_2)$	dim K^n	0	10	52	108	106	48	8
	dim H^n	0	2	6	8	6	2	0
$(0, -2\Lambda_1 - 5\Lambda_2)$	dim K^n	0	8	49	107	107	49	8
	dim H^n	0	0	1	4	6	4	1

Table D.5. $h = -2$

$(\Lambda^M, -i\Lambda^L)$	n	1	2	3	4	5	6	7
	dim C^n	1	12	39	56	39	12	1
$(\Lambda_1, -\Lambda_2)$	dim K^n	0	3	15	30	28	11	1
	dim H^n	0	2	6	6	2	0	0
$(\Lambda_1, \Lambda_1 - 3\Lambda_2)$	dim K^n	0	3	15	30	28	11	1
	dim H^n	0	2	6	6	2	0	0
$(\Lambda_2, \Lambda_1 - 4\Lambda_2)$	dim K^n	0	2	15	32	29	11	1
	dim H^n	0	1	5	8	5	1	0
$(\Lambda_2, -5\Lambda_2)$	dim K^n	0	2	15	32	29	11	1
	dim H^n	0	1	5	8	5	1	0
$(\Lambda_1, -\Lambda_1 - 5\Lambda_2)$	dim K^n	0	1	13	32	30	11	1
	dim H^n	0	0	2	6	6	2	0
$(\Lambda_1, -3\Lambda_1 - 4\Lambda_2)$	dim K^n	0	1	13	32	30	11	1
	dim H^n	0	0	2	6	6	2	0

Table D.6. $h = -3$

$(\Lambda^M, -i\Lambda^L)$	n	2	3	4	5	6
	dim C^n	2	8	12	8	2
$(0, -\Lambda_1 - \Lambda_2)$	dim K^n	1	5	8	6	2
	dim H^n	1	4	5	2	0
$(\Lambda_2, -2\Lambda_2)$	dim K^n	1	4	7	6	2
	dim H^n	1	3	3	1	0
$(0, -3\Lambda_2)$	dim K^n	1	5	8	6	2
	dim H^n	1	4	5	2	0
$(\Lambda_1, -4\Lambda_2)$	dim K^n	1	4	8	7	2
	dim H^n	1	3	4	3	1
$(0, -\Lambda_1 - 4\Lambda_2)$	dim K^n	0	4	9	7	2
	dim H^n	0	2	5	4	1
$(\Lambda_2, -2\Lambda_1 - 4\Lambda_2)$	dim K^n	0	3	8	7	2
	dim H^n	0	1	3	3	1
$(0, -3\Lambda_1 - 3\Lambda_2)$	dim K^n	0	4	9	7	2
	dim H^n	0	2	5	4	1

Table D.7. $h = 1$

$(\Lambda^M, -i\Lambda^L)$	n	3	4	5
	dim C^n	1	2	1
$(0, -2\Lambda_1 - 2\Lambda_2)$	dim K^n	1	2	1
	dim H^n	1	2	1
$(\Lambda_1, -\Lambda_1 - 2\Lambda_2)$	dim K^n	1	2	1
	dim H^n	1	2	1
$(\Lambda_2, -\Lambda_1 - 3\Lambda_2)$	dim K^n	1	2	1
	dim H^n	1	2	1
$(0, -\Lambda_1 - 4\Lambda_2)$	dim K^n	1	2	1
	dim H^n	1	2	1
$(\Lambda_1, -2\Lambda_1 - 3\Lambda_2)$	dim K^n	1	2	1
	dim H^n	1	2	1

Appendix E A Graphical Representation of $H_{\mathrm{pr}}(\mathcal{W}_3, \mathfrak{C})$

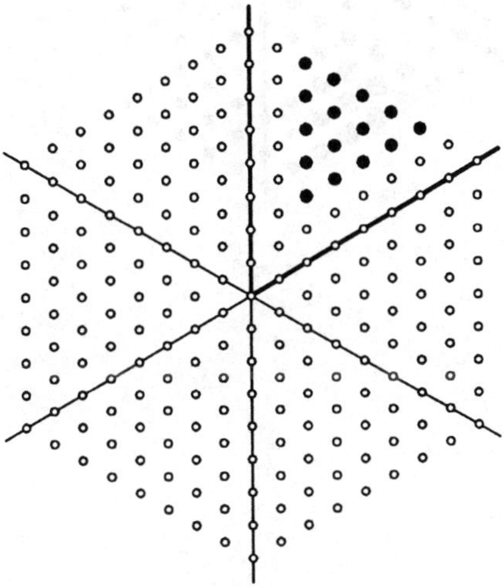

$n = 0$

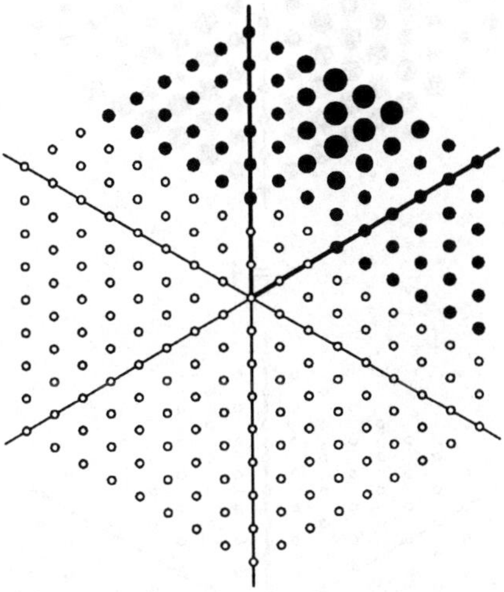

$n = 1$

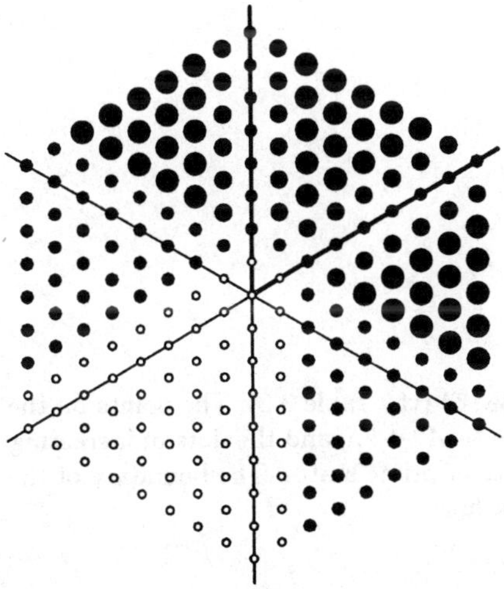

$n = 2$

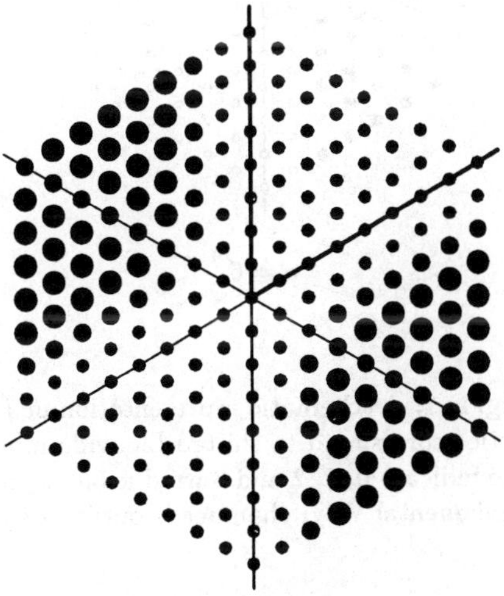

$n = 3$

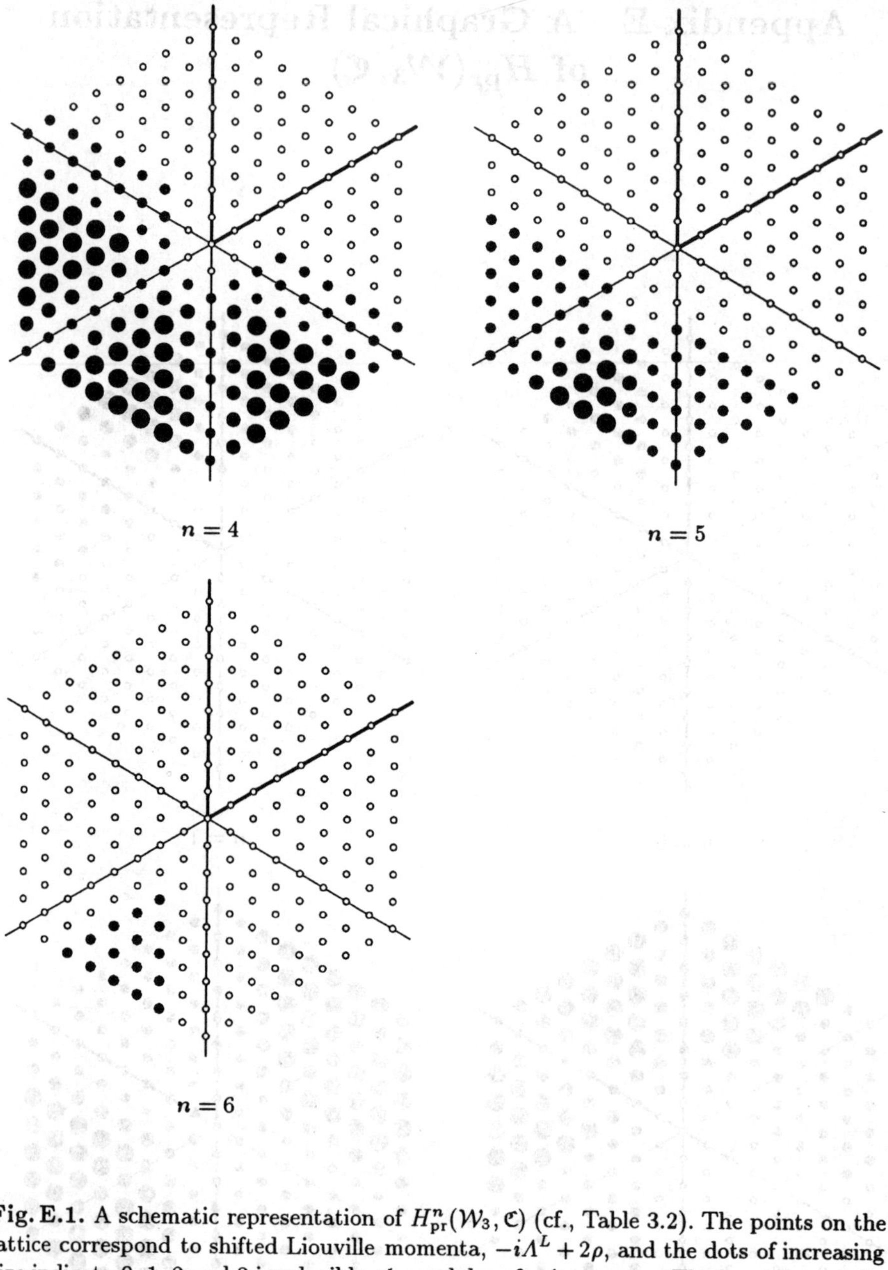

Fig. E.1. A schematic representation of $H^n_{\rm pr}(\mathcal{W}_3, \mathbb{C})$ (cf., Table 3.2). The points on the lattice correspond to shifted Liouville momenta, $-i\Lambda^L + 2\rho$, and the dots of increasing size indicate 0, 1, 2 and 3 irreducible $\mathfrak{sl}_3$ modules of prime states. The boundary of the fundamental Weyl chamber is outlined by thick lines.

Appendix F Polyderivations $\mathcal{P}(\mathcal{R}_N)$

In this appendix we derive additional results on the polyderivations $\mathcal{P}(\mathcal{R}_N)$ and, in particular, complete the proofs of Theorems 4.21 and 4.24.

F.1 Preliminary Results

Let

$$T_{i_1\ldots i_m,j_1\ldots j_n} = x_{i_1}\ldots x_{i_m} x^*_{j_1}\ldots x^*_{j_n}\,. \tag{F.1}$$

As an $\mathfrak{so}_{2N}$ tensor, $T_{i_1\ldots i_m,j_1\ldots j_m}$ is symmetric and traceless in $i_1,\ldots,i_m$, and antisymmetric in $j_1,\ldots,j_n$.

Lemma F.1 *Define the trace*

$$\widehat{T}_{i_1\ldots i_m,j_1,\ldots j_n} = g^{ij}\,T_{i\,i_1\ldots i_m,j\,j_1,\ldots j_n}\,. \tag{F.2}$$

Then

$$\begin{aligned} T^{i_1\ldots i_m}{}_{,j_1\ldots j_n} &= \widetilde{T}^{i_1\ldots i_m}{}_{,j_1,\ldots j_n} + a(m,n)\,\delta^{(i_1}{}_{[j_1}\widehat{T}^{i_2\ldots i_m)}{}_{,j_2\ldots j_n]} \\ &\quad + b(m,n)\,\delta^{(i_1}{}_{[j_1}\widehat{T}^{i_2\ldots i_{m-1}}{}_{j_2,}{}^{i_m)}{}_{j_3\ldots j_n]} \\ &\quad + c(m,n)\,g^{(i_1 i_2}\widehat{T}^{i_3\ldots i_m)}{}_{[j_1,j_2\ldots j_n]}\,, \end{aligned} \tag{F.3}$$

where

$$\begin{aligned} a(m,n) &= d(m,n)\,[(2N+2m-2)(2N+m-n-1)-2N]\,, \\ b(m,n) &= -2d(m,n)\,(m-1)(n-1)\,, \\ c(m,n) &= -d(m,n)\,(2N+m-n)(m-1)\,, \end{aligned} \tag{F.4}$$

and

$$d(m,n) = \frac{mn}{(2N+m-n)(2N+m-n-2)(2N+2m-2)}\,, \tag{F.5}$$

is the decomposition of T into its traceless and trace components $\widetilde{T}$ and $\widehat{T}$, respectively.

Proof. One verifies by explicit algebra that $\widetilde{T}$ defined by (F.3) is indeed traceless in all pairs of indices $i_1,\ldots,j_n$. □

Lemma F.2 *In the above notation we have*

$$P_{i_1\dots i_m, j_1\dots j_n} = \widetilde{T}_{i_1\dots i_{m-1}[i_m, j_1\dots j_n]}\,, \quad C_{i_1\dots i_m, j_1\dots j_n} = \widehat{T}_{i_1\dots[i_m, j_1\dots j_n]}\,. \quad (F.6)$$

Proof. See, the definitions in Sect. 4.2.3. □

To avoid confusion, let us denote

$$E^{i_1\dots i_m} = x^{i_1}\dots x^{i_m}\,. \quad (F.7)$$

From the decomposition (F.3) and Lemma F.2 we find

Lemma F.3

$$E^{i_1\dots i_m} P_{j_1,j_2\dots j_n} = P^{i_1\dots i_m}{}_{j_1,j_2\dots j_n} - \tfrac{m(n-1)}{2N+m-n}\,\delta^{(i_1}{}_{[j_1}C^{i_2\dots i_m)}{}_{j_2,j_3\dots j_n]}\,, \quad (F.8)$$

$$E^{i\,i_1\dots i_m} P_{i,j_1\dots j_n} = -\tfrac{n}{n+1}\,C^{i_1\dots i_m}{}_{,j_1\dots j_n}\,. \quad (F.9)$$

Clearly these relation allow a convenient construction of the entire basis in $\mathcal{P}(\mathcal{R}_N)$ in terms of products of the "generating elements" $E_{i_1\dots i_m}$, $P_{j_1,j_2\dots j_n}$, and C.

F.2 Proof of Theorem 4.21

We may now proceed with the proof of Theorem 4.21. The startegy is to first consider the products of generating elements, and then extend the result to arbitrary basis elements.

Case 1. Since

$$E_{i_1\dots i_m} E_{j_1\dots j_m} = E_{i_1\dots j_n}\,, \quad (F.10)$$

equation (4.73) is clearly satisfied in this case.

Case 2. Consider the Schouten bracket

$$[E^{i_1\dots i_m}, P_{j_1,j_2\dots j_n}]_S = m(n-1)\,\delta^{(i_1}{}_{[j_1}x^{i_2}\dots x^{i_m)}x_{j_2}x^*_{j_3}\dots x^*_{j_n]}\,. \quad (F.11)$$

Substituting the decomposition (F.3) in the rhs, we find that all the trace terms vanish upon symmetrization in $i_1,\dots,i_m$ and antisymmetrization in $j_1,\dots,j_n$, and we obtain

$$[E^{i_1\dots i_m}, P_{j_1,j_2\dots j_n}]_S = m(n-1)\,\delta^{(i_1}{}_{[j_1}P^{i_2\dots i_m)}{}_{j_2,j_3\dots j_n]}\,. \quad (F.12)$$

Hence, (F.8) can be rewritten as

$$E_{i_1\dots i_m} P_{j_1,j_2\dots j_n} = P_{i_1\dots i_m j_1,j_2\dots j_n} - \tfrac{1}{2N+m-n}\,C\,[E_{i_1\dots i_m}, P_{j_1,j_2\dots j_n}]_S\,, \quad (F.13)$$

which proves Theorem 4.21 in this case.

Case 3. The last special case follows easily from identities in Sect. 4.2.4. There we find

$$P_{i_1,i_2\dots i_m}P_{j_1,j_2\dots j_n} = (-1)^{m-1}\tfrac{m+n-1}{n}\,x_{[i_1}\,P_{i_2,i_3\dots i_m]j_1\dots j_n}\,, \tag{F.14}$$

which, combined with (F.8) and (4.70), yields

$$\begin{aligned} P_{i_1,i_2\dots i_m}P^{j_1,j_2\dots j_n} &= (-1)^{m-1}\tfrac{m+n-1}{n}\,P_{[i_1 i_2,i_3\dots i_m]}{}^{j_1\dots j_n} \\ &\quad - \tfrac{m+n-2}{2N-m-n+2}\,\delta_{[i_1}{}^{[j_1}C_{i_2,i_3\dots i_m]}{}^{j_2\dots j_n]}\,. \end{aligned} \tag{F.15}$$

However, we also have, see (4.71),

$$[P_{i_1,i_2\dots i_m}, P^{j_1,j_2\dots j_n}]_S = (-1)^{m-1}(m+n-2)\delta_{[i_1}{}^{[j_1}\,P_{i_2,i_3\dots i_m]}{}^{j_2\dots j_n]}\,. \tag{F.16}$$

This shows that (F.15) is indeed equivalent to (4.73).

Before we discuss the general case, let us simplify the notation, and write $E_{(m)}$ for $E_{i_1\dots i_m}$, $P_{(m,n)}$ for $P_{i_1\dots i_{m+1},i_{m+2}\dots i_{m+n}}$, and $C_{(m,n)}$ for $C_{i_1\dots i_{m+1},i_{m+2}\dots i_{m+n}}$. Also, for any $\mathfrak{so}_{2N}$ tensor T, let $(\!(T)\!)$ denotes its traceless component. Finally, let

$$g(m,n) = \tfrac{1}{2N+m-n}\,. \tag{F.17}$$

In this notation we may rewrite (F.13) as

$$E_{(m)}P_{(0,n)} = (\!(E_{(m)}P_{(0,n)})\!) - g(m,n)C[E_{(m)}, P_{(0,n)}]_S\,, \tag{F.18}$$

where $(\!(E_{(m)}P_{(0,n)})\!) = P_{(m,n)}$, and (F.15) as

$$P_{(0,m)}P_{(0,n)} = (\!(P_{(0,m)}P_{(0,n)})\!) + (-1)^m g(2,m+n)C\,[P_{(0,m)}, P_{(0,n)}]_S\,. \tag{F.19}$$

Lemma F.4

$$\begin{aligned} &E_{(m)}C\,[E_{(m')}, P_{(0,n')}]_S \\ &\quad = \tfrac{1}{1+mg(m',n')}\left(C\,[E_{(m+m')}, P_{(0,n')}]_S - C\,[E_{(m)}, P_{(m',n')}]\right)\,. \end{aligned} \tag{F.20}$$

Proof. Using (F.18), (F.10), (4.72), and the Leibnitz rule for the bracket, we obtain

$$\begin{aligned} E_{(m)}C\,[E_{(m')}, P_{(0,n')}]_S &= C\,[E_{(m+m')}, P_{(0,n')}]_S - C\,[E_{(m)}, E_{(m')}P_{(0,n')}]_S \\ &= C\,[E_{(m+m')}, P_{(0,n')}]_S - C\,[E_{(m)}, P_{(m',n')}] \\ &\quad - mg(m',n')C\,E_{(m)}[E_{(m')}, P_{(0,n')}]_S\,, \end{aligned} \tag{F.21}$$

which implies (F.20). □

Now, using Lemma F.4 and the identities above, we find

$$\begin{aligned} E_{(m)}P_{(m',n')} &= E_{(m)}\left(E_{(m')}P_{(0,n')} + g(m',n')C\,[E_{(m')}, P_{(0,n')}]_S\right) \\ &= P_{(m+m',n')} - g(m+m',n')C[E_{(m+m')}, P_{(0,n')}]_S \\ &\quad + g(m,n)C\,E_{(m)}[E_{(m')}, P_{(0,n')}]_S \\ &= P_{(m+m',n')} - g(m+m',n')C\,[E_{(m)}, P_{(m',n')}]_S\,, \end{aligned} \tag{F.22}$$

which agrees with (4.73).

Finally, in the general case, we find using (F.18), (F.19), (F.22) and (4.72),

$$\begin{aligned} P_{(m,n)}P_{(m',n')} &= (\!(P_{(m,n)}P_{(m',n')})\!) \\ &\quad + (-1)^n g(m+m'+1,n'-1)\, C\,[P_{(m,n)},P_{(m',n')}]_S\,. \end{aligned} \tag{F.23}$$

We omit the details of this somewhat lengthy, but otherwise completely straightforward algebra. This proves the first part of Theorem 4.21.

To show that the bracket on the left hand side in (F.23) is a linear combination of traceless elements we proceed similarly. Let us only illustrate the method on a simpler case of the bracket in (F.22). Using the same identities that led to (F.22), as well Theorem 4.21 in Case 2 above, and the Jacobi identity for the bracket, we find

$$\begin{aligned} [E_{(m)},P_{(m',n')}]_S &= E_{(m')}[E_{(m)},P_{(0,n')}]_S - mg(m',n')E_{(m)}[E_{(m')},P_{(0,n')}]_S \\ &\quad - g(m',n')C\,[E_{(m)},[E_{(m')},P_{(0,n')}]_S \\ &= (\!(E_{(m')}[E_{(m)},P_{(0,n')}]_S)\!) \\ &\quad - mg(m',n')(\!(E_{(m)}[E_{(m')},P_{(0,n')}]_S)\!) \\ &\quad + \big[g(m+m'-1,n'-1)\,(1+mg(m',n')) \\ &\quad - g(m',n')\big]C\,[E_{(m)},[E_{(m')},P_{(0,n')}]_S\,. \end{aligned} \tag{F.24}$$

Since the sum of the two terms inside the square bracket vanishes, we find that the bracket on the left hand side is indeed traceless. In the general case one reduces the bracket to a manifestly traceless expression using the same identities and, in addition, the already proven result in Case 3. This concludes the proof of Theorem 4.21. □

F.3 Proof of Theorem 4.24

The remaining two cases are, schematically, $\Delta_S(CP)$ and $\Delta_S(CC)$. In the first one, on the one hand we have

$$\begin{aligned} &\Delta_S(C_{(m,n)}P_{(m',n')}) - \Delta_S(C_{(m,n)})P_{(m',n')} - (-1)^n C_{(m,n)}\Delta_S(P_{(m',n')}) \\ &\quad = -(2N+m+m'-n-n'+2)(\!(P_{(m,n)}P_{(m',n')})\!) \\ &\qquad + (2N+m-n)P_{(m,n)}P_{(m',n')} \\ &\quad = -(m'-n'+2)(\!(P_{(m,n)}P_{(m',n')})\!) \\ &\qquad + (-1)^n(2N+m-n)g(m+m'+1,n+n'-1)\,C\,[P_{(m,n)},P_{(m',n')}]_S \end{aligned} \tag{F.25}$$

On the other hand,

$$\begin{aligned}
[C_{(m,n)}, P_{(m',n')}]_S \\
&= C[P_{(m,n)}, P_{(m',n')}]_S + (-1)^{n'(n-1)}[C, P_{(m',n')}]_S P_{(m,n)} \\
&= -(-1)^n (m' - n' + 2)(\!(P_{(m,n)} P_{(m',n')})\!) \\
&\quad + (1 - (m' - n' + 2) g(m + m' + 1, n + n' - 1)) \\
&\qquad \times C\,[P_{(m,n)}, P_{(m',n')}]_S\,.
\end{aligned} \tag{F.26}$$

Since

$$(2N+m-n)g(m+m'+1, n+n'-1) = 1-(m'-n'+2)g(m+m'+1, n+n'-1)\,, \tag{F.27}$$

the relation between Δ_S and $[-,-]_S$ also holds in this case. The last case is proved similarly. □

Appendix G BV Algebra $\mathcal{P}(A)$

In this appendix we prove some additional results for the algebra of polyvectors $\mathcal{P}(A)$ on the base affine space $A(G)$ and illustrate them with examples for the case $G = SL(3,\mathbb{C})$.

G.1 The Base Affine Space $A(SL(n,C))$

Let us first consider the space of function on the group $G = SL(n,\mathbb{C})$, and explicitly construct the decomposition of $\mathcal{E}(G)$ corresponding to (4.132). In terms of matrix elements $(g_{\dot\sigma\sigma})$, $g_{\dot\sigma\sigma} = \delta_{\dot\sigma\rho}g^{\rho}{}_{\sigma}$, the generators of $\mathfrak{g}_R$ and $\mathfrak{g}_L$ are[1]

$$\begin{aligned} R_{\sigma\dot\sigma} &= g_{\dot\rho\sigma}\frac{\partial}{\partial g_{\dot\rho}{}^{\dot\sigma}} - \tfrac{1}{n}\delta_{\sigma\dot\sigma}g_{\dot\rho}{}^{\dot\kappa}\frac{\partial}{\partial g_{\dot\rho}{}^{\dot\kappa}}\,, \\ L_{\sigma\dot\sigma} &= -g_{\dot\sigma\rho}\frac{\partial}{\partial g^{\sigma}{}_{\rho}} + \tfrac{1}{n}\delta_{\sigma\dot\sigma}g^{\kappa}{}_{\rho}\frac{\partial}{\partial g^{\kappa}{}_{\rho}}\,, \end{aligned} \tag{G.1}$$

with $\sigma, \dot\sigma = 1,\ldots,n$. We introduce the following elements $\Delta_k \in \mathcal{E}(G)$ by means of the minors of $g \in G$,

$$\Delta_k = \begin{vmatrix} g_{n-k+1 1} & \cdots & g_{n-k+1 k} \\ \vdots & & \vdots \\ g_{n1} & \cdots & g_{nk} \end{vmatrix}\,, \qquad k = 1,\ldots,n\,. \tag{G.2}$$

Since acting with $L_{\sigma\dot\sigma}$ replaces the σ-th row by the $\dot\sigma$-th row, and $R_{\sigma\dot\sigma}$ replaces the $\dot\sigma$-th column by the σ-th column, it is clear that all Δ_k are annihilated by $\mathfrak{n}_+^L \oplus \mathfrak{n}_+^R$. (Note that $\Delta_n = 1$.) The action of the Cartan subalgebra generators $\Pi_i^L = L_{ii} - L_{i+1i+1}$, $\Pi_i^R = R_{ii} - R_{i+1i+1}$, $i = 1,\ldots,n-1$, is given by

$$\Pi_i^L \Delta_k = \delta_{i,n-k}\Delta_k\,, \qquad \Pi_i^R \Delta_k = \delta_{i,k}\Delta_k\,. \tag{G.3}$$

Thus, in the decomposition (4.132), the highest weight vector corresponding to the weight $\Lambda = \sum_i s_i\Lambda_i$ is realized by the function $\prod_i \Delta_i^{s_i}$. By invariance with respect to N_+^L this is also the highest weight vector corresponding to Λ in the decomposition of $\mathcal{E}(A)$ in (4.133).

For $SL(3,\mathbb{C})$, the elements in $\mathcal{E}(A)$ corresponding to the fundamental representations are explicitly given by

---

[1] Strictly speaking these formulae define the action of $\mathfrak{g}_R$ and $\mathfrak{g}_L$ on the pull-back of $\mathcal{E}(G)$ to the functions on $GL(n,\mathbb{C})$.

$$x_\sigma = g_{3\sigma}\,, \quad x^\sigma = \epsilon^{\sigma\rho\kappa} g_{2\rho} g_{3\kappa}\,, \quad \sigma = 1,2,3\,, \tag{G.4}$$

with $x_1 = \Delta_1$ and $x^3 = \Delta_2$. Evidently, these functions satisfy the constraint

$$x^\sigma x_\sigma = 0\,. \tag{G.5}$$

Moreover, $\mathcal{E}(A)$ is spanned by the polynomials of those functions. Thus we have shown that $\mathcal{E}(A)$, for $A = SL(3,\mathbb{C})$, provides an explicit realization of the ground ring algebra $\mathcal{R}_3$. Of course this is also an immediate consequence of Theorems 4.28 and 4.38.

G.2 Polyvectors $\mathcal{P}(A)$

Here we collect some results referred to in Sect. 4.6.2.

Proof of Theorem 4.40. Recall that in Sect. 4.6.2 we have characterized $\mathcal{P}(A)$ as the space of invariants of the $\mathfrak{n}_+$ action on $\mathcal{E}(G, \bigwedge\mathfrak{b}_-)$ defined by $\Pi^L(x)+\Pi^{bc}(x)$, $x \in \mathfrak{n}_+$, where $\Pi^L(x)$ is the operator of the left regular action on $\mathcal{E}(G)$ and $\Pi^{bc}(x)$ is the operator of the adjoint action on $\bigwedge\mathfrak{b}_- \cong \bigwedge(\mathfrak{n}_+\backslash\mathfrak{g})$. Then the right regular action of $\mathfrak{g}$ on $\mathcal{E}(G)$ induces a $\mathfrak{g}$ module structure on $\mathcal{P}(A)$ and the complete reducibility of $\mathcal{P}(A)$ with respect to this $\mathfrak{g}$ action is a consequence of Theorem 4.38.

Under the right action of $\mathfrak{g}$, the space of polyvector fields $\mathcal{P}(A)$ decomposes as

$$\mathcal{P}(A) \cong \bigoplus_{\Lambda\in P_+} \mathrm{Hom}_{\mathfrak{g}}(\mathcal{L}(\Lambda), \mathcal{P}(A)) \otimes \mathcal{L}(\Lambda)\,. \tag{G.6}$$

A $\Phi \in \mathrm{Hom}_{\mathfrak{g}}(\mathcal{L}(\Lambda), \mathcal{P}(A))$ can be seen as a map $\Phi : G \to \mathrm{Hom}(\mathcal{L}(\Lambda), \bigwedge\mathfrak{b}_-)$ satisfying the equations

$$\begin{aligned} \Pi^R(x)\Phi(g) &= \Phi(g)D(x)\,, \quad x \in \mathfrak{g}\,, g \in G\,, \\ \Pi^L(x)\Phi(g) &= -\Pi^{bc}(x)\Phi(g)\,, \quad x \in \mathfrak{n}_+\,, g \in G\,, \end{aligned} \tag{G.7}$$

where $D(x)$ is the representative of $x \in \mathfrak{g}$ in the irreducible representation $\mathcal{L}(\Lambda)$. In particular, for $g = e$, where e is the identity element of G, we find

$$\Pi^{bc}(x)\Phi(e) = -\Pi^L(x)\Phi(e) = \Pi^R(x)\Phi(e) = \Phi(e)D(x)\,, \tag{G.8}$$

for all $x \in \mathfrak{n}_+$; i.e., $\Phi(e) \in \mathrm{Hom}_{\mathfrak{n}_+}(\mathcal{L}(\Lambda), \bigwedge\mathfrak{b}_-)$. Conversely, suppose we have a $\Phi \in \mathrm{Hom}_{\mathfrak{n}_+}(\mathcal{L}(\Lambda), \bigwedge\mathfrak{b}_-)$, then define $\widetilde{\Phi} : G \to \mathrm{Hom}(\mathcal{L}(\Lambda), \bigwedge\mathfrak{b}_-)$ by $\widetilde{\Phi}(g) = \Phi D(g)$, where, by slight abuse of notation, $D(g)$ also stands for the lift of the representation to G. One easily checks that $\widetilde{\Phi}(g)$ satisfies (G.7) and thus defines and element $\widetilde{\Phi} \in \mathrm{Hom}_{\mathfrak{g}}(\mathcal{L}(\Lambda), \mathcal{P}(A))$. □

An explicit computation of $\mathcal{P}(A)$. The space $\mathrm{Hom}_{\mathfrak{n}_+}(\mathcal{L}(\Lambda), \bigwedge\mathfrak{b}_-)$ is easy to characterize. Since $\mathcal{L}(\Lambda)$ is cyclic over $\mathfrak{n}_+$, a homomorphism Φ of $\mathcal{L}(\Lambda)$ into an arbitrary $\mathfrak{n}_+$ module is completely determined by the image of the lowest weight vector $v_{w_0\Lambda}$. Clearly, $\Phi(v_{w_0\Lambda})$ is restricted by $\Pi^{bc}(x)\Phi(v_{w_0\Lambda}) = 0$ for all $x \in U(\mathfrak{n}_+)$ such that $xv_{w_0\Lambda} = 0$.

Lemma G.1 *A pair $(v_{w_0\Lambda}, t)$, where $v_{w_0\Lambda}$ is the lowest weight vector of $\mathcal{L}(\Lambda)$ and $t \in \bigwedge\mathfrak{b}_-$, defines a homomorphism Φ of $\mathcal{L}(\Lambda)$ into $\bigwedge\mathfrak{b}_-$ by $\Phi(v_{w_0\Lambda}) = t$, provided t is a solution to the equations*

$$e_i^{(\Lambda^*+\rho,\alpha_i)}t = 0, \qquad i = 1, \ldots, \ell, \tag{G.9}$$

with $\Lambda^* = -w_0\Lambda$.

Proof. Any $\mathfrak{n}_+$ homomorphism can be extended to a unique $U(\mathfrak{n}_+)$ homomorphism, where $U(\mathfrak{n}_+)$ is the enveloping algebra of $\mathfrak{n}_+$. The elements of $U(\mathfrak{n}_+)$ that annihilate the lowest weight vector $v_{w_0\Lambda}$ of $\mathcal{L}(\Lambda)$ form a left ideal, $\mathcal{I}(\Lambda) \subset U(\mathfrak{n}_+)$, generated by the powers $e_i^{(\Lambda^*+\rho,\alpha_i)}$ of simple root generators (see, e.g., [Di]). This implies (G.9). □

The $\mathfrak{h}_L$ weight of an element $t \in \bigwedge\mathfrak{b}_-$ is of the form $\lambda = \sum_{\alpha\in\Delta_+} n_\alpha\alpha$, where $n_\alpha \in \{0, -1\}$. Thus, in particular, $\lambda + 2\rho \in P_+$. By (G.8), the resulting $\mathfrak{h}_L$ weight of the homomorphism defined by $(v_{w_0\Lambda}, t)$ is equal to $\lambda - w_0\Lambda = \lambda + \Lambda^*$.

Now consider homomorphisms into the trivial $\mathfrak{n}_+$ module $\bigwedge\mathfrak{b}_- \cong \mathbb{C}$. In this case (G.9) is satisfied for any weight $\Lambda \in P_+$. Obviously $\mathrm{Hom}_{\mathfrak{n}_+}(\mathcal{L}(\Lambda), \mathbb{C}) \cong \mathbb{C}_{\Lambda^*}$ as an $\mathfrak{h}_L$ module, and thus (4.137) reduces to (4.133) in agreement with Theorem 4.38.

For higher order polyvectors, with $n \geq 1$, we find that for a weight Λ sufficiently deep inside the fundamental Weyl chamber P_+ there are no restrictions on t due to (G.9), and hence $\mathrm{Hom}_{\mathfrak{n}_+}(\mathcal{L}(\Lambda), \bigwedge^n\mathfrak{b}_-) \cong \bigwedge^n\mathfrak{b}_-$ (cf. Theorem 4.42). However, if Λ lies close to the boundary of P_+ the constraint (G.9) becomes nontrivial. The following observation is quite useful in the computation of all polyvectors.

Lemma G.2 *Given $t \in \bigwedge\mathfrak{b}_-$, consider the set $\mathcal{C}(t)$ of all weights $(\Lambda, \Lambda') \in P_+ \otimes (P_+ - 2\rho)$ for which there exists a homomorphism that maps the lowest weight vector in $\mathcal{L}(\Lambda)$ to t and has the $\mathfrak{h}_L$ weight equal to $-w_0\Lambda'$. Then $\mathcal{C}(t)$ is a cone,*

$$\mathcal{C}(t) = \{S(t) + (\lambda, \lambda) \,|\, \lambda \in P_+\}, \tag{G.10}$$

with the "tip" $S(t)$.

Proof. From the computation of the $\mathfrak{h}_L$ weight of a homomorphism, it is clear that for fixed t, Λ' is determined by Λ, and that a shift $\Lambda \to \Lambda + \lambda$ induces the same shift $\Lambda' \to \Lambda' + \lambda$. So, it is sufficient to show that the set of Λ's is a cone in P_+. Now, for any $\Lambda, \lambda \in P_+$, we have $\mathcal{I}(\Lambda + \lambda) \subset \mathcal{I}(\Lambda)$ (where $\mathcal{I}(\Lambda)$ is the ideal introduced in the proof of Lemma G.1 above). Thus, if Λ corresponds to a nontrivial homomorphism, then so does $\Lambda + \lambda$. The lowest lying Λ weight with the required property is then determined by (G.9) and the minimal powers of simple root generators that annihilate t. □

Thus the problem of computing all homomorphisms is reduced to that of determining a finite set of cones. Later we will give the complete solution in the case of $SL(3,\mathbb{C})$.

To conclude, let us summarize the main steps of this construction of polyvector fields, and comment on the relation between $\mathcal{P}(A)$ and the polyderivations $\mathcal{P}(\mathcal{E}(A))$. Given a weight $\Lambda \in P_+$ and an element $t \in \bigwedge \mathfrak{b}_-$ of ghost number n satisfying (G.9), we first construct a homomorphism Φ by setting $\Phi(v_{w_0 \Lambda}) = t$. Using the exponentiation of the right G action in (G.7), having identified $\Phi = \Phi(e)$ we extend it to a function on G with values in $\mathrm{Hom}(\mathcal{L}(\Lambda), \bigwedge^n \mathfrak{b}_-)$. The components Φ_I of Φ, with respect to some basis in $\mathcal{L}(\Lambda)$, lie in $\mathcal{E}(G, \bigwedge^n \mathfrak{b}_-)$ and have an expansion of the form (in the notation of Sect. 4.6.3)

$$\Phi_I(g) \;=\; \Phi_I^{a_1 \ldots a_n}(g)\, c_{a_1} \ldots c_{a_n}\,, \qquad I = 1, \ldots, \dim \mathcal{L}(\Lambda). \tag{G.11}$$

Let $B_- = HN_-$ be the Borel subgroup in G, which, using the Gauss decomposition, may locally be identified with A. The vector fields $L_a = \Pi^L(e_a)$ on $B_- \subset G$ then provide a local trivialization of the tangent bundle of A. In the language of Definition 4.39, they correspond to sections $(nb, \Pi^{bc}(n) X_a)$, where $b \in B_-$, $n \in N_+$ and $X_a \in \mathfrak{n}_+ \backslash \mathfrak{g}$. Thus we identify $c_{a_1} \ldots c_{a_n}$ with the exterior product $L_{a_1} \wedge \ldots \wedge L_{a_n}$ of vector fields. One should remember that in general the vector fields L_a cannot be extended to the entire base affine space. However, if we identify a polyvector $\Phi = \Phi_I$ in (G.11) with

$$\Phi \;=\; \Phi^{a_1 \ldots a_n} L_{a_1} \wedge \ldots \wedge L_{a_n}\,, \tag{G.12}$$

then (4.135) implies that the this polyvector field is globally well defined on A, and thus defines a polyderivation of $\mathcal{E}(A)$.

G.3 Example of $SL(3,C)$

As an illustration for the above discussion, let us now determine polyvectors on the base affine space of $SL(3,\mathbb{C})$ and show that indeed they reproduce all polyderivations of the ground ring algebra $\mathcal{R}_3 \cong \mathcal{E}(A)$.

Consider $\mathcal{P}^1(A)$. The basis of $\mathfrak{b}_-$ consists of states obtained by acting with the ghost operators c_1, c_2, $c_{-\alpha_1}$, $c_{-\alpha_2}$, and $c_{-\alpha_3}$ on the vacuum. The $\mathfrak{n}_+$ module structure of $\mathfrak{b}_-$ is summarized by the diagram in Fig. G.1, in which each arrow corresponds to a nontrivial action of a given generator e_α. This must be compared with (G.9), which for a weight $\Lambda = s_1 \Lambda_1 + s_2 \Lambda_2$ reads

$$e_1^{s_2+1} t \;=\; 0\,, \quad e_2^{s_1+1} t \;=\; 0\,. \tag{G.13}$$

Clearly, depending on $t \in \mathfrak{b}_-$, the constraint (G.9) is satisfied provided $s_1 \geq 1$ for $c_{-\alpha_2}$, $s_2 \geq 1$ for $c_{-\alpha_1}$, and $s_1, s_2 \geq 1$ for $c_{-\alpha_3}$. There is no restriction on Λ for t equal to c_1 or c_2. These five cases correspond to five cones of vector fields on A, which we will now compute.

Using (G.4), we find the following identities

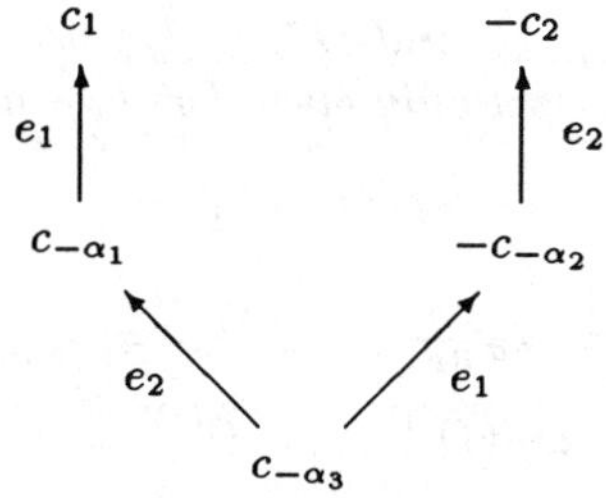

Fig. G.1. The $\mathfrak{n}_+$ module structure of $\mathfrak{b}_- \cong \mathfrak{n}_+\backslash\mathfrak{g}$.

$$\begin{aligned} \frac{\partial}{\partial g_{2\sigma}} &= \epsilon^{\sigma\rho\kappa} g_{3\rho} \frac{\partial}{\partial x^\kappa}, \\ \frac{\partial}{\partial g_{3\sigma}} &= \frac{\partial}{\partial x_\sigma} - \epsilon^{\sigma\rho\kappa} g_{2\rho} \frac{\partial}{\partial x^\kappa}. \end{aligned} \tag{G.14}$$

Substituting those identities in (G.1), it is straightforward, though somewhat laborious, to obtain an explicit formula for a polyvector if its "coordinates" $\Phi^{a_1\ldots a_n}$ in (G.12) are known. One should note that although intermediate steps of this calculation may explicitly involve the group elements (as in (G.14)), the final result for a polyvector field can always be expressed in terms of polynomials in the ground ring generators x^σ and x_σ, and the derivatives $\frac{\partial}{\partial x^\sigma}$ and $\frac{\partial}{\partial x_\sigma}$.

In the simplest nontrivial example we take $\Lambda = 0$ and $t = c_1$ or $t = c_2$. The corresponding vector fields are $\Pi_1^L = L_{11} - L_{22}$ and $\Pi_2^L = L_{22} - L_{33}$, respectively. Using (G.1) and (G.14), we find

$$L_{11} - L_{22} = x^\sigma \frac{\partial}{\partial x^\sigma}, \quad L_{22} - L_{33} = x_\sigma \frac{\partial}{\partial x_\sigma}, \tag{G.15}$$

which, as expected, coincide with C_- and C_+ in (4.97). Similarly, the vector fields at the tips of the other three cones – corresponding to the pairs $(\Lambda_1, c_{-\alpha_2})$, $(\Lambda_2, c_{-\alpha_1})$ and $(\Lambda_1 + \Lambda_2, c_{-\alpha_3})$ – reproduce the derivations $P_{\sigma,\rho}$, $P_{\dot\sigma,\dot\rho}$ and $\Lambda_{\sigma\dot\sigma}$ in (4.98).

For the higher ghost numbers the action of $\mathfrak{n}_+$ on $\bigwedge^n \mathfrak{b}_-$ follows from the diagram in Fig. G.1. The computation of the cones of polyvectors and representatives of the tips is essentially the same as above. The complete result may be summarized as follows:

Theorem G.3 *The space $\mathcal{P}(A)$ is isomorphic, as an $\mathfrak{sl}_3 \oplus (\mathfrak{u}_1)^2$ module, to the direct sum of irreducible modules $\mathcal{L}(\Lambda)\otimes\mathbb{C}_{\Lambda'}$ with weights $(\Lambda, \Lambda') \in P_+ \otimes (P_+ - 2\rho)$ lying in the set of disjoint cones $\{(\Lambda(t), \Lambda'(t)) + (\lambda, \lambda) \mid t \in \bigwedge \mathfrak{b}_-, \lambda \in P_+\}$; i.e.*

$$\mathcal{P}(A) \cong \bigoplus_{t \in \bigwedge \mathfrak{b}_-} \bigoplus_{\lambda \in P_+} \mathcal{L}(\Lambda(t) + \lambda) \otimes \mathbb{C}_{\Lambda'(t)+\lambda}, \tag{G.16}$$

where the tips of the cones, $(\Lambda(t), \Lambda'(t))$, $t \in \bigwedge\mathfrak{b}_-$, *satisfy* $\Lambda'(t) + 2\rho = \Lambda(t) + \rho - \sigma\rho$, $\sigma \in \widetilde{W}$. *They are listed in Table G.1, together with the corresponding polyvectors* ${D^{(n)}}_{\mu_1 \ldots \mu_n}$ *and* ${\widetilde{D}^{(n)}}_{\mu_1 \ldots \mu_n}$ *which are linear combinations of the generators in* $\mathcal{P}(\mathcal{R}_3)$ *explicitly given by*[2] *($\omega_{\mathcal{P}}$ as in Sect. 4.2.6)*

$$D^{(0)} = 1\,, \tag{G.17}$$

$$\begin{aligned} D^{(1)} &= x_\sigma \frac{\partial}{\partial x_\sigma}\,, \quad {D^{(1)}}_\sigma = \epsilon_{\sigma\mu\rho}\, x^\mu \frac{\partial}{\partial x_\rho}\,, \\ {D^{(1)}}_{\dot\sigma} &= \omega_{\mathcal{P}}({D^{(1)}}_\sigma)\,, \quad \widetilde{D}^{(1)} = \omega_{\mathcal{P}}(D^{(1)})\,, \end{aligned} \tag{G.18}$$

$${D^{(1)}}_{\sigma\dot\sigma} = x_\sigma \frac{\partial}{\partial x^{\dot\sigma}} - x_{\dot\sigma} \frac{\partial}{\partial x^\sigma} - \text{trace}\,, \tag{G.19}$$

$$D^{(2)} = \epsilon_{\sigma\mu\rho}\, x^\sigma \frac{\partial}{\partial x_\mu} \wedge \frac{\partial}{\partial x_\rho}\,, \quad \widetilde{D}^{(2)} = \omega_{\mathcal{P}}(D^{(2)})\,, \tag{G.20}$$

$${D^{(2)}}_\sigma = x_\sigma \frac{\partial}{\partial x_\rho} \wedge \frac{\partial}{\partial x^\rho} - x_\rho \frac{\partial}{\partial x_\rho} \wedge \frac{\partial}{\partial x^\sigma} + x^\rho \frac{\partial}{\partial x^\rho} \wedge \frac{\partial}{\partial x^\sigma}\,, \quad {D^{(2)}}_{\dot\sigma} = \omega_{\mathcal{P}}({D^{(2)}}_\sigma)\,, \tag{G.21}$$

$$\begin{aligned} {D^{(2)}}_{\sigma\rho} &= \epsilon_{\sigma\mu\nu}\Big(x^\mu \frac{\partial}{\partial x^\rho} - \tfrac{1}{2} x_\rho \frac{\partial}{\partial x_\mu}\Big) \wedge \frac{\partial}{\partial x_\nu} + (\sigma \leftrightarrow \rho)\,, \\ {D^{(2)}}_{\dot\sigma\dot\rho} &= \omega_{\mathcal{P}}({D^{(2)}}_{\sigma\rho})\,, \end{aligned} \tag{G.22}$$

$$D^{(3)} = x_\sigma \frac{\partial}{\partial x_\sigma} \wedge \frac{\partial}{\partial x_\rho} \wedge \frac{\partial}{\partial x^\rho} - x^\sigma \frac{\partial}{\partial x_\rho} \wedge \frac{\partial}{\partial x^\rho} \wedge \frac{\partial}{\partial x^\sigma}\,, \tag{G.23}$$

$${D^{(3)}}_\sigma = \epsilon_{\mu\nu\rho}\Big(x^\rho \frac{\partial}{\partial x^\sigma} - \tfrac{1}{3} x_\sigma \frac{\partial}{\partial x_\rho}\Big) \wedge \frac{\partial}{\partial x_\mu} \wedge \frac{\partial}{\partial x_\nu}\,, \quad {D^{(3)}}_{\dot\sigma} = \omega_{\mathcal{P}}({D^{(3)}}_\sigma)\,, \tag{G.24}$$

$${D^{(3)}}_{\sigma\dot\sigma} = \Big(x_\sigma \frac{\partial}{\partial x^\rho} - x_\rho \frac{\partial}{\partial x^\sigma}\Big) \wedge' \frac{\partial}{\partial x^{\dot\sigma}} \wedge \frac{\partial}{\partial x_\rho} + \Big(x_{\dot\sigma} \frac{\partial}{\partial x^{\dot\rho}} - x_{\dot\rho} \frac{\partial}{\partial x^{\dot\sigma}}\Big) \wedge' \frac{\partial}{\partial x^\sigma} \wedge \frac{\partial}{\partial x_{\dot\rho}}\,, \tag{G.25}$$

$$\begin{aligned} {D^{(4)}}_\sigma &= \epsilon_{\sigma\rho\pi}\epsilon^{\kappa\mu\nu}\Big(x_\kappa \frac{\partial}{\partial x_\rho} - \tfrac{1}{3} x^\rho \frac{\partial}{\partial x^\kappa}\Big) \wedge \frac{\partial}{\partial x_\pi} \wedge \frac{\partial}{\partial x^\mu} \wedge \frac{\partial}{\partial x^\nu}\,, \\ {D^{(4)}}_{\dot\sigma} &= \omega_{\mathcal{P}}({D^{(4)}}_\sigma)\,, \end{aligned} \tag{G.26}$$

$$D^{(5)} = \epsilon_{\sigma\rho\pi}\epsilon^{\kappa\mu\nu}\Big(x^\pi \frac{\partial}{\partial x^\nu} - x_\nu \frac{\partial}{\partial x_\pi}\Big) \wedge \frac{\partial}{\partial x_\sigma} \wedge \frac{\partial}{\partial x_\rho} \wedge \frac{\partial}{\partial x^\kappa} \wedge \frac{\partial}{\partial x^\mu}\,. \tag{G.27}$$

At this point it is interesting to ask whether the two spaces $\mathcal{P}(\mathcal{R}_3)$ and $\mathcal{P}(A)$ coincide. By decomposing the polyderivations $\mathcal{P}(\mathcal{R}_3)$ (see Theorem 4.16) with respect to $\mathfrak{sl}_3 \subset \mathfrak{so}_6$, and comparing with Table G.1, we conclude that indeed $\mathcal{P}(\mathcal{R}_3) \cong \mathcal{P}(A)$.

Remark. It is clear from the definition that $\bigwedge^n(\mathcal{P}^1(A)) \subset \mathcal{P}^n(A)$. In fact, from explicit computations as summarized above, we see that for $A = SL(3, \mathbb{C})$ it is a *proper* subset (unlike for the $SL(2, \mathbb{C})$ case).

---

[2] Here, and in Table G.1, $\wedge$ is the exterior product of vector fields, while $\wedge'$ implies, in addition, subtraction of all $\mathfrak{sl}_3$ invariant traces.

Table G.1. The decomposition of $\mathcal{P}(A)$ into cones of $\mathfrak{sl}_3 \oplus (\mathfrak{u}_1)^2$ modules.

t	$(\Lambda(t), \Lambda'(t))_\sigma$	D
1	$(0,0)_{r_3}$	$D^{(0)}$
c_1	$(0,0)_{r_3}$	$\widetilde{D}^{(1)}$
c_2	$(0,0)_{r_3}$	$D^{(1)}$
$c_{-\alpha_1}$	$(\Lambda_2, \Lambda_1 - \Lambda_2)_{r_{12}}$	${D^{(1)}}_{\dot\sigma}$
$c_{-\alpha_2}$	$(\Lambda_1, -\Lambda_1 + \Lambda_2)_{r_{21}}$	${D^{(1)}}_{\sigma}$
$c_{-\alpha_3}$	$(\Lambda_1 + \Lambda_2, 0)_{\sigma_2}$	${D^{(1)}}_{\sigma\dot\sigma}$
$c_1 c_2$	$(0,0)_{r_3}$	$\widetilde{D}^{(1)} \wedge D^{(1)}$
$c_1 c_{-\alpha_1}$	$(0, \Lambda_1 - 2\Lambda_2)_{r_{12}}$	$\widetilde{D}^{(2)}$
$c_2 c_{-\alpha_2}$	$(0, -2\Lambda_1 + \Lambda_2)_{r_{21}}$	$D^{(2)}$
$c_2 c_{-\alpha_1}$	$(\Lambda_2, \Lambda_1 - \Lambda_2)_{r_{12}}$	$D^{(1)} \wedge {D^{(1)}}_{\dot\sigma}$
$c_1 c_{-\alpha_2}$	$(\Lambda_1, -\Lambda_1 + \Lambda_2)_{r_{21}}$	$\widetilde{D}^{(1)} \wedge {D^{(1)}}_{\sigma}$
$c_{-\alpha_1} c_{-\alpha_2} + c_2 c_{-\alpha_3}$	$(\Lambda_2, -\Lambda_1)_{\sigma_2}$	${D^{(2)}}_{\dot\sigma}$
$c_{-\alpha_1} c_{-\alpha_2} + c_1 c_{-\alpha_3}$	$(\Lambda_1, -\Lambda_2)_{\sigma_2}$	${D^{(2)}}_{\sigma}$
$c_{-\alpha_1} c_{-\alpha_2}$	$(\Lambda_1 + \Lambda_2, 0)_{\sigma_1}$	${D^{(1)}}_{\sigma} \wedge' {D^{(1)}}_{\dot\sigma}$
$c_{-\alpha_1} c_{-\alpha_3}$	$(2\Lambda_2, -\Lambda_2)_{r_1}$	${D^{(2)}}_{\dot\sigma\dot\rho}$
$c_{-\alpha_2} c_{-\alpha_3}$	$(2\Lambda_1, -\Lambda_1)_{r_2}$	${D^{(2)}}_{\sigma\rho}$
$c_1 c_2 c_{-\alpha_1}$	$(0, \Lambda_1 - 2\Lambda_2)_{r_{12}}$	$D^{(1)} \wedge \widetilde{D}^{(2)}$
$c_1 c_2 c_{-\alpha_2}$	$(0, -2\Lambda_1 + \Lambda_2)_{r_{21}}$	$\widetilde{D}^{(1)} \wedge D^{(2)}$
$c_1 c_{-\alpha_1} c_{-\alpha_2} + c_1 c_2 c_{-\alpha_3} - c_2 c_{-\alpha_1} c_{-\alpha_2}$	$(0, -\Lambda_1 - \Lambda_2)_{\sigma_2}$	$D^{(3)}$
$c_1 c_{-\alpha_1} c_{-\alpha_2} + c_1 c_2 c_{-\alpha_3}$	$(\Lambda_2, -\Lambda_1)_{\sigma_1}$	$\widetilde{D}^{(1)} \wedge {D^{(2)}}_{\dot\sigma}$
$c_2 c_{-\alpha_1} c_{-\alpha_2} - c_1 c_2 c_{-\alpha_3}$	$(\Lambda_1, -\Lambda_2)_{\sigma_1}$	$D^{(1)} \wedge {D^{(2)}}_{\sigma}$
$c_1 c_{-\alpha_1} c_{-\alpha_3}$	$(\Lambda_2, -2\Lambda_2)_{r_1}$	${D^{(3)}}_{\dot\sigma}$
$c_2 c_{-\alpha_2} c_{-\alpha_3}$	$(\Lambda_1, -2\Lambda_1)_{r_2}$	${D^{(3)}}_{\sigma}$
$c_2 c_{-\alpha_1} c_{-\alpha_3}$	$(2\Lambda_2, -\Lambda_2)_{r_1}$	$D^{(1)} \wedge {D^{(2)}}_{\dot\sigma\dot\rho}$
$c_1 c_{-\alpha_2} c_{-\alpha_3}$	$(2\Lambda_1, -\Lambda_1)_{r_2}$	$\widetilde{D}^{(1)} \wedge {D^{(2)}}_{\sigma\rho}$
$c_{-\alpha_1} c_{-\alpha_2} c_{-\alpha_3}$	$(\Lambda_1 + \Lambda_2, -\Lambda_1 - \Lambda_2)_1$	${D^{(3)}}_{\sigma\dot\sigma}$
$c_1 c_2 c_{-\alpha_1} c_{-\alpha_2}$	$(0, -\Lambda_1 - \Lambda_2)_{\sigma_1}$	$\widetilde{D}^{(2)} \wedge D^{(2)}$
$c_1 c_2 c_{-\alpha_1} c_{-\alpha_3}$	$(\Lambda_2, -2\Lambda_2)_{r_1}$	$D^{(1)} \wedge {D^{(3)}}_{\dot\sigma}$
$c_1 c_2 c_{-\alpha_2} c_{-\alpha_3}$	$(\Lambda_1, -2\Lambda_1)_{r_2}$	$\widetilde{D}^{(1)} \wedge {D^{(3)}}_{\sigma}$
$c_1 c_{-\alpha_1} c_{-\alpha_2} c_{-\alpha_3}$	$(\Lambda_1, -\Lambda_1 - 2\Lambda_2)_1$	${D^{(4)}}_{\sigma}$
$c_2 c_{-\alpha_1} c_{-\alpha_2} c_{-\alpha_3}$	$(\Lambda_2, -2\Lambda_1 - \Lambda_2)_1$	${D^{(4)}}_{\dot\sigma}$
$c_1 c_2 c_{-\alpha_1} c_{-\alpha_2} c_{-\alpha_3}$	$(0, -2\Lambda_1 - 2\Lambda_2)_1$	$D^{(5)}$

G.4 The BV Algebra Structure of $\mathcal{P}(A)$

We have shown in Sect. 4.6.4 that $(\mathcal{P}(A), \cdot, \Delta_0)$ is a BV algebra. Given two polyvector fields $\Phi = \Phi^{a_1 \ldots a_m} c_{a_1} \ldots c_{a_m} |0\rangle$ and $\Psi = \Psi^{b_1 \ldots b_n} c_{b_1} \ldots c_{b_n} |0\rangle$, their product is

$$\Phi \cdot \Psi = \Phi^{a_1 \ldots a_m} \Psi^{b_1 \ldots b_n} c_{a_1} \ldots c_{a_m} c_{b_1} \ldots c_{b_n} |0\rangle . \tag{G.28}$$

The BV operator is defined by

$$\Delta_0' = -b^a \Pi^L(e_a) + \tfrac{1}{2} f_{ab}{}^c b^a b^b c_c \,. \tag{G.29}$$

Let us compute the bracket between two vector fields induced by Δ_0'.

Lemma G.4 *The bracket induced by Δ_0' on $\mathcal{P}(A)$ is the Schouten bracket between polyvector fields on A.*

Proof. For $\Phi = \Phi^a c_a$, $\Psi = \Psi^a c_a$ we find (as in (4.5)),

$$\begin{aligned} [\Phi, \Psi] &= -\Delta_0'(\Phi\Psi) + (\Delta_0'\Phi)\Psi - \Phi\Delta_0'(\Psi) \\ &= (\Phi^a L_a \Psi^c - \Psi^a L_a \Phi^c + f_{ab}{}^c \Phi^a \Psi^b)\, c_c \,, \end{aligned} \tag{G.30}$$

which is just the commutator between vector fields. The generalization to higher order polyvectors is essentially the same, and we omit the details. □

Clearly, Δ_0' commutes with the action of $\mathfrak{g}$ on $\mathcal{P}(A)$. It also commutes with the action of $\mathfrak{h}_L$, which is generated, up to a constant, by the operators $\Pi_i = [c_i, \Delta_0']$. To evaluate Δ_0' explicitly on the irreducible $\mathfrak{g}$ module of polyvectors corresponding to a homomorphism $\Phi(e)$ it is sufficient to determine the vector $\Delta_0'\Phi(e)(v_{w_0\Lambda})$. Using (G.8), (G.29) and that $xv_{w_0\Lambda} = 0$ for all $x \in \mathfrak{n}_-$, we obtain ($D_a = D(e_a)$)

$$\begin{aligned} \Delta_0'\Phi(e)(v_{w_0\Lambda}) &= b^a\Phi(e)(D_a v_{w_0\Lambda}) + \tfrac{1}{2} f_{ab}{}^c b^a b^b c_c \Phi(e)(v_{w_0\Lambda}) \\ &= b^i\Phi(e)(D_i v_{w_0\Lambda}) + \tfrac{1}{2} f_{ab}{}^c b^a b^b c_c \Phi(e)(v_{w_0\Lambda}) \\ &= -(\alpha_i^\vee, \Lambda^*) b^i \Phi(e)(v_{w_0\Lambda}) + \big(\tfrac{1}{2} f_{-\alpha-\beta}{}^{-\alpha-\beta} b^{-\alpha} b^{-\beta} c_{-\alpha-\beta} \\ &\qquad + f_{i-\alpha}{}^{-\alpha} b^i b^{-\alpha} c_{-\alpha}\big)\, \Phi(e)(v_{w_0\Lambda}) \,. \end{aligned} \tag{G.31}$$

In particular, on vector fields, this reduces to

$$\Delta_0'\Phi(e)(v_{w_0\Lambda}) = -(\alpha_i^\vee, \Lambda^* + 2\rho) b^i \Phi(e)(v_{w_0\Lambda}) \,, \quad \Phi \in \mathcal{P}(A) \,. \tag{G.32}$$

Finally, we have an analogue of Theorem 4.26.

Theorem G.5 *The homology of Δ_0' on the polyvector fields $\mathcal{P}(A)$ is given by*

$$H_n(\Delta_0', \mathcal{P}(A)) \cong \delta_{n,\dim(A)}\, \mathbb{C} \,. \tag{G.33}$$

The representative of the nontrivial homology is the polyvector corresponding to the homomorphism defined by $\Lambda = 0$ and $t = \prod_{i=1}^{\ell} c_i \prod_{\alpha\in\Delta_+} c_{-\alpha}$.

Proof. Note that c_i, $i = 1, \ldots, \ell$, are well-defined operators on $\mathcal{P}(A)$. Then, similarly as above, we find that for any $\Phi \in \mathcal{P}(A)$

$$\begin{aligned} [\Delta_0', c_i]\,\Phi(e)(v_{w_0\Lambda}) &= \Phi(e)(D_i v_{w_0\Lambda}) - (\alpha_i^\vee, 2\rho)\Phi(e)(v_{w_0\Lambda}) \\ &\qquad + \sum_{\alpha\in\Delta_+} (\alpha_i^\vee, \alpha) c_{-\alpha} b^{-\alpha} \Phi(e)(v_{w_0\Lambda}) \\ &= -(\alpha_i^\vee, \Lambda^* + 2\rho + \lambda)\Phi(e)(v_{w_0\Lambda}) \,, \end{aligned} \tag{G.34}$$

where λ is the weight of ghosts in $t = \Phi(e)(v_{w_0 \Lambda})$. In particular, for $C = \rho^i c_i$, where $\rho = \rho^i \alpha_i$, we get

$$[\Delta_0', C]\Phi(e)(v_{w_0\Lambda}) = -(\rho^\vee, \Lambda^* + 2\rho + \lambda)\Phi(e)(v_{w_0\Lambda}) \,. \tag{G.35}$$

Since $\lambda+2\rho \in P_+$, we find that the homology of Δ_0' is concentrated on polyvectors with $\Lambda = 0$ and $\lambda = -2\rho$. The space of those polyvectors is isomorphic with $\bigwedge^* \mathfrak{h}$, spanned by the products of c_i's. From (G.31) we find that on this subspace Δ_0' reduces to $2\sum_i b^i$, and the theorem follows by an elementary evaluation of the homology in the reduced case. □

Example: $G = SL(3,\mathbb{C})$. The vector fields on A that extend $\mathfrak{sl}_3$ symmetry to the $\mathfrak{so}_6$ symmetry in Sect. 4.2, are $P_{\sigma,\rho}$ and $P_{\dot\sigma,\dot\rho}$, corresponding to the homomorphisms $(\Lambda_1, c_{-\alpha_2})$ and $(\Lambda_2, c_{-\alpha_1})$, respectively, and $C_+ - C_-$ corresponding to $(0, c_2 - c_1)$. From (G.32) it follows easily that all are annihilated by Δ_0'. This, combined with invariance with respect to $\mathfrak{sl}_3$ and (4.6) yields

Theorem G.6 *The BV operators Δ_S and Δ_0' satisfy $\Delta_0' = -\Delta_S$, i.e., $\mathcal{P}(\mathcal{R}_3) \cong \mathcal{P}(A)$, as BV algebras.*

Proof. Given the $\mathfrak{so}_6$ invariance, we must only verify the overall normalizaton of both operators. Since $C = C_+ + C_-$ we find using (G.32), $\Delta_0' C = 4$, which thus agrees with (4.88). □

Appendix H Free Modules of $\mathfrak{P}_\pm$

In this appendix we outline an explicit construction of a free G module on one generator of the chiral subalgebras, $\mathfrak{P}_+$ and $\mathfrak{P}_-$. An immediate application of this result is to prove Theorem 4.35.

Consider the holomorphic subalgebra $\mathfrak{P}_+$. As a dot algebra it is generated by

$$x_\sigma\,, \quad D^\sigma = \epsilon^{\sigma\rho\pi} P_{\rho,\pi}\,, \quad P_{\sigma,\rho\pi} = \epsilon_{\sigma\rho\pi} P\,, \tag{H.1}$$

subject to relations, see (4.63) and (4.65),

$$x_\sigma \cdot D^\sigma = 0\,, \tag{H.2}$$

$$D^\sigma \cdot D^\rho = -\tfrac{3}{2}\epsilon^{\sigma\rho\pi} x_\pi \cdot P\,, \quad D^\alpha \cdot P = 0\,, \quad P \cdot P = 0\,. \tag{H.3}$$

Since the bracket (and the BV operator) on $\mathfrak{P}$ vanishes when restricted to $\mathfrak{P}_+$, the free G module, $\mathfrak{M}_\Gamma$, is spanned by elements of the form[1]

$$\Phi_0 \cdot [\Phi_1, [\ldots, [\Phi_n, \Gamma] \ldots]]\,, \quad \Phi_0, \ldots, \Phi_n \in \mathfrak{P}_+\,. \tag{H.4}$$

In fact, given (4.102), it is sufficient that the Φ_i, $i \geq 1$, run over the set of generators (H.1). Denote by

$$\partial_\sigma = [x_\sigma, -]\,, \quad \mathcal{D}^\sigma = [D^\sigma, -]\,, \quad \mathcal{P} = [P, -]\,, \tag{H.5}$$

the generators of the bracket action of $\mathfrak{P}_+$ on $\mathfrak{M}_\Gamma$. From (4.102) and (4.103), and the vanishing of the bracket on $\mathfrak{P}_+$, it follows that – together with the operators 1, x_σ, D^σ and P, corresponding to the dot action of $\mathfrak{P}_+$ on $\mathfrak{M}_\Gamma$ – they generate a graded commutative algebra, $\mathfrak{Q}_+ = \bigoplus_{n\in\mathbb{Z}} \mathfrak{Q}_+^n$.

Table H.1. The generators of $\mathfrak{Q}_+$

n	-1	0	1	2
Φ_I	∂_σ	$x_\sigma, \mathcal{D}^\sigma$	$D^\sigma, \mathcal{P}$	P

[1] We will omit here the subscripts on the bracket and the BV operator.

In addition to (H.2) and (H.3),

$$x_\sigma \mathcal{D}^\sigma + D^\sigma \partial_\sigma = 0\,, \tag{H.6}$$

$$D^\sigma \mathcal{D}^\rho - D^\rho \mathcal{D}^\sigma = -\tfrac{3}{2}\epsilon^{\sigma\rho\pi}\left(x_\pi \mathcal{P} + P\partial_\pi\right), \tag{H.7}$$

$$D^\sigma \mathcal{P} + P\mathcal{D}^\sigma = 0\,, \quad P\mathcal{P} = 0\,, \tag{H.8}$$

exhaust the defining relations between the generators of $\mathfrak{Q}_+$.

Since $\mathfrak{M}_\Gamma$ is freely generated by $\mathfrak{Q}_+$ acting on Γ, the problem of determining the free module is equivalent to that of computing $\mathfrak{Q}_+$, i.e., the quotient of a free graded commutative algebra by the ideal generated by the relations (H.2), (H.3) and (H.6)-(H.8).

Consider the subalgebra $\mathfrak{Q}'_+$ generated by the operators D^σ, P, ∂_σ and $\mathcal{P}$. Clearly it is finite dimensional and nonvanishing for $n = -3, \ldots, 2$. The quotient algebra $\mathfrak{Q}_+/(\mathfrak{Q}_+\mathfrak{Q}'_+)$, is spanned by monomials

$$x_{\sigma_1} \ldots x_{\sigma_{s_1}} \mathcal{D}^{\rho_1} \ldots \mathcal{D}^{\rho_{s_2}}\,, \quad s_1, s_2 \geq 0\,, \tag{H.9}$$

of order zero, so that the $\mathfrak{Q}^n_+$ are nonzero for the same range of n as $\mathfrak{Q}'_+$. Moreover, all monomials with the same s_1 and s_2 comprise a single $\mathfrak{sl}_3 \oplus (\mathfrak{u}_1)^2$ module with the highest weight $(\Lambda, r_2\Lambda)$, where $\Lambda = s_1\Lambda_1 + s_2\Lambda_2$, the reason being that by (H.6) all modules spanned by the trace components in (H.9) vanish in the quotient. This heuristically shows that $\mathfrak{Q}_+$ can be decomposed into a direct sum of disjoint r_2-twisted cones of $\mathfrak{sl}_3 \oplus (\mathfrak{u}_1)^2$ modules, although to determine the set of those cones we must study products of the monomials (H.9) with $\mathfrak{Q}'_+$. Given the finite number of cases to be considered, this can be done explicitly.

In the case $n = 2$ we must consider all expressions of the form

$$x_{\sigma_1} \ldots x_{\sigma_{s_1}} \mathcal{D}^{\rho_1} \ldots \mathcal{D}^{\rho_{s_2}} P\,, \quad s_1, s_2 \geq 0\,, \tag{H.10}$$

whose trace components still vanish because of (H.3) and (H.6). This yields a single cone of $\mathfrak{sl}_3 \oplus (\mathfrak{u}_1)^2$ modules with the tip, $(0, \Lambda_1 - 2\Lambda_2)$, at the weights of P.

For $n = 1$ all terms can be reduced, using (H.2), (H.3), (H.7) and (H.8), to just three types:

$$x_{\sigma_1} \ldots x_{\sigma_{s_1}} \mathcal{D}^{\rho_1} \ldots \mathcal{D}^{\rho_{s_2}} P\,\partial_\pi\,, \quad s_1, s_2 \geq 0\,, \tag{H.11}$$

$$x_{\sigma_1} \ldots x_{\sigma_{s_1}} \mathcal{D}^{\rho_1} \ldots \mathcal{D}^{\rho_{s_2}} \mathcal{P}\,, \quad s_1, s_2 \geq 0\,, \tag{H.12}$$

$$x_{\sigma_1} \ldots x_{\sigma_{s_1}} \mathcal{D}^{\rho_1} \ldots \mathcal{D}^{\rho_{s_2}} D^{\rho_{s_2+1}}\,, \quad s_1, s_2 \geq 0\,. \tag{H.13}$$

As for $n = 2$, we have no trace in (σ, ρ) in (H.11). Thus the decomposition into $\mathfrak{sl}_3$ modules is that of the tensor product $(s_1, s_2) \otimes (1, 0)$, giving rise to three cones, $(\Lambda_1, 2\Lambda_1 - 2\Lambda_2)$, $(\Lambda_2, 3\Lambda_1 - 2\Lambda_2)$ and $(0, 3\Lambda_1 - 3\Lambda_2)$. The traceless component in (H.12) yields a cone at $(0, \Lambda_1 - 2\Lambda_2)$, while the trace component, using (H.6) and (H.8), is equivalent the last cone in (H.11). Finally, we may assume complete symmetry in $\rho_1, \ldots, \rho_{s_2+1}$ of (H.13), as all terms with mixed symmetry can be expressed in terms of (H.11) and (H.12). Also, the trace terms either vanish due

Table H.2. The r_2-twisted cone decomposition of $\mathfrak{Q}_+$

n	$\mathfrak{Q}_+$	(Λ, Λ')
2	P	$(0, \Lambda_1 - 2\Lambda_2)$
1	D^ρ	$(\Lambda_2, \Lambda_1 - \Lambda_2)$
	$\mathcal{P}$	$(0, \Lambda_1 - 2\Lambda_2)$
	$P\,\partial_\sigma$	$(\Lambda_1, 2\Lambda_1 - 2\Lambda_2)$, $(\Lambda_2, 3\Lambda_1 - 2\Lambda_2)$, $(0, 3\Lambda_1 - 3\Lambda_2)$
0	1	$(0,0)$
	$D^\mu\partial_\sigma$	$(\Lambda_1 + \Lambda_2, 2\Lambda_1 - \Lambda_2)$, $(2\Lambda_2, 3\Lambda_1 - \Lambda_2)$, $(0, 2\Lambda_1 - \Lambda_2)$
	$\partial_\sigma\,\mathcal{P}$	$(\Lambda_1, 2\Lambda_1 - 2\Lambda_2)$, $(0, 3\Lambda_1 - 3\Lambda_2)$, $(\Lambda_2, 3\Lambda_1 - 2\Lambda_2)$
	$P\,\partial_\sigma\partial_\rho$	$(0, 4\Lambda_1 - 2\Lambda_2)$, $(\Lambda_1, 4\Lambda_1 - 3\Lambda_2)$, $(\Lambda_2, 3\Lambda_1 - 2\Lambda_2)$
−1	∂_σ	(Λ_1, Λ_1), $(\Lambda_2, 2\Lambda_1)$, $(0, 2\Lambda_1 - \Lambda_2)$
	$D^\mu\partial_\sigma\partial_\pi$	$(\Lambda_1, 3\Lambda_1 - \Lambda_2)$, $(\Lambda_2, 4\Lambda_1 - \Lambda_2)$, $(2\Lambda_2, 3\Lambda_1 - \Lambda_2)$
	$\partial_\sigma\partial_\rho\mathcal{P}$	$(0, 4\Lambda_1 - 2\Lambda_2)$, $(\Lambda_1, 4\Lambda_1 - 3\Lambda_2)$, $(\Lambda_2, 3\Lambda_1 - 2\Lambda_2)$
	$\partial_\sigma\partial_\rho\partial_\pi P$	$(0, 4\Lambda_1 - 2\Lambda_2)$
−2	$\partial_\sigma\partial_\rho$	$(0, 3\Lambda_1)$, $(\Lambda_1, 3\Lambda_1 - \Lambda_2)$, $(\Lambda_2, 2\Lambda_1)$
	$D^\mu\partial_\sigma\partial_\rho\partial_\pi$	$(\Lambda_2, 4\Lambda_1 - \Lambda_2)$
	$\partial_\sigma\partial_\rho\partial_\pi\mathcal{P}$	$(0, 4\Lambda_1 - 2\Lambda_2)$
−3	$\partial_\sigma\partial_\rho\partial_\pi$	$(0, 3\Lambda_1)$

to (H.2) or are reduced to (H.11) and (H.12) using (H.6) and (H.3). This leaves just a single cone $(\Lambda_2, \Lambda_1 - \Lambda_2)$.

All the remaining cases can be analyzed similarly. The complete decomposition of $\mathfrak{Q}_+$ into r_2-twisted cones of $\mathfrak{sl}_3 \oplus (\mathfrak{u}_1)^2$ modules is given in Table H.2. We list there both the weights, (Λ, Λ'), of the tips of the cones as well as elements of $\mathfrak{Q}'_+$ that give rise to them upon multiplication with monomials (H.9).

To obtain the $\mathfrak{sl}_3 \oplus (\mathfrak{u}_1)^2$ decomposition of a free module $\mathfrak{M}_\Gamma$, where an $\mathfrak{sl}_3$ singlet Γ has order m and the $(\mathfrak{u}_1)^2$ weight Λ_Γ, one must merely shift $n \to n+m$ and $(\Lambda, \Lambda') \to (\Lambda, \Lambda'+\Lambda_\Gamma)$. The corresponding result for $\mathfrak{P}_-$ modules is obtained by interchanging the fundamental weights Λ_1 and Λ_2.

Proof of Theorem 4.35. There is a unique G homomorphism $\iota : \mathfrak{M}_{\Gamma_1} \to \mathfrak{P}_{r_1}$ of $\mathfrak{P}_-$ modules defined by $\iota(\Gamma_1) = \Gamma_1$. Since the twisted cone decompositions of both modules are identical, we first verify that the images of all tips of the cones do not vanish. To conclude the proof we must show that the entire $\mathfrak{P}_{r_1}$ is generated from those tips by the action of $x_{\dot\sigma}$ and $[\,D^{\dot\sigma}, -\,]$. Since those operators generate the underlying $\mathcal{R}_3$ module M_{r_1}, the required extension to higher order polyderivations seems rather obvious. □

Appendix I Computation of $H(\mathfrak{n}_+, \mathcal{L}(\Lambda) \otimes \bigwedge \mathfrak{b}_-)$ for $\mathfrak{sl}_3$

In Sect. 4.6.2 (Theorem 4.42) we have computed the cohomology $H(\mathfrak{n}_+, \mathcal{L}(\Lambda) \otimes \bigwedge \mathfrak{b}_-)$ for dominant integral weights Λ in the bulk, i.e., Λ sufficiently deep inside the fundamental Weyl chamber. In this appendix we compute this cohomology for all $\Lambda \in P_+$ in the cases of interest, namely for $\mathfrak{sl}_2$ and $\mathfrak{sl}_3$.

In principle, the cohomology for Λ away from the bulk can be computed by explicitly going through the spectral sequence discussed in the proof of Theorem 4.42. For $\mathfrak{sl}_2$ this is particularly easy since only the singlet weight $\Lambda = 0$ is not in the bulk. In this case one quickly verifies that the states $c_{-\alpha}|bc\rangle \otimes |\sigma\omega\rangle$ and $c_1|bc\rangle \otimes \sigma^\alpha|\sigma\omega\rangle$ in E_1 are eliminated in E_2. In other words, for $\mathfrak{sl}_2$ one finds that $H(\mathfrak{n}_+, \mathcal{L}(\Lambda) \otimes \bigwedge \mathfrak{b}_-)$ for $\Lambda = 0$ contains six states (organized in three doublets at ghost numbers $0, 1$ and 2 and $\mathfrak{h}$ weights $0, -\alpha, -2\alpha$, respectively, where a doublet at ghost number n is a pair of states of the same weight at ghost numbers n and $n+1$), as opposed to eight states (Theorem 4.42) for $\Lambda \neq 0$.

For algebras other than $\mathfrak{sl}_2$, going through the spectral sequence becomes rather cumbersome. Instead we will present an independent calculation, based on free field techniques, for the other case of special interest – namely, $\mathfrak{sl}_3$. Here we will determine the cohomology $H(\mathfrak{n}_+, \mathcal{L}(\Lambda) \otimes \bigwedge \mathfrak{b}_-)$, for arbitrary $\Lambda \in P_+$, through a spectral sequence associated with the resolution $(\mathcal{F}, \delta)$ of the irreducible module $\mathcal{L}(\Lambda)$ in terms of free field Fock spaces which are co-free over $\mathfrak{n}_+$, i.e., isomorphic to contragredient Verma modules. The reader should consult [BMP1,FeFr1] for a review of such techniques, but for completeness we will recall the little of this theory which is required here.

Introduce the free oscillators $\beta_\alpha, \gamma^\alpha$, $\alpha \in \Delta_+$, with nontrivial commutators $[\gamma^\alpha, \beta_\beta]_+ = \delta^\alpha{}_\beta$ and associated Fock space $F_\Lambda^{\beta\gamma}$ with vacuum $|\Lambda\rangle$, $\Lambda \in P$, satisfying $\beta_\alpha|\Lambda\rangle = 0$. Then $F_\Lambda^{\beta\gamma}$ can be given the structure of a $\mathfrak{g}$ module in a natural way, with highest weight Λ. For the example of $\mathfrak{sl}_3$, the positive root generators are realized as

$$\begin{aligned} e_{\alpha_1} &= \beta_{\alpha_1}\,, \\ e_{\alpha_2} &= \beta_{\alpha_2} - \gamma^{\alpha_1}\beta_{\alpha_3}\,, \\ e_{\alpha_3} &= \beta_{\alpha_3}\,. \end{aligned} \tag{I.1}$$

For $\Lambda \in P_+$ there exists a complex of such Fock modules

$$0 \longrightarrow \mathcal{F}_\Lambda^{(0)} \stackrel{\delta^{(0)}}{\longrightarrow} \mathcal{F}_\Lambda^{(1)} \stackrel{\delta^{(1)}}{\longrightarrow} \ldots \stackrel{\delta^{(s-1)}}{\longrightarrow} \mathcal{F}_\Lambda^{(s)} \longrightarrow 0\,, \tag{I.2}$$

where $s = |\Delta_+|$ and

$$\mathcal{F}_{\Lambda}^{(i)} = \bigoplus_{\{w \in W \,|\, \ell(w)=i\}} F_{w(\Lambda+\rho)-\rho}\,, \tag{I.3}$$

which gives a resolution of the irreducible module $\mathcal{L}(\Lambda)$. The differential of the complex is constructed from the so-called "screeners." For $\mathfrak{sl}_3$ the screeners are given by

$$\begin{aligned} s_{\alpha_1} &= -\beta_{\alpha_1} + \gamma^{\alpha_2}\beta_{\alpha_3}\,, \\ s_{\alpha_2} &= -\beta_{\alpha_2}\,, \\ s_{\alpha_3} &= -\beta_{\alpha_3}\,. \end{aligned} \tag{I.4}$$

Applying this resolution, we may proceed with the usual manipulations on the ensuing double complex $(\mathcal{F} \otimes F^{bc} \otimes F^{\sigma\omega}, d, \delta)$. The first spectral sequence associated with this double complex collapses at the second term,

$$\begin{aligned} E_{\infty}^{p,q} \cong E_2^{p,q} &\cong H^p(\mathfrak{n}_+, H^q(\delta, \mathcal{F}) \otimes \textstyle\bigwedge \mathfrak{b}_-) \\ &\cong \delta^{q,0}\, H^p(\mathfrak{n}_+, \mathcal{L}(\Lambda) \otimes \textstyle\bigwedge \mathfrak{b}_-)\,, \end{aligned} \tag{I.5}$$

and produces the cohomology that we want to compute, while the E_2' term for the second spectral sequence is given by

$$E_2'^{p,q} \cong H^q(\delta, H^p(\mathfrak{n}_+, \mathcal{F} \otimes \textstyle\bigwedge \mathfrak{b}_-))\,. \tag{I.6}$$

Let us now restrict to the case of $\mathfrak{sl}_3$. To proceed it is convenient to first make a similarity transformation on d as follows: Introduce the operators

$$\begin{aligned} X &= -\gamma^{\alpha_1} c_1 b^{-\alpha_1} - \gamma^{\alpha_2} c_2 b^{-\alpha_2} - \gamma^{\alpha_3}(c_1 + c_2) b^{-\alpha_3} - \gamma^{\alpha_1}\gamma^{\alpha_2} c_2 b^{-\alpha_3}\,, \\ Y &= \gamma^{\alpha_1} c_{-\alpha_2} b^{-\alpha_3} - \gamma^{\alpha_2} c_{-\alpha_1} b^{-\alpha_3}\,, \end{aligned} \tag{I.7}$$

then

$$e^Y e^X\, d\, e^{-X} e^{-Y} = \sigma^{\alpha} e_{\alpha} - \sigma^{\alpha_1}\sigma^{\alpha_2}\omega_{\alpha_3}\,. \tag{I.8}$$

This shows

$$\begin{aligned} E_1'^{p,q} &\cong H^p(\mathfrak{n}_+, \mathcal{F}^{(q)} \otimes \textstyle\bigwedge \mathfrak{b}_-) \cong H^p(\mathfrak{n}_+, \mathcal{F}^{(q)}) \otimes \textstyle\bigwedge \mathfrak{b}_- \\ &\cong \delta^{p,0} \bigoplus_{\{w \in W \,|\, \ell(w)=q\}} \mathbb{C}_{w(\Lambda+\rho)-\rho} \otimes \textstyle\bigwedge \mathfrak{b}_-\,. \end{aligned} \tag{I.9}$$

Having done this, at each Fock space in the original resolution we simply have a copy of $\bigwedge \mathfrak{b}_-$, and we must calculate the cohomology of the similarity-transformed δ on such. The operators $\delta^{(i)}$, in turn, are made up of similarity-transformed screeners. Since the screeners now operate on states with no γ's, we drop their β dependence in writing the transformed result below ($s_i = e^Y e^X s_{\alpha_i} e^{-X} e^{-Y}$).

$$\begin{aligned} s_1 &= c_1 b^{-\alpha_1} - c_{-\alpha_2} b^{-\alpha_3}\,, \\ s_2 &= c_2 b^{-\alpha_2} + c_{-\alpha_1} b^{-\alpha_3}\,, \\ s_3 &= (c_1 + c_2) b^{-\alpha_3}\,. \end{aligned} \tag{I.10}$$

The resolution is given in Fig. I.1. In this figure, $F_w \equiv F_{w(\Lambda+\rho)-\rho}$ and the intertwiners $Q_{w,w'} : F_w \to F_{w'}$ are given by

$$\begin{aligned} Q_{w,r_iw} &= s_i^{(\Lambda+\rho,w\alpha_i)}, \qquad \text{if} \quad \ell(r_iw) = \ell(w)+1\,, \\ Q_{r_1,r_1r_2} &= \sum_{0\le j\le l_2} b(l_2, l_1+l_2; j)(s_2)^{l_2-j}(s_3)^j(s_1)^{l_2-j}\,, \\ Q_{r_2,r_2r_1} &= \sum_{0\le j\le l_1} b(l_1, l_1+l_2; j)(s_1)^{l_1-j}(-s_3)^j(s_2)^{l_1-j}\,, \end{aligned} \tag{I.11}$$

where $l_i = (\Lambda+\rho, \alpha_i)$ and

$$b(m,n;j) \;=\; \frac{m\,!\,n\,!}{j\,!\,(m-j)\,!\,(n-j)\,!}\,. \tag{I.12}$$

$$\begin{array}{ccccccc} & & F_{r_1} & \longrightarrow & F_{r_1r_2} & & \\ & \nearrow & & & & \searrow & \\ F_\Lambda & & & \times & & & F_{r_1r_2r_1} \\ & \searrow & & & & \nearrow & \\ & & F_{r_2} & \longrightarrow & F_{r_2r_1} & & \end{array}$$

Fig. I.1. Fock space resolution of $\mathcal{L}(\Lambda)$.

It is now clear that since $s_i^n = 0$ for $i = 1,2$, $n \ge 3$, and $s_3^n = 0$ for $n \ge 2$ as well as $s_i^n s_3 = s_3 s_i^n = 0$ for $i = 1,2$, $n \ge 2$, all the differentials $\delta^{(i)}$ on E_1' vanish if $(\Lambda, \alpha_i) \ge 2$ for $i = 1,2$. Thus the spectral sequence collapses at E_1', and leads to a result consistent with that of Theorem 4.42. In the remaining cases the spectral sequence collapses at E_2' and the result is obtained by straightforward algebra. To formulate the answer in an elegant way let us introduce an extension $\widetilde{W}$ of the Weyl group W of $\mathfrak{sl}_3$ by $\widetilde{W} \equiv W \cup \{\sigma_1, \sigma_2\}$, and extend the length function on W by assigning $\ell(\sigma_1) = 1$ and $\ell(\sigma_2) = 2$. Let $\widetilde{W}$ act on $\mathfrak{h}^*$ by defining $\sigma_i\lambda = 0$, $i = 1,2$. We can now parametrize the weights in $\bigwedge \mathfrak{n}_-$ by $\sigma \in \widetilde{W}$ as follows

$$P(\textstyle\bigwedge^n \mathfrak{n}_-) \;=\; \{\sigma\rho - \rho \,|\, \sigma \in \widetilde{W}, \ell(\sigma) = n\}\,. \tag{I.13}$$

Then we have

Theorem I.1 *For $\mathfrak{g} \cong \mathfrak{sl}_3$, the cohomology $H(\mathfrak{n}_+, \mathcal{L}(\Lambda) \otimes \bigwedge \mathfrak{b}_-)_{\Lambda'}$ is nontrivial only if there exists a $w \in W$ and $\sigma \in \widetilde{W}$ such that $\Lambda' = (w(\Lambda+\rho)-\rho)+(\sigma\rho-\rho)$. The set of of allowed pairs (w,σ) depends on Λ and is given in Table I.1. For each allowed pair (w,σ) there is a quartet of cohomology states at ghost numbers $n, n+1, n+1$ and $n+2$ where $n = \ell(w)+\ell(\sigma)$.*

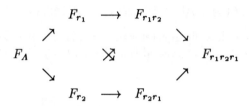

Table I.1. Condition on Λ for the pair (w, σ) to be allowed ($m_i = (\Lambda, \alpha_i)$ and "–" means there is no condition on $\Lambda \in P_+$).

$w \backslash \sigma$	1	r_1	r_2	σ_1	$r_1 r_2$	$r_2 r_1$	σ_2	$r_1 r_2 r_1$
1	–	$m_1 \geq 1$	$m_2 \geq 1$	$m_1 \geq 1$, $m_2 \geq 1$	$m_1 \geq 2$	$m_2 \geq 2$	–	$m_1 \geq 1$, $m_2 \geq 1$
r_1	–	–	$m_1 \geq 1$	$m_2 \geq 1$	$m_2 \geq 1$	$m_2 \geq 1$	$m_1 \geq 1$	–
r_2	–	$m_2 \geq 1$	–	$m_1 \geq 1$	$m_1 \geq 1$	$m_1 \geq 1$	$m_2 \geq 1$	–
$r_1 r_2$	–	–	$m_1 \geq 1$	$m_1 \geq 1$	–	$m_1 \geq 1$	$m_2 \geq 1$	–
$r_2 r_1$	–	$m_2 \geq 1$	–	$m_2 \geq 1$	$m_2 \geq 1$	–	$m_1 \geq 1$	–
$r_1 r_2 r_1$	–	$m_1 \geq 2$	$m_2 \geq 2$	–	$m_1 \geq 1$	$m_2 \geq 1$	$m_1 \geq 1$, $m_2 \geq 1$	–

For Λ in the bulk, the quartet structure of $H(\mathfrak{n}_+, \mathcal{L}(\Lambda) \otimes \bigwedge \mathfrak{b}_-)$ corresponds to the decomposition

$$H(\mathfrak{n}_+, \mathcal{L}(\Lambda) \otimes \textstyle\bigwedge \mathfrak{b}_-) \;\cong\; (H(\mathfrak{n}_+, \mathcal{L}(\Lambda)) \otimes \textstyle\bigwedge \mathfrak{n}_-) \otimes \textstyle\bigwedge \mathfrak{h}\,. \tag{I.14}$$

It is quite remarkable that the quartet structure persists even away from the bulk because, e.g., it is not true in general that $H(\mathfrak{n}_+, \mathcal{L}(\Lambda) \otimes \bigwedge \mathfrak{b}_-) \cong H(\mathfrak{n}_+, \mathcal{L}(\Lambda) \otimes \bigwedge \mathfrak{n}_-) \otimes \bigwedge \mathfrak{h}$.

Appendix J Some Explicit Cohomology States

In this appendix we give a complete list of explicit representatives of the cohomology classes that are required for the calculations discussed in Chap. 5. We have listed only operators corresponding to the highest weights in $\mathfrak{sl}_3 \oplus (\mathfrak{u}_1)^2$ modules. Other operators in those modules can be obtained through the action of the $\mathfrak{sl}_3$ currents (3.48). The normalization has been chosen to simplify formulae in Chap. 5.

J.1 The Ground Ring Generators

$$
\begin{aligned}
\varPsi^{(0)}_{\Lambda_1,\Lambda_1} = \Big(& - 4b^{[2]}b^{[3]}c^{[2]}c^{[3]} - \tfrac{10}{\sqrt{3}} b^{[2]}b^{[3]}\partial c^{[3]}c^{[3]} + \sqrt{3} b^{[2]}\partial^2 c^{[3]} - 2\sqrt{3} b^{[3]}c^{[2]} \\
& + 4b^{[3]}\partial c^{[3]} - 2i\sqrt{6}\partial\phi^{M,1}b^{[2]}\partial c^{[3]} - 3i\sqrt{2}\partial\phi^{M,1}b^{[3]}c^{[3]} \\
& - 4\partial\phi^{M,1}\partial\phi^{M,2}b^{[2]}c^{[3]} - 2\sqrt{3}\partial\phi^{M,1}\partial\phi^{M,2} - i\sqrt{3}\partial\phi^{M,1}\partial\phi^{L,1}b^{[2]}c^{[3]} \\
& + 3i\partial\phi^{M,1}\partial\phi^{L,1} - i\partial\phi^{M,1}\partial\phi^{L,2}b^{[2]}c^{[3]} + i\sqrt{3}\partial\phi^{M,1}\partial\phi^{L,2} \\
& - i\sqrt{6}\partial\phi^{M,1}\partial b^{[2]}c^{[3]} - 2i\sqrt{2}\partial\phi^{M,2}b^{[2]}\partial c^{[3]} - i\sqrt{6}\partial\phi^{M,2}b^{[3]}c^{[3]} \\
& + \tfrac{4}{\sqrt{3}}\partial\phi^{M,2}\partial\phi^{M,2}b^{[2]}c^{[3]} + 2\partial\phi^{M,2}\partial\phi^{M,2} - i\partial\phi^{M,2}\partial\phi^{L,1}b^{[2]}c^{[3]} \\
& + i\sqrt{3}\partial\phi^{M,2}\partial\phi^{L,1} - \tfrac{i}{\sqrt{3}}\partial\phi^{M,2}\partial\phi^{L,2}b^{[2]}c^{[3]} + i\partial\phi^{M,2}\partial\phi^{L,2} \\
& - i\sqrt{2}\partial\phi^{M,2}\partial b^{[2]}c^{[3]} - 3\sqrt{2}\partial\phi^{L,1}b^{[2]}c^{[2]} - 2\sqrt{6}\partial\phi^{L,1}b^{[2]}\partial c^{[3]} \\
& - 3\sqrt{2}\partial\phi^{L,1}b^{[3]}c^{[3]} - \sqrt{3}\partial\phi^{L,1}\partial\phi^{L,1}b^{[2]}c^{[3]} + 3\partial\phi^{L,1}\partial\phi^{L,1} \\
& - 2\partial\phi^{L,1}\partial\phi^{L,2}b^{[2]}c^{[3]} + 2\sqrt{3}\partial\phi^{L,1}\partial\phi^{L,2} - \sqrt{6}\partial\phi^{L,2}b^{[2]}c^{[2]} \\
& - 2\sqrt{2}\partial\phi^{L,2}b^{[2]}\partial c^{[3]} - \sqrt{6}\partial\phi^{L,2}b^{[3]}c^{[3]} - \tfrac{1}{\sqrt{3}}\partial\phi^{L,2}\partial\phi^{L,2}b^{[2]}c^{[3]} \\
& + \partial\phi^{L,2}\partial\phi^{L,2} + \tfrac{2}{\sqrt{3}}\partial b^{[2]}b^{[2]}c^{[2]}c^{[3]} + \tfrac{20}{3}\partial b^{[2]}b^{[2]}\partial c^{[3]}c^{[3]} \\
& + 2\partial b^{[2]}c^{[2]} + \tfrac{8}{\sqrt{3}}\partial b^{[2]}\partial c^{[3]} + 6\partial b^{[3]}c^{[3]} + \sqrt{6}\partial^2\phi^{L,1}b^{[2]}c^{[3]} - 3\sqrt{2}\partial^2\phi^{L,1} \\
& + \sqrt{2}\partial^2\phi^{L,2}b^{[2]}c^{[3]} - \sqrt{6}\partial^2\phi^{L,2} + \sqrt{3}\partial^2 b^{[2]}c^{[3]} \Big) \mathcal{V}_{\Lambda_1,\Lambda_1} ,
\end{aligned} \tag{J.1}
$$

$$
\begin{aligned}
\Psi^{(0)}_{\Lambda_2,\Lambda_2} = {} & \Big(4b^{[2]}b^{[3]}c^{[2]}c^{[3]} - \tfrac{10}{\sqrt{3}} b^{[2]}b^{[3]}\partial c^{[3]}c^{[3]} + \sqrt{3}b^{[2]}\partial^2 c^{[3]} - 2\sqrt{3}b^{[3]}c^{[2]} \\
& - 4b^{[3]}\partial c^{[3]} + 2\sqrt{3}\partial\phi^{M,1}\partial\phi^{M,1}b^{[2]}c^{[3]} - 3\partial\phi^{M,1}\partial\phi^{M,1} \\
& - 4i\sqrt{2}\partial\phi^{M,2}b^{[2]}\partial c^{[3]} + 2i\sqrt{6}\partial\phi^{M,2}b^{[3]}c^{[3]} - \tfrac{2}{\sqrt{3}}\partial\phi^{M,2}\partial\phi^{M,2}b^{[2]}c^{[3]} \\
& + \partial\phi^{M,2}\partial\phi^{M,2} - \tfrac{4i}{\sqrt{3}}\partial\phi^{M,2}\partial\phi^{L,2}b^{[2]}c^{[3]} - 4i\partial\phi^{M,2}\partial\phi^{L,2} \\
& - 2i\sqrt{2}\partial\phi^{M,2}\partial b^{[2]}c^{[3]} + 2\sqrt{6}\partial\phi^{L,2}b^{[2]}c^{[2]} - 4\sqrt{2}\partial\phi^{L,2}b^{[2]}\partial c^{[3]} \\
& + 2\sqrt{6}\partial\phi^{L,2}b^{[3]}c^{[3]} - \tfrac{4}{\sqrt{3}}\partial\phi^{L,2}\partial\phi^{L,2}b^{[2]}c^{[3]} - 4\partial\phi^{L,2}\partial\phi^{L,2} \\
& + \tfrac{2}{\sqrt{3}}\partial b^{[2]}b^{[2]}c^{[2]}c^{[3]} - \tfrac{20}{3}\partial b^{[2]}b^{[2]}\partial c^{[3]}c^{[3]} - 2\partial b^{[2]}c^{[2]} + \tfrac{8}{\sqrt{3}}\partial b^{[2]}\partial c^{[3]} \\
& - 6\partial b^{[3]}c^{[3]} + 2\sqrt{2}\partial^2\phi^{L,2}b^{[2]}c^{[3]} + 2\sqrt{6}\partial^2\phi^{L,2} + \sqrt{3}\partial^2 b^{[2]}c^{[3]}\Big)\,\mathcal{V}_{\Lambda_2,\Lambda_2}\,.
\end{aligned}
\tag{J.2}
$$

J.2 The Identity Quartet

$$
\mathbf{1}(z)\,, \tag{J.3}
$$

$$
\begin{aligned}
C^{[2]}(z) = {} & -4(\partial c^{[2]} + \partial b^{[2]}\partial c^{[3]}c^{[3]} + b^{[2]}\partial^2 c^{[3]}c^{[3]}) \\
& - \tfrac{1}{\sqrt{2}}(\partial\phi^{L,1} + \sqrt{3}\partial\phi^{L,2})(c^{[2]} + b^{[2]}\partial c^{[3]}c^{3]}) \\
& - \tfrac{1}{\sqrt{2}}(\sqrt{3}\partial\phi^{L,1} - \partial\phi^{L,2})\partial c^{[3]} + \sqrt{2}(\sqrt{3}\partial^2\phi^{L,1} + \partial^2\phi^{L,2})c^{[3]} \\
& - \partial\phi^{L,1}\partial\phi^{L,2}c^{[3]} - \tfrac{\sqrt{3}}{2}(\partial\phi^{L,1}\partial\phi^{L,1} - \partial\phi^{L,2}\partial\phi^{L,2})c^{[3]}\,,
\end{aligned}
\tag{J.4}
$$

$$
\begin{aligned}
C^{[3]}(z) = {} & -4\sqrt{3}\partial^2 c^{[3]} - \sqrt{\tfrac{3}{2}}(\sqrt{3}\partial\phi^{L,1} - \partial\phi^{L,2})(c^{[2]} + b^{[2]}\partial c^{[3]}c^{[3]}) \\
& - \sqrt{3}(\sqrt{2}\partial^2\phi^{L,1} + 3\partial^2\phi^{L,2})c^{[3]} - 3\sqrt{\tfrac{3}{2}}(\partial\phi^{L,1} + \sqrt{3}\partial\phi^{L,2})\partial c^{[3]} \\
& - \tfrac{\sqrt{3}}{2}(\partial\phi^{L,1}\partial\phi^{L,1} + 2\sqrt{3}\partial\phi^{L,1}\partial\phi^{L,2} + \partial\phi^{L,2}\partial\phi^{L,2})c^{[3]}\,,
\end{aligned}
\tag{J.5}
$$

$$
C^{[23]}(z) = (C^{[2]} \cdot C^{[3]})(z)\,. \tag{J.6}
$$

J.3 Generators of $\mathfrak{H}^n_1$, $n \geq 1$

The following is a complete list of operators generating $\mathfrak{H}^n_1$, $n \geq 1$, under the dot product action of $\mathfrak{H}^0$ and the bracket action of $\mathfrak{sl}_3$. They are obtained by explicitly evaluating the multiple commutators (5.40). The normalization of $\check{X}$ in (J.30) is chosen such that (5.41) holds, with the ground ring generators normalized as in (J.1) and (J.2).

$n = 1$:

$$\widehat{P}_{\dot{2},\dot{3}} = \Psi^{(1)}_{\Lambda_1,-\Lambda_1+\Lambda_2}\,, \quad \widehat{P}_{1,\dot{3}} = \tfrac{1}{2}\Psi^{(1)}_{\Lambda_1+\Lambda_2,0}\,, \quad \widehat{P}_{1,2} = \Psi^{(1)}_{\Lambda_2,\Lambda_1-\Lambda_2}\,, \tag{J.7}$$

$$\widehat{P}_{\sigma,}{}^{\sigma} = \widehat{C}_+ - \widehat{C}_-\,, \quad \widehat{C} = \widehat{C}_+ + \widehat{C}_-\,. \tag{J.8}$$

$$\begin{aligned}\Psi^{(1)}_{\Lambda_1,-\Lambda_1+\Lambda_2} = \tfrac{1}{36}\big(&12\sqrt{3}b^{[2]}c^{[2]}\partial c^{[3]} + 15b^{[2]}\partial^2 c^{[3]}c^{[3]} - 18b^{[3]}c^{[2]}c^{[3]}\\ &- 24\sqrt{3}b^{[3]}\partial c^{[3]}c^{[3]} - 3i\sqrt{6}\partial\phi^{M,1}b^{[2]}c^{[2]}c^{[3]}\\ &- 21i\sqrt{2}\partial\phi^{M,1}b^{[2]}\partial c^{[3]}c^{[3]} - 9i\sqrt{2}\partial\phi^{M,1}c^{[2]}\\ &- 18i\partial\phi^{M,1}\partial\phi^{L,2}c^{[3]} - 12i\sqrt{6}\partial\phi^{M,1}\partial c^{[3]}\\ &- 3i\sqrt{2}\partial\phi^{M,2}b^{[2]}c^{[2]}c^{[3]} - 7i\sqrt{6}\partial\phi^{M,2}b^{[2]}\partial c^{[3]}c^{[3]}\\ &- 3i\sqrt{6}\partial\phi^{M,2}c^{[2]} - 6i\sqrt{3}\partial\phi^{M,2}\partial\phi^{L,2}c^{[3]}\\ &- 12i\sqrt{2}\partial\phi^{M,2}\partial c^{[3]} + 12\sqrt{2}\partial\phi^{L,2}b^{[2]}c^{[2]}c^{[3]}\\ &- 2\sqrt{6}\partial\phi^{L,2}b^{[2]}\partial c^{[3]}c^{[3]} - 6\sqrt{6}\partial\phi^{L,2}c^{[2]}\\ &- 12\sqrt{3}\partial\phi^{L,2}\partial\phi^{L,2}c^{[3]} - 24\sqrt{2}\partial\phi^{L,2}\partial c^{[3]} + 36\partial b^{[2]}\partial c^{[3]}c^{[3]}\\ &+ 18\sqrt{2}\partial^2\phi^{L,2}c^{[3]} + 6\sqrt{3}\partial^2 c^{[3]}\big)\mathcal{V}_{\Lambda_1,-\Lambda_1+\Lambda_2}\,,\end{aligned} \tag{J.9}$$

$$\Psi^{(1)}_{\Lambda_1+\Lambda_2,0}(z) = \Big(c^{[2]} + \sqrt{\tfrac{3}{2}}i\partial\phi^{M,1}c^{[3]} - \tfrac{1}{\sqrt{2}}i\partial\phi^{M,2}c^{[3]} - b^{[2]}\partial c^{[3]}c^{[3]}\Big)\mathcal{V}_{\Lambda_1+\Lambda_2,0}\,. \tag{J.10}$$

$$\begin{aligned}\Psi^{(1)}_{\Lambda_2,\Lambda_1-\Lambda_2} = \tfrac{1}{36}\big(&12\sqrt{3}b^{[2]}c^{[2]}\partial c^{[3]} - 15b^{[2]}\partial^2 c^{[3]}c^{[3]} + 18b^{[3]}c^{[2]}c^{[3]}\\ &- 24\sqrt{3}b^{[3]}\partial c^{[3]}c^{[3]} - 6i\sqrt{2}\partial\phi^{M,2}b^{[2]}c^{[2]}c^{[3]}\\ &+ 14i\sqrt{6}\partial\phi^{M,2}b^{[2]}\partial c^{[3]}c^{[3]} + 6i\sqrt{6}\partial\phi^{M,2}c^{[2]} - 18i\partial\phi^{M,2}\partial\phi^{L,1}c^{[3]}\\ &- 6i\sqrt{3}\partial\phi^{M,2}\partial\phi^{L,2}c^{[3]} - 24i\sqrt{2}\partial\phi^{M,2}\partial c^{[3]} + 6\sqrt{6}\partial\phi^{L,1}b^{[2]}c^{[2]}c^{[3]}\\ &+ 3\sqrt{2}\partial\phi^{L,1}b^{[2]}\partial c^{[3]}c^{[3]} + 9\sqrt{2}\partial\phi^{L,1}c^{[2]} - 9\sqrt{3}\partial\phi^{L,1}\partial\phi^{L,1}c^{[3]}\\ &- 18\partial\phi^{L,1}\partial\phi^{L,2}c^{[3]} - 12\sqrt{6}\partial\phi^{L,1}\partial c^{[3]} + 6\sqrt{2}\partial\phi^{L,2}b^{[2]}c^{[2]}c^{[3]}\\ &+ \sqrt{6}\partial\phi^{L,2}b^{[2]}\partial c^{[3]}c^{[3]} + 3\sqrt{6}\partial\phi^{L,2}c^{[2]} - 3\sqrt{3}\partial\phi^{L,2}\partial\phi^{L,2}c^{[3]}\\ &- 12\sqrt{2}\partial\phi^{L,2}\partial c^{[3]} - 36\partial b^{[2]}\partial c^{[3]}c^{[3]} + 9\sqrt{6}\partial^2\phi^{L,1}c^{[3]}\\ &+ 9\sqrt{2}\partial^2\phi^{L,2}c^{[3]} + 6\sqrt{3}\partial^2 c^{[3]}\big)\,\mathcal{V}_{\Lambda_2,\Lambda_1-\Lambda_2}\,,\end{aligned} \tag{J.11}$$

$n = 2$:

$$\widehat{P}_{1,23} = \Psi^{(2)}_{0,-2\Lambda_1+\Lambda_2}\,, \quad \widehat{P}_{1,\sigma}{}^{\sigma} = \Psi^{(2)}_{\Lambda_1,-\Lambda_2}\,, \quad \widehat{P}_{1,2\dot{3}} = \Psi^{(2)}_{2\Lambda_2,-\Lambda_2}\,, \tag{J.12}$$

$$\widehat{P}_{\dot{1},\dot{2}\dot{3}} = \Psi^{(2)}_{0,\Lambda_1-2\Lambda_2}\,, \quad \widehat{P}_{\dot{3},\dot{\sigma}}{}^{\dot{\sigma}} = \Psi^{(2)}_{\Lambda_2,-\Lambda_1}\,, \quad \widehat{P}_{\dot{3},12} = \Psi^{(2)}_{2\Lambda_1,-\Lambda_1}\,. \tag{J.13}$$

$$\begin{aligned}\Psi^{(2)}_{0,-2\Lambda_1+\Lambda_2} = \tfrac{1}{108}\big(&6b^{[2]}c^{[2]}\partial^2c^{[3]}c^{[3]} - 6b^{[2]}\partial c^{[2]}\partial c^{[3]}c^{[3]} + 12\sqrt{3}b^{[2]}\partial^2c^{[3]}\partial c^{[3]}c^{[3]}\\&+ 6\sqrt{3}c^{[2]}\partial^2c^{[3]} + 3\sqrt{6}\partial\phi^{L,1}c^{[2]}\partial c^{[3]} + 9\partial\phi^{L,1}\partial\phi^{L,1}\partial c^{[3]}c^{[3]}\\&+ 6\sqrt{3}\partial\phi^{L,1}\partial\phi^{L,2}\partial c^{[3]}c^{[3]} - 3\sqrt{6}\partial\phi^{L,1}\partial c^{[2]}c^{[3]}\\&+ 9\sqrt{2}\partial\phi^{L,1}\partial^2c^{[3]}c^{[3]} + 3\sqrt{2}\partial\phi^{L,2}c^{[2]}\partial c^{[3]} + 3\partial\phi^{L,2}\partial\phi^{L,2}\partial c^{[3]}c^{[3]}\\&- 3\sqrt{2}\partial\phi^{L,2}\partial c^{[2]}c^{[3]} + 3\sqrt{6}\partial\phi^{L,2}\partial^2c^{[3]}c^{[3]} + 6\partial b^{[2]}c^{[2]}\partial c^{[3]}c^{[3]}\\&+ 6\partial c^{[2]}c^{[2]} - 6\sqrt{3}\partial c^{[2]}\partial c^{[3]} + 3\sqrt{6}\partial^2\phi^{L,1}c^{[2]}c^{[3]}\\&- 9\sqrt{2}\partial^2\phi^{L,1}\partial c^{[3]}c^{[3]} + 3\sqrt{2}\partial^2\phi^{L,2}c^{[2]}c^{[3]} - 3\sqrt{6}\partial^2\phi^{L,2}\partial c^{[3]}c^{[3]}\\&+ 21\partial^2c^{[3]}\partial c^{[3]} + \partial^3c^{[3]}c^{[3]}\big)\,\mathcal{V}_{0,-2\Lambda_1-\Lambda_2}\,,\end{aligned} \tag{J.14}$$

$$\begin{aligned}\Psi^{(2)}_{0,\Lambda_1-2\Lambda_2} = \tfrac{1}{108}\big(&-6b^{[2]}c^{[2]}\partial^2c^{[3]}c^{[3]} + 6b^{[2]}\partial c^{[2]}\partial c^{[3]}c^{[3]} + 12\sqrt{3}b^{[2]}\partial^2c^{[3]}\partial c^{[3]}c^{[3]}\\&+ 6\sqrt{3}c^{[2]}\partial^2c^{[3]} + 6\sqrt{2}\partial\phi^{L,2}c^{[2]}\partial c^{[3]} - 12\partial\phi^{L,2}\partial\phi^{L,2}\partial c^{[3]}c^{[3]}\\&- 6\sqrt{2}\partial\phi^{L,2}\partial c^{[2]}c^{[3]} - 6\sqrt{6}\partial\phi^{L,2}\partial^2c^{[3]}c^{[3]} - 6\partial b^{[2]}c^{[2]}\partial c^{[3]}c^{[3]}\\&- 6\partial c^{[2]}c^{[2]} - 6\sqrt{3}\partial c^{[2]}\partial c^{[3]} + 6\sqrt{2}\partial^2\phi^{L,2}c^{[2]}c^{[3]}\\&+ 6\sqrt{6}\partial^2\phi^{L,2}\partial c^{[3]}c^{[3]} - 21\partial^2c^{[3]}\partial c^{[3]} - \partial^3c^{[3]}c^{[3]}\big)\,\mathcal{V}_{0,-\Lambda_1-2\Lambda_2}\,,\end{aligned} \tag{J.15}$$

$$\begin{aligned}\Psi^{(2)}_{\Lambda_1,-\Lambda_2} = \tfrac{1}{108}\big(&12b^{[2]}c^{[2]}\partial c^{[3]}c^{[3]} + 12\sqrt{3}c^{[2]}\partial c^{[3]} + 3i\sqrt{6}\partial\phi^{M,1}c^{[2]}c^{[3]}\\&- 6i\sqrt{2}\partial\phi^{M,1}\partial c^{[3]}c^{[3]} + 3i\sqrt{2}\partial\phi^{M,2}c^{[2]}c^{[3]} - 2i\sqrt{6}\partial\phi^{M,2}\partial c^{[3]}c^{[3]}\\&+ 6\sqrt{6}\partial\phi^{L,1}c^{[2]}c^{[3]} + 6\sqrt{2}\partial\phi^{L,1}\partial c^{[3]}c^{[3]} + 6\sqrt{2}\partial\phi^{L,2}c^{[2]}c^{[3]}\\&+ 2\sqrt{6}\partial\phi^{L,2}\partial c^{[3]}c^{[3]} + 12\partial^2c^{[3]}c^{[3]}\big)\,\mathcal{V}_{\Lambda_1,-\Lambda_2}\,,\end{aligned} \tag{J.16}$$

$$\begin{aligned}\Psi^{(2)}_{\Lambda_2,-\Lambda_1} = \tfrac{1}{54}\big(&-6b^{[2]}c^{[2]}\partial c^{[3]}c^{[3]} + 6\sqrt{3}c^{[2]}\partial c^{[3]} + 3i\sqrt{2}\partial\phi^{M,2}c^{[2]}c^{[3]}\\&+ 2i\sqrt{6}\partial\phi^{M,2}\partial c^{[3]}c^{[3]} + 6\sqrt{2}\partial\phi^{L,2}c^{[2]}c^{[3]}\\&- 2\sqrt{6}\partial\phi^{L,2}\partial c^{[3]}c^{[3]} - 6\partial^2c^{[3]}c^{[3]}\big)\,\mathcal{V}_{\Lambda_2,-\Lambda_1}\,,\end{aligned} \tag{J.17}$$

$$\Psi^{(2)}_{2\Lambda_2,-\Lambda_2} = \tfrac{1}{18\sqrt{3}}\big(3c^{[2]}c^{[3]} - 2\sqrt{3}\partial c^{[3]}c^{[3]}\big)\,\mathcal{V}_{2\Lambda_2,-\Lambda_2}\,, \tag{J.18}$$

$$\Psi^{(2)}_{2\Lambda_1,-\Lambda_1} = -\tfrac{1}{18\sqrt{3}}\big(3c^{[2]}c^{[3]} + 2\sqrt{3}\partial c^{[3]}c^{[3]}\big)\,\mathcal{V}_{2\Lambda_1,-\Lambda_1}\,, \tag{J.19}$$

$n = 3$:

$$\widehat{P}_{1,23\dot{3}} = \Psi^{(3)}_{\Lambda_2,-2\Lambda_2}\,,\quad \widehat{P}_{1,\dot{2}\dot{3}1} = \Psi^{(3)}_{\Lambda_1,-2\Lambda_1}\,, \tag{J.20}$$

$$\widehat{P}_{1,2\dot{2}\dot{3}} = \Psi^{(3)}_{\Lambda_1+\Lambda_2,-\Lambda_1-\Lambda_2}\,,\quad \widehat{P}_{\sigma,}{}^{\sigma}{}_{\rho}{}^{\rho} = \Psi^{(3)}_{0,0}\,. \tag{J.21}$$

$$\begin{aligned}\Psi^{(3)}_{\Lambda_2,-2\Lambda_2} = & -\tfrac{1}{864}\big(12c^{[2]}\partial^2 c^{[3]}c^{[3]} + 9\sqrt{2}\partial\phi^{L,1}c^{[2]}\partial c^{[3]}c^{[3]} \\ & + 3\sqrt{6}\partial\phi^{L,2}c^{[2]}\partial c^{[3]}c^{[3]} + 6\sqrt{3}\partial c^{[2]}c^{[2]}c^{[3]} - 6\partial c^{[2]}\partial c^{[3]}c^{[3]} \\ & + 7\sqrt{3}\partial^2 c^{[3]}\partial c^{[3]}c^{[3]}\big)\,\mathcal{V}_{\Lambda_2,-2\Lambda_2}\,,\end{aligned} \tag{J.22}$$

$$\begin{aligned}\Psi^{(3)}_{\Lambda_1,-2\Lambda_1} = & \tfrac{1}{864}\big(-12c^{[2]}\partial^2 c^{[3]}c^{[3]} - 6\sqrt{6}\partial\phi^{L,2}c^{[2]}\partial c^{[3]}c^{[3]} + 6\sqrt{3}\partial c^{[2]}c^{[2]}c^{[3]} + \\ & 6\partial c^{[2]}\partial c^{[3]}c^{[3]} + 7\sqrt{3}\partial^2 c^{[3]}\partial c^{[3]}c^{[3]}\big)\,\mathcal{V}_{\Lambda_1,-2\Lambda_1}\,,\end{aligned} \tag{J.23}$$

$$\Psi^{(3)}_{\Lambda_1+\Lambda_2,-\Lambda_1-\Lambda_2} = \tfrac{1}{48}c^{[2]}\partial c^{[3]}c^{[3]}\,\mathcal{V}_{\Lambda_1+\Lambda_2,-\Lambda_1-\Lambda_2}\,, \tag{J.24}$$

$$\Psi^{(3)}_{0,0} = \tfrac{1}{32\sqrt{3}}\big(2\partial c^{[2]}c^{[2]}c^{[3]} - \partial^2 c^{[3]}\partial c^{[3]}c^{[3]}\big)\,\mathcal{V}_{0,0}\,. \tag{J.25}$$

$n = 4$:

$$\widehat{P}_{1,23\dot{2}\dot{3}} = \Psi^{(4)}_{\Lambda_1,-\Lambda_1-2\Lambda_2}\,, \quad \widehat{P}_{\dot{1},\dot{2}\dot{3}12} = \Psi^{(4)}_{\Lambda_2,-2\Lambda_1-\Lambda_2}\,. \tag{J.26}$$

$$\Psi^{(4)}_{\Lambda_1,-\Lambda_1-2\Lambda_2} = -\tfrac{1}{360\sqrt{3}}\big(c^{[2]}\partial^2 c^{[3]}\partial c^{[3]}c^{[3]} + \sqrt{3}\partial c^{[2]}c^{[2]}\partial c^{[3]}c^{[3]}\big)\,\mathcal{V}_{\Lambda_1,-\Lambda_1-2\Lambda_2}\,, \tag{J.27}$$

$$\Psi^{(4)}_{\Lambda_2,-2\Lambda_1-\Lambda_2} = -\tfrac{1}{360\sqrt{3}}\big(c^{[2]}\partial^2 c^{[3]}\partial c^{[3]}c^{[3]} - \sqrt{3}\partial c^{[2]}c^{[2]}\partial c^{[3]}c^{[3]}\big)\,\mathcal{V}_{\Lambda_2,-2\Lambda_1-\Lambda_2}\,. \tag{J.28}$$

$n = 5$:

$$\widehat{X} = \Psi^{(5)}_{0,-2\Lambda_1-2\Lambda_2}\,. \tag{J.29}$$

$$\Psi^{(5)}_{0,-2\Lambda_1-2\Lambda_2} = \tfrac{1}{1728\sqrt{3}}\partial c^{[2]}c^{[2]}\partial^2 c^{[3]}\partial c^{[3]}c^{[3]}\,\mathcal{V}_{0,-2\Lambda_1-2\Lambda_2}\,. \tag{J.30}$$

J.4 Twisted Modules of the Ground Ring

$$\widehat{\Omega}_1(z) = \mathbf{1}(z)\,, \tag{J.31}$$

$$\widehat{\Omega}_{r_1}(z) = 2\big(b^{[2]}\partial c^{[3]}c^{[3]} + c^{[2]} + \sqrt{2}\partial\phi^{L,2}c^{[3]} + \sqrt{3}\partial c^{[3]}\big)\,\mathcal{V}_{0,-2\Lambda_1+\Lambda_2}\,, \tag{J.32}$$

$$\widehat{\Omega}_{r_2}(z) = \big(-2b^{[2]}\partial c^{[3]}c^{[3]} - 2c^{[2]} + \sqrt{6}\partial\phi^{L,1}c^{[3]} + \sqrt{2}\partial\phi^{L,2}c^{[3]} + 2\sqrt{3}\partial c^{[3]}\big)\,\mathcal{V}_{0,\Lambda_1-2\Lambda_2}\,, \tag{J.33}$$

$$\widehat{\Omega}_{r_{12}}(z) = c^{[2]}c^{[3]}\,\mathcal{V}_{0,-3\Lambda_2}\,, \tag{J.34}$$

$$\widehat{\Omega}_{r_{21}}(z) = c^{[2]}c^{[3]}\,\mathcal{V}_{0,-3\Lambda_1}\,, \tag{J.35}$$

$$\widehat{\Omega}_{r_3}(z) = c^{[2]}\partial c^{[3]}c^{[3]}\,\mathcal{V}_{0,-2\Lambda_1-2\Lambda_2}\,. \tag{J.36}$$

References

[AGSY] Aharony, O., Ganor, O., Sonnenschein, J., Yankielowicz, S.: $C = 1$ string as a topological G/G model. Phys. Lett. **305B**, 35-42 (1993), hep-th/9302027

[ASY] Aharony, O., Sonnenschein, J., Yankielowicz, S.: G/G models and $\mathcal{W}_N$ strings. Phys. Lett. **289B**, 309-316 (1992), hep-th/9206063

[BBSS] Bais, F.A., Bouwknegt, P., Schoutens, K., Surridge, M.: Extensions of the Virasoro algebra constructed from Kac-Moody algebras using higher order Casimir invariants. Nucl. Phys. **B304**, 348-370 (1988)

[BaVi] Batalin, I., Vilkovisky, G.: Quantization of gauge theories with linearly dependent generators. Phys. Rev. **D28**, 2567-2582 (1983)

[BPZ] Belavin, A.A., Polyakov, A.M., Zamolodchikov, A.B.: Infinite conformal symmetry in two dimensional quantum field theory. Nucl. Phys. **B241**, 333-380 (1984)

[BBRT] Bergshoeff, E., de Boer, J., de Roo, M., Tjin, T.: The cohomology of the noncritical $\mathcal{W}$-string. Nucl. Phys. **B420**, 379-408 (1994), hep-th/9312185

[BSS] Bergshoeff, E., Sevrin, A., Shen, S.: A derivation of the BRST operator for non-critical $\mathcal{W}$ strings. Phys. Lett. **296B**, 95-103 (1992), hep-th/9209037

[BGG1] Bernstein, I.N., Gel'fand, I.M., Gel'fand, S.I.: Structure of representations generated by vectors of highest weight. Funct. Anal. Appl. **5**, 1-8 (1971)

[BGG2] Bernstein, I.N., Gel'fand, I.M., Gel'fand, S.I.: Differential operators on the base affine space and a study of $\mathfrak{g}$-modules. In: Gel'fand, I.M. (ed.) Lie groups and their representations. Proceedings, Budapest 1971, pp. 21-64. New York: Halsted 1975

[BLNW1] Bershadsky, M., Lerche, W., Nemeschansky, D., and Warner, N.P.: A BRST operator for noncritical W strings. Phys. Lett. **292B**, 35-41 (1992), hep-th/9207067

[BLNW2] Bershadsky, M., Lerche, W., Nemeschansky, D., and Warner, N.P.: Extended $N = 2$ superconformal structure of gravity and W gravity coupled to matter. Nucl. Phys. **B401**, 304-347 (1993), hep-th/9211040

[BeOo] Bershadsky, M., Ooguri, H.: Hidden $SL(n)$ symmetry in conformal field theories. Comm. Math. Phys. **126**, 49-83 (1989)

[BiFl] Biedenharn, L.C., Flath, D.E.: On the structure of tensor operators in SU(3). Comm. Math. Phys. **93**, 143-169 (1984)

[Bi] Bilal, A.: What is W geometry? Phys. Lett. **249B**, 56-62 (1990)

[Br] Borcherds, R.E.: Vertex algebras, Kac-Moody algebras, and the monster. Proc. Natl. Acad. Sci. USA **83**, 3068-3071 (1986)

[Bt] Bott., R.: Homogeneous vector bundles. Ann. Math. **66**, 203-248 (1957)

[BoTu] Bott, R., Tu, L.W.: Differential forms in algebraic topology. Berlin, Heidelberg, New York: Springer 1982

[Bo] Bouwknegt, P.: Extended conformal algebras from Kac-Moody algebras. In: Kac, V.G. (ed.) Infinite-dimensional Lie algebras and Lie groups. Advanced Series in Mathematical Physics, Vol. **7**, pp. 527-555. Singapore: World Scientific 1989

[BMP1] Bouwknegt, P., McCarthy, J., Pilch, K.: Quantum group structure in the Fock space resolutions of $\widehat{sl}(n)$ representations. Comm. Math. Phys. **131**, 125-155 (1990)

[BMP2] Bouwknegt, P., McCarthy, J., Pilch, K.: Some spects of free field resolutions in $2D$ CFT with application to the quantum Drinfel'd-Sokolov reduction. In: Berkovits, N. et al. (eds.) Strings and Symmetries 1991. Proceedings, Stony Brook 1991, pp. 407-422. Singapore: World Scientific 1991, hep-th/9110007

[BMP3] Bouwknegt, P., McCarthy, J., Pilch, K.: BRST analysis of physical states for 2d gravity coupled to $c \leq 1$ matter. Comm. Math. Phys. **145**, 541-560 (1992)

[BMP4] Bouwknegt, P., McCarthy, J., Pilch, K.: Semi-infinite cohomology in conformal field theory and $2D$ gravity. In: Gielerak, R., Borowiec, A. (eds.) Infinite dimensional geometry in physics. Proceedings, Karpacz 1992. J. Geom. Phys. **11**, 225-249 (1993)

[BMP5] Bouwknegt, P., McCarthy, J., Pilch, K.: Semi-infinite cohomology of $\mathcal{W}$-algebras. Lett. Math. Phys. **29**, 91-102 (1993)

[BMP6] Bouwknegt, P., McCarthy, J., Pilch, K.: On the BRST structure of $\mathcal{W}_3$ gravity coupled to $c = 2$ matter. In: Penner, R., Yau, S.T. (eds.) Perspectives in Mathematical Physics. Proceedings, Los Angeles 1993, pp. 77-89. Boston: International Press 1994, hep-th/9303164

[BMP7] Bouwknegt, P., McCarthy, J., Pilch, K.: On the $\mathcal{W}$-gravity spectrum and its G-structure. In: Baulieu, L. et al. (eds.) Quantum field theory and string theory. Proceedings, Cargese 1994, pp. 59-70. New York: Plenum Press 1995, hep-th/9311137

[BMP8] Bouwknegt, P., McCarthy, J., Pilch, K.: Operator algebra of the $4D$ $\mathcal{W}_3$ string. In: Bars, I. et al (eds.) Strings '95: Future Perspectives in String Theory. Proceedings, Los Angeles 1995. Singapore: World Scientific (to appear), hep-th/9509121

[BoPi] Bouwknegt, P., Pilch, K.: The BV-algebra structure of $\mathcal{W}_3$ cohomology. In: Aktaş, G. et al. (eds.) Gürsey Memorial Conference I: Strings and Symmetries. Lecture Notes in Physics, Vol. **447**, pp. 283-291. Berlin, Heidelberg, New York: Springer 1995, hep-th/9509126

[BoSc1] Bouwknegt, P., Schoutens, K.: $\mathcal{W}$ symmetry in conformal field theory. Phys. Rep. **223**, 183-276 (1993), hep-th/9210010

[BoSc2] Bouwknegt, P., Schoutens, K.: $\mathcal{W}$-symmetry. Advanced Series in Mathematical Physics, Vol. **22**. Singapore: World Scientific 1995

[Da] David, F.: Conformal field theories coupled to 2-d gravity in the conformal gauge. Mod. Phys. Lett. **A3**, 1651-1656 (1988)

[dBGo1] de Boer, J., Goeree, J.: $\mathcal{W}$ gravity from Chern-Simons theory. Nucl. Phys. **B381**, 329-359 (1992), hep-th/9112060

[dBGo2] de Boer, J., Goeree, J.: KPZ analysis for $\mathcal{W}_3$ gravity. Nucl. Phys. **B405**, 669-694 (1993), hep-th/9211108

[DSTS] Deckmyn, A., Siebelink, R., Troost, W., Sevrin, A.: On the Lagrangian realization of noncritical $\mathcal{W}$ strings. Phys. Rev. **D51**, 6970-6980 (1995), hep-th/9411221

[dVvD1] de Vos, K., van Driel, P.: The Kazhdan-Lusztig conjecture for finite $\mathcal{W}$ algebras. Lett. Math. Phys. **35**, 333-344 (1995), hep-th/9312016.

[dVvD2] de Vos, K., van Driel, P.: The Kazhdan-Lusztig conjecture for $\mathcal{W}$ algebras. Preprint, hep-th/9508020

[DiKa] Distler, J., Kawai, H.: Conformal field theory and 2-D quantum gravity. Nucl. Phys. **B321**, 509-527 (1989)

[Di] Dixmier, J.: Enveloping algebras. Amsterdam: North Holland 1977

[FaLu1] Fateev, V.A., Lukyanov, S.L.: The models of two dimensional conformal quantum field theory with $\mathbb{Z}_n$ symmetry. Int. J. Mod. Phys. **A3**, 507-520 (1988)

[FaLu2] Fateev, V.A., Lukyanov, S.L.: Additional symmetries and exactly soluble models in two-dimensional conformal field theory. Sov. Sci. Rev. A. Phys. **15**, 1-117 (1990)

[FaZa] Fateev, V.A., Zamolodchikov, A.B.: Conformal quantum field theory models in two dimensions having $\mathbb{Z}_3$ symmetry. Nucl. Phys. **B280**, 644-660 (1987)

[Fe] Feigin, B.L.: The semi-infinite cohomology of the Virasoro and Kac-Moody Lie algebras. Usp. Mat. Nauk **39**, 195-196 (1984)

[FeFr1] Feigin, B.L., Frenkel, E.V.: Affine Kac-Moody algebras and semi-infinite flag manifolds. Comm. Math. Phys. **128**, 161-189 (1990)

[FeFr2] Feigin, B.L., Frenkel, E.V.: Quantization of the Drinfeld-Sokolov reduction. Phys. Lett. **246B**, 75-81 (1990)

[FeFr3] Feigin, B.L., Frenkel, E.V.: Affine Kac-Moody algebras at the critical level and Gelfand-Dikii algebras. Int. J. Mod. Phys. **A7** (Suppl. A1) 197-215 (1992)

[FeFu] Feigin, B.L., Fuchs, D.B.: Representations of the Virasoro algebra. In: Vershik, A.M., Zhelobenko, D.P. (eds.) Representation of Lie groups and related topics. Advanced Studies in Contemporary Mathematics, Vol. **7**, pp. 465-554. New York: Gordon and Breach 1990

[Fi] Figueroa-O'Farrill, J.M.: On the homological construction of Casimir algebras. Nucl. Phys. **B343**, 450-466 (1990)

[FKW] Frenkel, E.V, Kac, V.G., Wakimoto, M.: Characters and fusion rules for $\mathcal{W}$-algebras via quantized Drinfeld-Sokolov reductions. Comm. Math. Phys. **147** 295-328 (1992)

[FGZ] Frenkel, I.B., Garland, Zuckerman, G.J.: Semi-infinite cohomology and string theory. Proc. Natl. Acad. Sci. USA **83**, 8442-8446 (1986)

[FHL] Frenkel, I.B., Huang, Y.-Z., Lepowsky, J.: On axiomatic approaches to vertex operator algebras and modules. Mem. Amer. Math. Soc. **104**, no. 494 (1993)

[FrKa] Frenkel, I.B., Kac, V.G.: Basic representations of affine Lie algebras and dual resonance models. Inv. Math. **62**, 23-66 (1980)

[FLM] Frenkel, I.B., Lepowsky, J., Meurman, A.: Vertex operator algebras and the monster. Pure and Applied Mathematics, Vol. **134**. San Diego: Academic Press 1988

[FrNi] Fröhlicher, A., Nijenhuis, A.: Theory of vector-valued differential forms I. Indag. Math., **18**, 338-359 (1956)

[GeKi1] Gel'fand, I.M., Kirillov, A.A.: On the structure of the field of quotients of the enveloping algebra of a semisimple Lie algebra. Sov. Math. Dokl. **9**, 669-671 (1968)

[GeKi2] Gel'fand, I.M., Kirillov, A.A.: The structure of the Lie field connected with a split semisimple Lie algebra. Funct. Anal. Appl. **3**, 6-21 (1969)

[GeZl] Gel'fand, I.M., Zelevinskij, A.V.: Models of representations of classical groups and their hidden symmetries. Funct. Anal. Appl. **18**, 183-198 (1984)

[Gs] Gerstenhaber, M.: The cohomology structure of an associative ring. Ann. Math. **78** 267-288 (1962). On the deformation of rings and algebras. Ann. Math. **79** 59-103 (1964)

[Gv] Gervais, J.-L.: *W* geometry from chiral embeddings. In: Gielerak, R., Borowiec, A. (eds.) Infinite dimensional geometry in physics. Proceedings, Karpacz 1992. J. Geom. Phys. **11**, 293-304 (1993)

[GvMa] Gervais, J.-L., Matsuo, Y.: $\mathcal{W}$ geometries. Phys. Lett. **274B**, 309-316 (1992), hep-th/9110028. Classical A_n $\mathcal{W}$ geometry. Comm. Math. Phys. **152**, 317-368 (1993), hep-th/9201026

[Gt] Getzler, E.: Batalin-Vilkovisky algebras and two-dimensional topological field theories. Comm. Math. Phys. **159**, 265-285 (1994), hep-th/9212043

[Go] Goddard, P.: Meromorphic conformal field theory. In: Kac, V.G. (ed.) Infinite dimensional Lie algebras and Lie groups. Advanced Series in Mathematical Physics, Vol. **7**, pp. 556-587. Singapore: World Scientific 1989

[GNOS] Goddard, P., Nahm, W., Olive, D., Schwimmer, A.: Vertex operators for non-simply-laced algebras. Comm. Math. Phys. **107**, 179-212 (1986)

[GoVa] Ghoshal, D., Vafa, C.: $c = 1$ string as the topological theory of the conifold. Nucl. Phys. **B453**, 121-128 (1995), hep-th/9506122

[Gu] Gurarie, V.: Logarithmic operators in conformal field theory. Nucl. Phys. **B410**, 535-549 (1993), hep-th/9303160

[Hu1] Hull, C.: The geometry of *W* gravity. Phys. Lett. **269B**, 257-263 (1991). *W* geometry. Comm. Math. Phys. **156**, 245-275 (1993)

[Hu2] Hull, C.: Lectures on *W* gravity, *W* geometry and *W* strings. In: E. Gava et al. (eds.) High energy physics and cosmology 1992. Proceedings, Trieste 1992, pp. 76-142. Singapore: World Scientific 1993, hep-th/9302110

[Ja] Jantzen, J.C.: Moduln mit einem höchsten Gewicht. Lecture Notes in Mathematics, Vol. **750**. Berlin, Heidelberg, New York: Springer 1979

[Ka] Kac, V.G.: Infinite dimensional Lie algebras. Cambridge: Cambridge University Press 1985

[KaPe] Kac, V.G., Peterson, D.H.: Infinite dimensional Lie algebras, theta functions and modular forms. Adv. Math. **53**, 125-264 (1984)

[KaRa] Kac, V.G., Raina, A.K.: Highest weight representations of infinite dimensional Lie algebras. Advanced Series in Mathematical Physics, Vol. **2**. Singapore: World Scientific 1987

[Kl] Klebanov, I.: Ward identities in two-dimensional string theory. Mod. Phys. Lett. **A7**, 723-732 (1992), hep-th/9201005

[Kn] Knapp, A.W.: Lie groups, Lie algebras, and cohomology. Princeton: Princeton University Press 1988

[K-S] Kosmann-Schwarzbach, Y.: Exact Gerstenhaber algebras and Lie bialgebroids. Preprint, U.R.A. 169 CNRS (1994)

[Kt] Kostant, B.: Lie algebra cohomology and the generalized Borel-Weil theorem. Ann. Math. **74**, 329-387 (1961)

[Ko] Koszul, J.-L.:Crochet de Schouten-Nijenhuis et cohomologie. Astérisque (hors série), 257-271 (1985)

[Kr] Krasil'shchik, I.S.: Hamiltonian cohomology of canonical algebras. Sov. Math. Dokl. **21**, 625-629 (1980). Schouten bracket and canonical algebras. Lecture Notes in Mathematics, Vol. **1334**, pp. 79-110. Berlin, Heidelberg, New York: Springer 1988

[KMS] Kutasov, D., Martinec, E., Seiberg, N.: Ground rings and their modules in 2-d gravity with $c < 1$. Phys. Lett. **276B**, 437-444 (1992), hep-th/9111048

[LeSe] Lerche, W., Sevrin, A.: On the Landau-Ginzburg realization of topological gravities. Nucl. Phys. **B428**, 259-281 (1994), hep-th/9403183

[LiZu1] Lian, B.H., Zuckerman, G.J.: New selection rules and physical states in 2-d gravity; conformal gauge. Phys. Lett. **254B**, 417-423 (1991). 2-d gravity with $c = 1$ matter. Phys. Lett. **266B**, 21-28 (1991). Semi-infinite homology and 2-d gravity. Comm. Math. Phys. **145**, 561-593 (1992)

[LiZu2] Lian, B.H., Zuckerman, G.J.: New perspectives on the BRST-algebraic structure of string theory. Comm. Math. Phys. **154**, 613-646 (1993), hep-th/9211072

[LiZu3] Lian, B.H., Zuckerman, G.J.: From string backgrounds to topological field theories. hep-th/9512039

[LPWX] Lu, H., Pope, C.N., Wang, X.J., Xu, K.W.: The complete cohomology of the $\mathcal{W}_3$ string. Class. Quant. Grav. **11**, 967-982 (1994), hep-th/9309041

[MMMO] Marshakov, A., Mironov, A., Morozov, A., Olshanetsky, M.: $c = r_G$ theories of W_G-gravity: the set of observables as a model of simply laced G. Nucl. Phys. **B404**, 427-456 (1993), hep-th/9203044

[Ma] Matsuo, Y.: Classical W geometry in conformal and light cone gauges. Prog. Theor. Phys. Suppl. **114**, 243-247 (1993)

[McC] McCarthy, J.: Operator algebra of the $D = 2 + 2$ $\mathcal{W}_3$ string. In: Carey, A.L. et al. (eds.) Confronting the infinite. Proceedings, Adelaide 1994, pp. 123-135, Singapore: World Scientific 1995, hep-th/9509134

[PeSc] Penkava, M., Schwarz, A.: On some algebraic structures arising in string theory. In: Penner, R., Yau, S.T. (eds.) Perspectives in Mathematical Physics. Proceedings, Los Angeles 1993, pp. 219-227. Boston: International Press 1994, hep-th/9212072

[Po] Pope, C.N.: $\mathcal{W}$ strings 93. In: M. Halpern et al. (eds.) Strings '93. Proceedings, Berkeley 1993. Singapore: World Scientific 1995, hep-th/9309125

[PSSW] Pope, C.N., Sezgin, E., Stelle, K.S., Wang, X.J.: Discrete states in the W_3 string. Phys. Lett. **299B**, 247-254 (1993), hep-th/9209111

[RSS] Ragoucy, E., Sevrin, A., Sorba, P.: Strings from $N = 2$ gauged Wess-Zumino-Witten models. hep-th/9511049

[RoSa] Rozansky, L., Saleur, H.: Quantum field theory for the multivariable Alexander-Conway polynomial. Nucl. Phys. **B376**, 461-509 (1992)

[Sa] Sadov, V.: On the spectra of $SL(N)_k/SL(N)_k$ cosets and $\mathcal{W}_N$ gravities, I. Int. J. Mod. Phys. **A8**, 5115-5128 (1993), hep-th/9302060

[Sc] Schouten, J.A.: Über differentialkomitanten zweier kontravarianter grössen. Konink. Nederl. Akad. Wetensch., Proc. Ser. **A43**, 449-452 (1940)

[SSvN] Schoutens, K., Sevrin, A., van Nieuwenhuizen, P.: Induced gauge theories and $\mathcal{W}$-gravity. In: Berkovits, N. et al. (eds.) Strings and Symmetries 1991. Proceedings, Stony Brook 1991, pp. 558-590. Singapore: World Scientific 1991, hep-th/9110007

[Th] Thielemans, C.: A Mathematica package for computing operator product expansions. Int. J. Mod. Phys. **C2**, 787-798 (1991)

[TM] Thierry-Mieg, J.: BRS analysis of Zamolodchikov's spin two and three current algbera. Phys. Lett. **197B**, 368-372 (1987)

[Ve] Verlinde, E.: The master equation of 2-D string theory. Nucl. Phys. **B381**, 141-157 (1992), hep-th/9202021

[Wa1] Watts, G.M.T.: Determinant formulae for extended algebras in two dimensional conformal field theory. Nucl. Phys. **B326**, 648-672 (1989). Erratum, Nucl. Phys. **B336**, 720 (1990)

[Wa2] Watts, G.M.T.: W-algebras and coset models. Phys. Lett. **245B**, 65-71 (1990)

[Wi1] Witten, E.: A note on the antibracket formalism. Mod. Phys. Lett. **A5**, 487-494 (1990)

[Wi2] Witten, E.: Ground ring of two-dimensional string theory. Nucl. Phys. **B373**, 187-213 (1992), hep-th/9108004

[WiZw] Witten, E., Zwiebach, B.: Algebraic structures and differential geometry in 2d string theory. Nucl. Phys. **B377**, 55-112 (1992), hep-th/9201056

[WuZh] Wu, Y.S., Zhu, C.J.: The complete structure of the cohomology ring and associated symmetries in $d = 2$ string theory. Nucl. Phys. **B404**, 245-287 (1993), hep-th/9209011

[Za] Zamolodchikov, A.B.: Infinite additional symmetries in two dimensional conformal quantum field theory. Theor. Math. Phys. **65**, 1205-1213 (1985)

[Zh] Zhu, C.J.: The structure of the ground ring in critical $\mathcal{W}_3$ gravity. Preprint, hep-th/9508125

Glossary of Notation

$\mathcal{O}$	category of modules, see Section 2.2.1
$\mathrm{JH}(V)$	Jordan-Hölder composition series for $V \in \mathcal{O}$.
$U(\cdot)$	universal enveloping algebra functor
$U(\cdot)_{\mathrm{loc}}$	corresponding local completion
$\mathfrak{g}$	complex simple Lie algebra
$\mathfrak{g} \cong \mathfrak{n}_+ \oplus \mathfrak{h} \oplus \mathfrak{n}_-$	Cartan decomposition
$\mathfrak{h}$	Cartan subalgebra with dual $\mathfrak{h}^*$
$(\, , \,)$	bilinear form on $\mathfrak{h}$ or $\mathfrak{h}^*$, sometimes also denoted by $\cdot$
ρ	element of $\mathfrak{h}^*$ such that $\rho(h_i) = 1, i = 1, \ldots, \ell$
ℓ	rank of $\mathfrak{g}$
Δ, $\Delta_\pm$	roots, positive/negative roots of $\mathfrak{g}$
$Q = \mathbb{Z} \cdot \Delta_+$	root lattice of $\mathfrak{g}$
$Q_+ = \mathbb{Z}_{\geq 0} \cdot \Delta_+$	
P, P_+, P_{++}	set of integral, dominant integral, strictly dominant integral weights, respectively
$\alpha_i, i = 1, \ldots, \ell$	simple roots of $\mathfrak{g}$
$\Lambda_i, i = 1, \ldots, \ell$	fundamental weights of $\mathfrak{g}$
D_+	fundamental Weyl chamber, i.e., $D_+ = \{\lambda \in \mathfrak{h}^*_{\mathbb{R}} \mid (\lambda, \alpha_i) \geq 0, i = 1, \ldots, \ell\}$
$\mathcal{L}(\Lambda)$	finite dimensional irreducible representation of $\mathfrak{g}$ with highest weight $\Lambda \in P_+$
$m_\Lambda^{\Lambda'}$	multiplicity of the weight Λ in $\mathcal{L}(\Lambda')$
W	Weyl group of $\mathfrak{g}$
w_0	Coxeter element of W, i.e., longest element in W
r_i	reflection in simple root α_i; for $\mathfrak{sl}_3$ $i = 1, 2$, $r_{ij} = r_i r_j$, $r_3 = r_1 r_2 r_1 = r_2 r_1 r_2 = w_0$
$\widetilde{W} = W \cup \{\sigma_1, \sigma_2\}$	where σ_i, $i = 1, 2$, act by zero on all weights $\lambda \in \mathfrak{h}^*$
$\ell(\sigma)$	the length of $\sigma \in \widetilde{W}$
$\ell_w(\sigma)$, $w \in W$	twisted length of $\sigma \in \widetilde{W}$
$\sigma \circ \Lambda = \Lambda + \rho - \sigma\rho$, $\sigma \in \widetilde{W}$	
$\widehat{\mathfrak{g}}$	affine Lie algebra with underlying finite-dimensional Lie algebra $\mathfrak{g}$
$\Lambda_i, i = 0, \ldots, \ell$	fundamental weights of $\widehat{\mathfrak{g}}$
$\widehat{W} \cong W \ltimes T$	Weyl group of $\widehat{\mathfrak{g}}$
$T \equiv \{t_\alpha \mid \alpha \in Q\}$	translation subgroup of $\widehat{W}$
G, N_+, H, N_-	complex groups generated by $\mathfrak{g}$, $\mathfrak{n}_+$, $\mathfrak{h}$, $\mathfrak{n}_-$ respectively
$A = N_+ \backslash G$	base affine space
$\mathcal{E}(G)$	regular functions on G
$\mathcal{E}(A)$	regular functions on A

$\mathcal{A}$	Heisenberg algebra
$F(\Lambda, \alpha_0)$	Fock space ($\mathcal{A}$-module) with weight Λ and background charge $\alpha_0 \in \mathbb{C}$, see Section 2.2.3
F^{gh}	ghost Fock space, see Section 3.1.1
$w \cdot \Lambda w(\Lambda + \alpha_0\rho) - \alpha_0\rho,\ w \in W$	
$M(h, w, c)$	Verma module of $\mathcal{W}_3$, see Definition 2.9
$M^{(\kappa)}(h, w, c)$	generalized Verma module of $\mathcal{W}_3$, see Definition 2.21
$M^{(\kappa)}(\Lambda, \alpha_0) \equiv M(h(\Lambda, \alpha_0), w(\Lambda, \alpha_0), c(\alpha_0))$	as specified in (2.32)
$M^{(\kappa)}[s_1, s_2] \equiv M^{(\kappa)}(s_1\Lambda_1 + s_2\Lambda_2, 0)$	as found below Theorem 2.33
$M(S) \equiv M(v_1, v_2, \ldots)$	submodule of $M(\Lambda, \alpha_0)$ generated by $S = \{v_1, v_2, \ldots\}$
$[\ ,\ ]$	graded commutator (e.g., anti-commutator for ghost fields)
$\mathfrak{C}$	chiral algebra specified in Theorem 3.5
$H(\mathcal{W}_3, \mathfrak{C})$	cohomology of the complex $(\mathfrak{C}, d)$ with differential d given by (3.10) acting as in (3.12)
$\mathfrak{H} = H(\mathcal{W}_3, \mathfrak{C})$	considered as an operator algebra
$\mathfrak{H}_w = H(\mathcal{W}_3, \mathfrak{C}_w)$	cohomology of the subcomplex $(\mathfrak{C}_w, d)$ of operators with $-i\Lambda^L + 2\rho \in w^{-1}P_+$
$(\mathfrak{A}, \cdot, [-,-])$	Gerstenhaber algebra, see Definition 4.1
$(\mathfrak{A}, \cdot, \Delta)$	BV-algebra with BV-operator Δ, see Definition 4.2
$\mathcal{P}^n(\mathcal{R}, M)$	order n polyderivations of an algebra $\mathcal{R}$ with values in M
$\mathcal{D}(\mathcal{R}, M) \equiv \mathcal{P}^1(\mathcal{R}, M)$	
$\mathcal{P}^n(\mathcal{R}) \equiv \mathcal{P}^n(\mathcal{R}, \mathcal{R})$	
$\mathfrak{P}_w$	twisted polyderivations, see Section 4.5.2
$\mathrm{BV}[\mathfrak{g}]$	the BV algebra $H(\mathfrak{n}_+, \mathcal{E}(G) \otimes \bigwedge \mathfrak{b}_-)$